T0245288

COMPUTATIONAL INTELLIGENCE TECHNIQUES FOR SUSTAINABLE SUPPLY CHAIN MANAGEMENT

Uncertainty, Computational Techniques and Decision Intelligence Book Series

Series Editors

Tofigh Allahviranloo, PhD
Faculty of Engineering and Natural Sciences, Istinye University, Istanbul, Turkey

Narsis A. Kiani, PhD
Algorithmic Dynamics Lab, Department of Oncology-Pathology & Center of Molecular Medicine, Karolinska Institute, Stockholm, Sweden

Witold Pedrycz, PhD
Department of Electrical and Computer Engineering, University of Alberta, Canada

For more information about the UCTDI series, please visit:
https://www.elsevier.com/books-and-journals/book-series/uncertainty-computational-techniques-and-decision-intelligence

Uncertainty, Computational
Techniques and Decision Intelligence

Computational Intelligence Techniques for Sustainable Supply Chain Management

Volume Editors

SANJOY KUMAR PAUL

UTS Business School, University of Technology Sydney, Sydney, NSW, Australia

SANDEEP KAUTISH

LBEF Campus (APU Malaysia) Kathmandu, Nepal; Model Institute of Engineering and Technology, Jammu, Jammu and Kashmir, India

ACADEMIC PRESS

An imprint of Elsevier

ISBN: 978-0-443-18464-2

For Information on all Academic Press publications
visit our website at https://www.elsevier.com/books-and-journals

Publisher: Mara Conner
Editorial Project Manager: Manisha Rana
Production Project Manager: Surya Narayanan Jayachandran
Cover Designer: Vicky Pearson Esser

Typeset by MPS Limited, Chennai, India

Working together
to grow libraries in
developing countries
www.elsevier.com • www.bookaid.org

Contents

10. Computational techniques for sustainable green procurement and production   275

Bhakti Parashar, Sandeep Kautish and Amrita Chaurasia

11. Predictive big data analytics for supply chain demand forecasting   301

Supriyo Ahmed, Ripon K. Chakrabortty and Daryl L. Essam

List of contributors

Imtiaz Ahmed
Department of Industrial and Management Systems Engineering, West Virginia University, Morgantown, WV, United States

Supriyo Ahmed
School of Systems & Computing, UNSW Canberra at ADFA, Campbell, ACT, Australia; Department of Electrical and Electronic Engineering, BRAC University, Dhaka, Bangladesh

Sharfuddin Ahmed Khan
Industrial Systems Engineering, Faculty of Engineering and Applied Science, University of Regina, Canada

Yusuf Jibrin Alkali
Federal Inland Revenue Service, Abuja, Nigeria

Pradeep Bedi
Department of Computer Science, Regional Campus Manipur, Indira Gandhi National Tribal University, Amarkantak, Madhya Pradesh, India; Department of Computer Science and Engineering, Galgotias University, Greater Noida, Uttar Pradesh, India

Malini Mittal Bishnoi
School of Humanities, Arts and Applied Sciences, Amity University, Dubai, United Arab Emirates

Ripon K. Chakrabortty
School of Systems & Computing, UNSW Canberra at ADFA, Campbell, ACT, Australia

Amrita Chaurasia
School of Commerce, Finance and Accountancy Christ (Deemed to be University), Ghaziabad, Uttar Pradesh, India

Sanjoy Das
Department of Computer Science, Indira Gandhi National Tribal University, Regional Campus Manipur, Makhan, India

Kavitha Desai
SVKM's Narsee Monjee Institute of Management Studies, Bangalore, Karnataka, India

Ashish Dwivedi
Jindal Global Business School, O.P. Jindal Global University, Sonipat, Haryana, India

Idiano D'Adamo
Department of Computer, Control and Management Engineering, Sapienza University of Rome, Rome, Italy

Daryl L. Essam
School of Systems & Computing, UNSW Canberra at ADFA, Campbell, ACT, Australia

S.B. Goyal
Faculty of Information Technology, City University, Petaling Jaya, Malaysia

Bharath H.
CMS Business School, JAIN (Deemed-to-Be-University), Bangalore, Karnataka, India

Muhammad Ikram
School of Business Administration, Al Akhawayn University in Ifrane, Ifrane, Morocco

Charbel Jose Chiappetta Jabbour
Department of Information Systems, Supply Chain Management & Decision Support, NEOMA Business School, Mont-Saint-Aignan, France

Mustafa Jahangoshai Rezaee
Faculty of Industrial Engineering, Urmia University of Technology, Urmia, Iran

Sandeep Kautish
LBEF Campus (APU Malaysia) Kathmandu, Nepal; Model Institute of Engineering and Technology, Jammu, Jammu and Kashmir, India

Gholamreza Khojasteh
Faculty of Industrial Engineering, Urmia University of Technology, Urmia, Iran

S. Mahalakshmi
Faculty of Management Studies, CMS Business School, JAIN (Deemed-to-Be-University), Bangalore, Karnataka, India

Rahul Reddy Nadikattu
Department of IT, University of Cumbersome, Williamsburg, KY, United States

Anitha Nallasivam
Faculty of Management Studies, CMS Business School, JAIN (Deemed-to-Be-University), Bangalore, Karnataka, India

Bhakti Parashar
VIT-Bhopal University, Bhopal, Madhya Pradesh, India

Sanjoy Kumar Paul
UTS Business School, University of Technology Sydney, Sydney, NSW, Australia

Towfique Rahman
UTS Business School, University of Technology Sydney, Sydney, NSW, Australia; Department of Business Strategy and Innovation, Griffith University, Gold Coast, QLD, Australia

Ahmed Shoyeb Raihan
Department of Industrial and Management Systems Engineering, West Virginia University, Morgantown, WV, United States

Anand Singh Rajawat
School of Computer Sciences & Engineering, Sandip University, Nashik, Maharashtra, India

Swamynathan Ramakrishnan
Amity Business School, Amity University, Dubai, United Arab Emirates

A. Reyana
Department of Computer Science and Engineering, Karunya Institute of Technology and Sciences, Coimbatore, Tamilnadu, India

Morteza Saberi
School of Computer Science, University of Technology Sydney, Ultimo, NSW, Australia

Muhammad Shujaat Mubarik
Department of Marketing and Operations, Edinburgh Business School, Heriot-Watt University, Edinburgh, United Kingdom

Arpit Singh
Jindal Global Business School, O.P. Jindal Global University, Sonipat, Haryana, India

Amin Vafadarnikjoo
Sheffield University Management School, The University of Sheffield, Sheffield, United Kingdom

Pawan Whig
Vivekananda Institute of Professional Studies, New Delhi, India

About the editors

Sanjoy Kumar Paul, PhD, is an Associate Professor in operations and supply chain management at the University of Technology Sydney (UTS), Sydney, Australia. He has published more than 140 articles in top-tier journals. He is also an associate editor, area editor, editorial board member, and active reviewer of several reputed journals. Dr. Paul has received several awards, including the *ASOR Rising Star Award* from the Australian Society for Operations Research, the *Excellence in Early Career Research Award* from the UTS Business School, and the *Stephen Fester Prize* for most outstanding thesis from UNSW. Based on his citation records in 2020, 2021, and 2022, he was included in the top 2% of scientists in author databases of standardized citation indicators. His research interests include sustainable supply chain management, supply chain resilience, applied operations research, modeling and simulation, and intelligent decision-making.

Sandeep Kautish, PhD, is an academician by choice and has more than 19 years of full-time experience in teaching and research. He earned his doctorate in computer science on intelligent systems in social networks. He has over 80 publications, and his research works have been published in highly reputed journals, i.e., *IEEE Transaction of Industrial Informatics*, *IEEE Access, Multimedia Tools and Applications*, etc. Dr. Kautish has edited more than 20 books with leading publishers, i.e., Elsevier, Springer, Emerald, and IGI Global, and is an editorial member/reviewer of various reputed journals. His research interests include healthcare analytics, business analytics, machine learning, data mining, and information systems.

Preface

This book covered various computational intelligence techniques that can play a vital role in achieving sustainable supply chain objectives. By leveraging technology, that is, automation and robotics, artificial intelligence, machine learning, data analytics, tracing and mapping technologies, and transportation innovations such as electric vehicles—businesses can achieve transparency, energy efficiency, and waste minimization—just to name a few—across the entire supply chain operations. Computational intelligence techniques, that is, artificial intelligence and machine learning, are growing faster than ever before, permeating the supply chain industry. These technologies bring new automation power, such as facilitating planning activities, demand forecasting, predictive maintenance, synchromodality, and collaborative shipping, using scenario analysis and numerical analytics. As a result, intelligence capabilities can help reduce error rates significantly, decrease operational costs, optimize supply chain flow, and improve sustainable performance.

We focus on presenting the state of the art using computational intelligence in supply chain sustainability issues and logistic problems. In addition, the chapters of this book address different problems in advanced topics in sustainable supply chains, such as sustainable logistics, sustainable procurement, sustainable manufacturing, sustainable inventory and production management, sustainable scheduling, sustainable transportation, and sustainable network design.

This book is an attempt to fill the gap between general textbooks on sustainable supply chain management and more specialized literature dealing with methods for computational intelligence. It is often difficult for readers to proceed from introductory texts on sustainable supply chain management to sophisticated literature dealing with advanced computational intelligence methods. This book fills the gaps by providing the state-of-the-art descriptions of the corresponding problems in sustainable supply chains and computational intelligence techniques for solving them.

Chapter 1 presents a state-of-the-art review of computational tools, techniques, and methods related to sustainable supply chains. It also uncovers the latest technological advancements, exploring how computational techniques shape the future of supply chain sustainability, from predictive modeling to real-time optimization.

Chapter 2 discusses how big data analytics uses computational techniques in construction-related projects and how it can improve project outcomes. It also explores the synergy between big data analytics and computational techniques in construction projects.

Chapter 3 explores the role of Unmanned Aerial Vehicles (UAVs) in sustainable supply chains while using queuing and ant colony optimization approaches.

Chapter 4 highlights the advantages of machine learning in supply chains and investigates new paradigms offered by machine learning in sustainable supply chains. It also discovers the novel paradigms introduced by machine learning, showcasing its potential to drive sustainable practices, optimize resource allocation, and adapt to evolving market demands.

Chapter 5 presents a case study in which scenario analysis is conducted for long-term planning of supply chains. This chapter also reveals decision-making models used in sustainable supply chains. It also provides insights into the decision-making models employed, providing a comprehensive view of how scenario analysis contributes to sustainable supply chain strategies.

Chapter 6 is exemplary of machine learning techniques used in industrial process control. It examines real-world examples illustrating how machine learning enhances efficiency, reliability, and adaptability in industrial processes, setting new standards for control systems.

Chapter 7 presents how computational intelligence techniques can be utilized to develop intelligent and sustainable infrastructure for procurement and distribution. It uncovers the innovative methods that pave the way for smarter procurement and distribution systems, contributing to sustainable practices and optimizing resource utilization.

Chapter 8 discusses how machine learning can be used for route optimization and logistics management. It also explores how machine learning algorithms streamline logistics operations, minimize costs, reduce environmental impact, and maximize overall supply chain logistics efficiency.

Chapter 9 discusses robotics in action for improving sustainable supply chains. It dives into real-world applications showcasing how robotics optimizes supply chain processes, from warehouse automation to last-mile delivery, contributing to a greener and more efficient ecosystem.

Chapter 10 advocates the uses of computational intelligence for green procurement and production. It uncovers how computational intelligence techniques drive eco-friendly procurement decisions and sustainable production practices, aligning businesses with environmental conservation goals.

Chapter 11 explores predictive big data analytics techniques for supply chain demand forecasting. It delves into the applications of predictive analytics in anticipating market demands, minimizing uncertainties, and enhancing the accuracy of demand forecasts, which is crucial for sustainable supply chain planning.

Chapter 12 is a unique chapter that discusses the Bayesian network based on cross bow-tie to analyze the differential effects of internal and external risks on sustainable supply chains. It gains insights into the innovative cross bow-tie methodology, providing a comprehensive understanding of internal and external risks and their differential effects on sustainable supply chain operations.

Chapter 13 presents a case study of BigBasket.com, which explains how the company uses digital technologies to become more customer-centric and optimize supply chain operations. It explores how digital technologies have enabled BigBasket.com to enhance customer experience, optimize operations, and stay competitive in the dynamic landscape of online grocery retail.

Chapter 14 is another case study that explores the applications of artificial intelligence in Echo Global Logistics. It advocates the use of artificial intelligence in supply chain operations optimization. This chapter also discovers how Echo Global Logistics leverages artificial intelligence to maximize profits, improve operational efficiency, and navigate the complexities of modern supply chain management, offering valuable insights for industry practitioners.

Sanjoy Kumar Paul
UTS Business School, University of Technology Sydney,
Sydney, NSW, Australia

Sandeep Kautish
LBEF Campus (APU Malaysia) Kathmandu, Nepal;
Model Institute of Engineering and Technology, Jammu,
Jammu and Kashmir, India

Acknowledgments

We are delighted to extend a warm welcome to the readers of our book, "Computational Intelligence Techniques for Sustainable Supply Chain Management." We sincerely congratulate all authors for their invaluable contributions and patience throughout the rigorous review process. We would like to extend our sincere thanks to all the reviewers who generously dedicated their precious time to ensure the quality of the content. Our gratitude extends to the Elsevier editorial team, particularly Ms. Manisha Rana, for her unwavering support and guidance at every stage—from proposal submission to the review process and online support during submissions. Special thanks are also due to the series editors of the "Uncertainty, Computational Techniques and Decision Intelligence" book series for their valuable insights during the review of the book proposal.

We would also like to thank the UTS Business School, University of Technology Sydney, Sydney, Australia and LBEF Campus, Kathmandu, Nepal, for their support during the reviewing and editing process of the book.

Finally, we express our heartfelt thanks to the Almighty for blessing us with a wonderful life and guiding us through the various challenges and triumphs encountered in the journey of completing this book.

Sanjoy Kumar Paul
Sandeep Kautish

A review of computational tools, techniques, and methods for sustainable supply chains

Towfique Rahman[1,2] and Sanjoy Kumar Paul[1]
[1]UTS Business School, University of Technology Sydney, Sydney, NSW, Australia
[2]Department of Business Strategy and Innovation, Griffith University, Gold Coast, QLD, Australia

1.1 Introduction

Research into managing sustainable supply chains is expanding rapidly as businesses and organizations worldwide explore ways to reduce their environmental impact while improving their sustainability performance (Sonar et al., 2022). One significant area of focus in this discipline is the application of computational tools, methodologies, and strategies to resolve various sustainability-related supply chain challenges.

The use of computational intelligence techniques, such as artificial intelligence (AI), machine learning (ML), and optimization, to solve various sustainability concerns in the supply chain has gained popularity in recent years (Modgil et al., 2021), including advanced topics like sustainable logistics, sustainable purchasing, sustainable manufacturing, sustainable inventory and production management, sustainable scheduling, sustainable transportation, and sustainable network design.

For instance, researchers have applied ML algorithms to schedule business operations as efficiently as possible and reduce energy use and carbon emissions (Liu et al., 2020). Others have studied sustainable transportation networks to reduce the environmental effect of product delivery using optimization algorithms (Moktadir et al., 2019). Still, others have developed decision-support tools that can help business decision-makers use AI approaches to choose more environmentally friendly suppliers (Marshall et al., 2015).

Modeling and simulation are other computational methods widely used in the literature to enhance sustainable practices in managing supply

Computational Intelligence Techniques for Sustainable Supply Chain Management.
DOI: https://doi.org/10.1016/B978-0-443-18464-2.00008-X

chains (Ivanov, 2017). Moreover, future supply chain behavior, and the environmental effects of various strategies and policies, can be forecast by modeling and simulation methods (Tan et al., 2020).

Therefore, the applications of computational tools, techniques, and approaches may, by increasing the sustainability of supply chains, hugely help companies make better decisions more quickly. Given it is difficult to accomplish sustainable development goals through these strategies and approaches alone (Macdonald et al., 2018), it is crucial to integrate teamwork, stakeholder involvement, and good governance to fulfill sustainable development goals while developing supply chain strategies (Longo et al., 2022).

One way to accomplish sustainable supply chains is to apply computational tools, methodologies, and procedures to traditional supply chains (Bui et al., 2021). These computational techniques can help decision-makers identify and address sustainability-related issues in several different industries, including manufacturing, logistics, purchasing, inventory and production management, scheduling, transportation, and supply chain network optimization. It is necessary to understand that these methods and approaches should be combined with other sustainability-related practices to achieve the desired results in terms of fulfilling sustainable development goals (Liu et al., 2020).

Section 1.2 of this chapter explores recent studies on sustainable supply chains. Section 1.3 elaborates on computational intelligence for managing sustainable supply chains. Section 1.4 elaborates further on the different types of computational tools, techniques, and methods for sustainable supply chains. Finally, Section 1.5 brings the chapter to a conclusion and offers several research directions for the future.

1.2 A brief description of sustainable supply chains

Sustainable management of supply chains involves designing, implementing, and improving supply chain operations using sustainability ideas and practices (de Vargas Mores et al., 2018). By addressing the economic, social, and environmental aspects of sustainability, a sustainable supply chain in a business model attempts to provide long-term value for all stakeholders engaged in producing and delivering goods and services (Moktadir et al., 2018).

Environmental, social, and economic factors can be grouped as the three fundamental forces behind the management of sustainable supply chains, as presented in Fig. 1.1. Environmental sustainability seeks to minimize environmental damage and preserve natural resources (Fernández–Miguel et al., 2022). The welfare of employees, the health of local communities, and the defense of human rights are all related to social sustainability (Tseng et al., 2022). Economic sustainability involves generating long-term value through cost reduction, risk management, and innovation for businesses and their stakeholders (Nasir et al., 2022).

Responsible sourcing is one of the essential elements of a sustainable supply chain (Iakovou et al., 2010). It entails implementing supplier assessment, monitoring, and evaluation methods and choosing suppliers based on environmental and social performance (Ang et al., 2017). This can involve using environmentally friendly products and ethical labor practices and complying with environmental laws (Mehrjerdi & Shafiee, 2021). Companies can guarantee they manufacture their products ecologically and with social responsibility by selecting suppliers that share their sustainability principles (Sonar et al., 2022).

Minimizing the environmental impact of logistical operations is one of the key components of managing a sustainable supply chain (Mandal, 2014). Industries can do this by utilizing energy-efficient modes of transportation, such as trains or electric cars, thus reducing their use of fossil fuels (Modgil et al., 2021). Moreover, businesses may attempt to lessen their overall transportation requirements by streamlining their supply chain processes and minimizing the distance that goods must travel (Yavari & Ajalli, 2021). This may be accomplished by utilizing digital technologies like the Internet of Things (IoT), big data analytics, and AI to enhance logistical operations, boost efficiency, and reduce total supply chain costs (Rehman & Ali, 2021).

Another vital part of managing a sustainable supply chain is waste reduction (Vijayan et al., 2014), including implementing recycling and

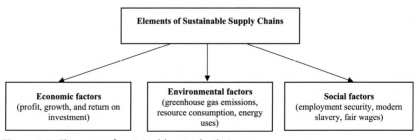

Figure 1.1 Elements of sustainable supply chains.

reuse initiatives and cooperating with suppliers to decrease packaging and related waste. This practice not only decreases the harmful effects on the environment of a company's activities but also reduces total supply chain costs (Mackay et al., 2019). To develop more effective and sustainable production processes, businesses may also try to apply circular economy mechanisms, such as product design for recycling, closed-loop supply chains, and the utilization of renewable resources (Hultberg & Pal, 2021).

Promoting fair labor standards (de Vargas Mores et al., 2018) is also crucial. This might involve ensuring suppliers abide by regional labor regulations and introducing initiatives to advance worker safety and well-being. Companies may guarantee that the items they offer are created ethically by collaborating with suppliers to improve working conditions. To build a more inclusive workplace, businesses should also incorporate sustainable human resource management principles, such as diversity and inclusion, employee engagement, and work-life balance (Karmaker et al., 2021).

Transparency in all aspects of the supply chain plays a significant role in managing a sustainable supply chain (Karmaker et al., 2021). This can include giving specific details about a business's supply chain activities, such as the sources of the items, the production methods, and the actions taken to improve sustainability performance across the supply chain (Papadopoulos et al., 2017). Companies may increase their accountability and responsibility by being transparent with their stakeholders, investors, and consumers about how they run their business (Munny et al., 2019).

Sustainable supply chain management is multidimensional and complex, requiring the integration of sustainability principles and practices into all aspects of supply chain activities. This involves collaborating and engaging all stakeholders, including suppliers, customers, and governments, to achieve shared sustainability goals (Prost et al., 2017). Companies that adopt sustainable supply chain practices can improve their reputation and social license to operate and benefit from cost savings, risk management, and innovation opportunities.

1.3 Computational intelligence for sustainable supply chain management

Computational intelligence includes a collection of AI methods designed to resemble human intellect and provide machines or software with the ability to analyze data, make conclusive decisions, and acquire new skills (Mostert et al., 2021). Examples of computational intelligence methods are AI, ML, data mining, optimization algorithms, and expert systems, often recognized in the literature for developing sustainable supply chains (Dwivedi et al., 2019). These computational methods and strategies can help supply chain decision-makers make timely decisions to manage risks and disruptions within supply chains by modeling complex supply chain systems, analyzing data, making accurate anticipations, etc. (Belhadi et al., 2022).

The capacity of computational intelligence to analyze large amounts of data simultaneously is one of the key reasons this technology can develop sustainable practices for managing supply chains (Dohale et al., 2021). Data in the supply chain are frequently dynamic, complicated, and full of interdependencies, making them challenging to manage manually (Modgil et al., 2021). By automating data gathering and analysis, computational intelligence helps streamline this process and helps identify trends and patterns that can guide risk assessment (Bianco et al., 2023).

Optimizing supply chain processes is one of the advanced tools of computational intelligence. Optimization considers diverse variables, including cost, time, and sustainability, and helps determine the best ways to deliver products from suppliers to customers (Paul et al., 2016). In determining the best routes, modes of transportation, and inventory levels, computational intelligence methods (like genetic algorithms and neural networks) optimize supply chain operations (Razavian et al., 2021).

Moreover, supply chain management choices can benefit from predictive models using computational intelligence (Mohamadi & Yaghoubi, 2017). Predictive modeling uses past data when forecasting future occurrences, such as product demand or anticipated supply chain interruptions (Choi et al., 2021). As well as creating more precise demand projections, computational intelligence may help businesses anticipate and reduce possible supply chain hazards by studying previous data.

Despite the many advantages of computational intelligence, issues still require resolution for supply chain management to be sustainable, such as the need for high-quality data (Um & Han, 2021). Organizations must ensure their data are reliable, thorough, and up-to-date since computational intelligence approaches rely on data to predict and guide decision-making (Taghikhah et al., 2022). Another difficulty is the requirement for qualified employees to handle and analyze the output of computational intelligence. If employees are to use computational intelligence approaches successfully, they must receive training and development (Njomane & Telukdarie, 2022).

Computational intelligence has become an integrated part of managing a sustainable supply chain because it enables businesses to improve their supply chain processes, develop predictive models, and make better decisions. The importance of computational intelligence in sustainable supply chains is enormous and increasing day by day as businesses continue to prioritize sustainability in their operations (Cavalcante et al., 2019). To utilize the capabilities of these methods, companies must address issues related to computational intelligence, such as data quality, for better design (Sahu et al., 2016).

The following sub-sections elaborate on different aspects of sustainable supply chains—namely, reverse logistics, intelligent infrastructure, and green procurement—and the application of computational intelligence in these areas.

1.3.1 Reverse logistics

Reverse logistics means managing the flow of products from their end-use back to their point of origin to recapture their value or dispose of them properly (Dheeraj & Vishal, 2012). This process is becoming increasingly important as environmental concerns, regulatory requirements, and cost-saving opportunities drive companies to optimize their supply chains (Islam et al., 2017). Computational intelligence is a branch of AI that develops algorithms and models to solve complex problems (Um & Han, 2021). The integration of computational intelligence with reverse logistics can potentially improve the efficiency and effectiveness of reverse logistics operations (Ali et al., 2018). Product disposition, routing, and inventory management choices can be improved through computational intelligence techniques, including artificial neural networks, fuzzy logic, evolutionary algorithms, and swarm intelligence (Govindan et al., 2014). These strategies may also be used to anticipate returns, estimate demand, and spot possible waste and inefficiency sources (Sarker et al., 2018). Recent studies have demonstrated that computer intelligence in reverse logistics may significantly reduce costs, minimize harmful environmental effects, and boost customer satisfaction (Islam et al., 2017). However, a thorough analysis of the unique problems and possibilities of various markets, goods, and supply chains is necessary to successfully use computational intelligence in reverse logistics (Sabouhi et al., 2021).

1.3.2 Intelligent infrastructure

A new topic of study—"intelligent sustainable infrastructure for procurement and distribution"—strives to create cutting-edge answers for effective and eco-friendly procurement and distribution systems (Alhalalmeh, 2022). To enhance procurement and distribution operations while reducing their environmental impact, the infrastructure for these activities combines various cutting-edge technologies, such as AI, IoT, and cloud computing (Chopra et al., 2021). The use of AI and ML algorithms to evaluate data on procurement and distribution processes, detect inefficiencies, and suggest ideal alternatives is a crucial component of smart infrastructure (Chowdhury et al., 2021). Predictive models, for instance, may be used to estimate demand and manage inventory levels, cutting down on waste and boosting effectiveness (Paul et al., 2021). Purchasing and distribution support systems incorporate IoT and blockchain to track and monitor products and vehicles in real-time, delivering useful information on their whereabouts, conditions, and usage (Pavlov et al., 2019). The distribution and procurement operations may be made more environmentally friendly by using this information to enhance delivery routes, lower transportation emissions, and decrease waste (Paul et al., 2021). Another benefit of cloud computing is the development of intelligent decision-making systems that can automate different procurement and distribution operations, including order placing, inventory management, and transportation planning (Hultberg & Pal, 2021). Consequently, decisions are quicker and more accurate, reducing costs and increasing productivity. Infrastructure for procurement and distribution is a significant area of study that aims to use the latest technology to develop effective and sustainable systems for procurement and distribution (Pettit et al., 2019). Its advantages include less environmental impact, fewer expenses, and more customer satisfaction.

1.3.3 Green procurement

Computational methods for green manufacturing and procurement aim to enhance and maximize sustainability in supply chains (Dohale et al., 2021). Several methods, including simulation, optimization, and ML, can be deployed to recognize and manage environmental consequences across the supply chain as part of computational strategies for eco-friendly manufacturing and procurement (Grzybowska & Tubis, 2022). Using life-cycle assessment (LCA) to examine how items and processes within supply chains impact the environment is one of the most important parts of computational tools for environmentally responsible manufacturing and purchasing. A detailed analysis of the environmental impact of a product or service can be obtained by the LCA process, which considers its carbon footprint, water use, and other aspects that can harm the environment (Lee et al., 2021). LCA data analysis may be used to find opportunities to increase sustainability using computational methods like optimization and ML. The creation of sustainable buying strategies is a significant computational tool for green purchasing and production (Mostert et al., 2021). This entails assessing suppliers' social and environmental performance in addition to their financial viability (Taghikhah et al., 2020). Models may be developed using computational approaches to examine supplier data and assist sustainable purchasing decisions. Production processes for sustainability may be optimized using computational methods for green manufacturing and procurement (Ivanov & Dolgui, 2021). The environmental effects of various industrial processes may be assessed using simulation models, and opportunities for improvement may be found (Choi, 2021). Moreover, using ML techniques, energy use may be optimized with waste reduction and general efficiency enhancement (Vali-Siar & Roghanian, 2022). In summary, computational methods for green production and procurement constitute a significant area of study that aims to use cutting-edge computational methods to enhance sustainable production and buying practices (Chen et al., 2022). Some of its advantages are less environmental impact, enhanced effectiveness, and improved economic viability.

1.4 Computational tools, techniques, and methods for sustainable supply chains

Managing supply chains sustainably is a vast area of operations management. It requires integrating sustainability practices and ideas into all areas of supply chain operations to fulfill sustainable development goals (Rajesh, 2020). Many computational tools, approaches, and procedures have been developed in the literature to help businesses design, implement, and improve their supply chain operations to integrate sustainable management practices effectively (Vijayan et al., 2014).

Simulation is a popular computational tool for managing sustainable supply chains (Kamalahmadi et al., 2021). The economic and environmental effects of various supply chain methods may be analyzed and evaluated using simulation models (Dolgui & Ivanov, 2021). A business may, for instance, use a simulation model to examine the energy savings that could result from instituting a recycling support system or to compare the carbon footprints of various modes of transportation (Schätter et al., 2019). These technologies may be used to forecast the outcomes of various situations, enhance decision-making processes, and optimize supply chain networks (Zhao et al., 2019).

Optimization is another frequently used computational technique (Heidari-Fathian & Pasandideh, 2018). The optimal solution to a certain issue, such as lowering costs, boosting revenues, or lessening environmental concerns, can be found using optimization models (Zhao et al., 2017). An organization may use an optimization model, for instance, to determine the most economical transportation routes or pinpoint the suppliers that provide the highest environmental performance (Mohammed et al., 2021). Under complicated and dynamic circumstances, these methods can be utilized to identify the most effective and long-lasting solution to a problem (Al-Haidous, Govindan, et al., 2022).

Among computational tools, data analytics is another important part of a sustainable supply chain (Singh & Singh, 2019). Data analytics can be used to collect, examine, and understand large amounts of data to help decision-making processes (Njomane & Telukdarie, 2022). For example, a company could use data analytics to track the environmental performance of its suppliers, identify behavioral patterns in customers, or anticipate future demands (Cavalcante et al., 2019). These methods can help companies identify potential areas for improvement, develop their supply chain operations better, and make more timely decisions to manage all risks (Dubey et al., 2019).

Several other strategies and procedures may be applied in addition to these computational tools to enhance the management of sustainable supply chains (Hsu et al., 2022). For instance, businesses may use social life-cycle assessment to examine the social consequences of their supply chain activities or use LCA to assess the environmental impacts of their goods and services (Mostert et al., 2021). Businesses may also develop and enhance their sustainability practices using environmental management systems or social accountability management systems (Vali-Siar & Roghanian, 2022).

The use of computational methods, tools, and techniques greatly supports the management of sustainable supply chains (Um & Han, 2021). By utilizing these tools, firms may better understand the economic, social, and environmental ramifications of their supply chain operations and make more sustainable decisions (Fernández-Miguel et al., 2022). Businesses may improve the design of their supply chains, make data-driven decisions, and reduce total supply chain costs by applying these computational tools while limiting undesirable environmental and social implications (Aurisano et al., 2021). Fig. 1.2 gives an overview of the types of computational tools, techniques, and methods for sustainable supply chains.

The following sections provide a quick overview of the many computational tools, techniques, and approaches that may be applied to improve the sustainability of supply networks.

1.4.1 Big data, artificial intelligence, and machine learning for sustainable supply chains

The following sub-sections will provide a quick overview of the many computational approaches, such as big data, AI, and ML, that may be applied to improve the sustainability of supply networks.

1.4.1.1 Big data and artificial intelligence for green supply chain and digital logistics

Big data and AI integration in digital logistics and green supply chain management have grown significantly as a field of study and application (Liu et al., 2020). By empowering workers to make informed decisions that improve supply chain operations while reducing waste and emissions, big data and AI may help firms decrease their environmental impact (Zhao et al., 2017).

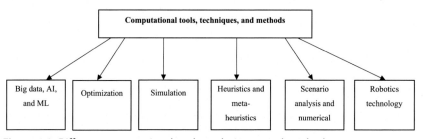

Figure 1.2 Different computational tools, techniques, and methods.

Big data analytics may assist in identifying patterns and trends that can be utilized to improve supply chain performance using the enormous volumes of data created by supply chain activities (Bag, 2016). Using AI, businesses may create predictive models that can aid in identifying possible supply chain interruptions, enabling proactive management and optimization (Papadopoulos et al., 2017).

By improving route design, optimal vehicle routing, and mode selection, AI-enabled logistic systems can help reduce transportation emissions to the environment and improve route efficiency (Singh & Singh, 2019). AI-enabled logistics enhance inventory management and reduce waste by allowing just-in-time inventory management and decreasing excess inventory and related expenses (Hsu et al., 2022). Big data and AI integration in green supply chain management and digital logistics offer excellent opportunities to improve supply chain operations while reducing overall environmental impacts (Kamble et al., 2020).

1.4.1.2 Predictive big data analytics for demand forecasting

With predictive big data analytics, supply chains can operate better, and projections about demand may be made more accurately (Khan et al., 2021). Demand forecasting is crucial to managing supply chains since it involves predicting future demand for a good or service (Sazvar et al., 2021). Large amounts of data, statistical algorithms, and ML methods are used in predictive big data analytics to examine historical demand patterns and forecast future needs (Choi, 2021).

Demand forecasting using predictive big data analytics has several advantages, including the capacity to estimate demand more accurately, identify current trends, and enhance inventory management. By utilizing predictive analytics, organizations can understand patterns and trends in the behavior of customers, market conditions, and other external factors that affect demand (Lohmer et al., 2020). This helps businesses estimate demand more accurately, thus increasing customer satisfaction, improving inventory management, and reducing total supply chain costs.

Moreover, predictive big data analytics can help businesses develop sustainable practices in their supply chains. Organizations can identify new customer trends and preferences and develop new products and services to fulfill those needs by analyzing data from various sources, including social media, web browsing habits, and purchasing history (Katsikopoulos et al., 2022).

Predictive big data analytics for demand forecasting may increase customer satisfaction, reduce total supply chain costs, and make supply chain operations more efficient. By identifying emerging consumer and industry trends, businesses may make strategic decisions that will increase their growth and fulfill sustainable development goals (Ivanov, 2021a).

1.4.1.3 Artificial intelligence-enabled solutions for plant locations and equipment efficiency

AI-enabled solutions are increasingly in demand for improving plant locations and equipment efficiency (Modgil et al., 2021). By utilizing ML algorithms and AI techniques, these systems assess large amounts of data to determine the optimal locations for plants and efficient equipment layouts (Dwivedi et al., 2019).

To determine the optimal sites for plants, ML algorithms can be used to analyze data on transportation costs, labor costs, workforce availability, resource availability, and similar aspects (Modgil et al., 2021). AI-enabled solutions can also improve equipment productivity by figuring out how to spend less energy while producing more output (Belhadi et al., 2022). To do this, AI-enabled solutions analyze performance data and the utilization of machines and other equipment (Rajesh, 2020).

These technologies can help companies reduce total supply chain costs, improve efficiency, and increase productivity by optimizing plant locations and equipment arrangements within supply chains (Choi et al., 2021). Moreover, AI-powered systems can provide analytical data on the relationships between various aspects, allowing organizations to make decisions based on real-time data (Rehman & Ali, 2021).

It is important to analyze the data, the algorithms, and the trade-offs between various optimization targets before using AI-enabled solutions for plant locations and equipment efficiency (Ivanov, 2021c). However, employing these solutions can considerably increase operational effectiveness and reduce total supply chain costs, eventually giving organizations a competitive edge in today's challenging and increasingly complicated business climate (Longo et al., 2022).

1.4.1.4 Machine learning for route optimization and logistics management

Route optimization and logistics management benefit greatly from ML, which effectively solves the challenges of both fields (Modgil et al., 2021). Massive volumes of data on routes, traffic patterns, and other important factors may be analyzed by ML algorithms to improve route planning, shorten delivery times, and lower transportation costs (Peng et al., 2021). Reinforcement learning—a form of ML system that can learn from experience and adapt to changing contexts—is commonly used in route optimization (Pound & Campbell, 2015). It trains agents to make route-planning decisions that depend on the traffic circumstances and delivery needs currently in effect (Lohmer et al., 2020). Other ML methods, such as neural networks and decision trees, can be easily used to forecast demand and achieve optimal route scheduling (Dwivedi et al., 2019). Moreover, inventory optimization, transportation routing, and demand forecasting can be done using ML-based methods (Handfield et al., 2020) by examining data from historical sales information, customer preferences, and other important sources. Moreover, the use of ML for logistics management and route optimization has the potential to increase operational effectiveness, reduce total supply chain costs, and increase customer satisfaction (Ivanov, 2020). However, for ML algorithms to be successful, the data sources, suitable algorithms, and essential metrics must be analyzed to assess the model's validity (Bastas & Garza-Reyes, 2022).

1.4.2 Optimization techniques for sustainable supply chain operations

Sustainability has become a critical issue for businesses, governments, and society at large. Integrating sustainability practices into supply chain operations has become necessary as companies seek ways to minimize their environmental impact, reduce waste, and improve social conditions along the supply chain network (Moktadir et al., 2021). Optimization techniques can play a vital role in achieving these objectives by improving the efficiency and effectiveness of supply chain operations (Al-Haidous, Govindan, et al., 2022). Mathematical programming techniques, such as linear and nonlinear programming, can be used to formulate supply chain decision problems as optimization problems (Wu et al., 2020). This allows decision-makers to identify the best possible solution given a set of constraints and objectives (Dolgui & Ivanov, 2021).

Simulation models can also be used to represent supply chain dynamics and test different scenarios, allowing decision-makers to evaluate the impact of different decisions on sustainability outcomes (Ivanov, 2017). Heuristic algorithms, such as genetic algorithms and ant colony optimization, use rule-based approaches to find near-optimal solutions to complex supply chain problems (Aouad et al., 2019). These algorithms can handle large and complex datasets, making them suitable for supply chain optimization (Wang & Webster, 2021). The use of these optimization techniques can help reduce greenhouse gas emissions, minimize waste, and improve social conditions along the supply chain, leading to a more sustainable future (Katsikopoulos et al., 2022).

1.4.3 Simulation-based design for sustainable supply chains

Simulation-based designs of sustainable supply chains try to include sustainability issues in the design and operation of supply chains (Zalitis et al., 2022). This method uses simulation models to assess how various design decisions and operational methods would affect sustainability goals, including lowering carbon emissions, cutting waste, and enhancing social conditions along the supply chain (Belhadi et al., 2021). Using simulation models, decision-makers may test various hypotheses and assess how various actions would affect sustainability results (Werner et al., 2021). These models may depict the supply chain dynamics and capture interactions among the many supply chain participants, including suppliers, manufacturers, distributors, and customers (Ghufran et al., 2022). Decision-makers may optimize design decisions and operational strategies to accomplish sustainability goals by modeling supply chain activities (Gholami-Zanjani et al., 2021). The definition of sustainability goals, modeling of the supply chain, selection and validation of the simulation model, and analysis of the simulation findings are some of the phases in the simulation-based design of sustainable supply chains (Ivanov, 2021b). This method may be used throughout the supply chain to find opportunities to reduce environmental impact, enhance social circumstances, and boost economic performance (Golan et al., 2020). In the end, simulation-based design can result in more robust and sustainable supply chains that can handle the demands of a constantly shifting business environment.

1.4.4 Heuristics and meta-heuristics for supply chain planning

Heuristics and meta-heuristics are two established mathematical optimization methods widely used in the literature for supply chain planning (Wang & Webster, 2021). These methods aim to find optimal solutions to the complex problems often found in the supply chain planning of businesses (Peng et al., 2021). Heuristics are rule-based methods using problem-specific knowledge to guide the search for solutions.

Meta-heuristics are general-purpose algorithms for solving different problems within supply chains (Ade Irawan et al., 2022).

Heuristics can be used to find better solutions quickly, which is essential for supply chain planning, where decisions must be made quickly (Ivanov & Rozhkov, 2020). Examples of heuristics that are commonly used in supply chain planning are the nearest-neighbor, insertion, and sweep algorithms (Al-Haidous, Govindan, et al., 2022). These heuristics are used to solve problems such as facility location, inventory routing, and vehicle routing (Paul et al., 2016). Meta-heuristics are more flexible than heuristics and can be used to solve a wider range of problems (Dohale et al., 2021). Meta-heuristics are based on iterative procedures that search for solutions by modifying and improving a set of candidate solutions (Aldrighetti et al., 2021). Examples of meta-heuristics commonly used in supply chain planning are genetic algorithms, simulated annealing, and tabu search (Katsikopoulos et al., 2022).

When used in supply chain planning, heuristics and meta-heuristics can significantly increase operational effectiveness and reduce total supply chain costs (Rahman et al., 2022). These methods can significantly improve delivery times, reduce transportation costs, and optimize inventory levels (Wang & Yao, 2021). In using heuristics and meta-heuristics, one must carefully consider the trade-offs between the quality of the solution and the amount of time needed to obtain it (Cheramin et al., 2021). However, heuristics and meta-heuristics can be effective supply chain planning tools, especially when combined with other optimization approaches (Ade Irawan et al., 2022).

1.4.5 Scenario analysis and numerical analytics in sustainable supply chain management

Scenario analysis and numerical analytics are essential tools used in supply chain management to manage risk and make better-informed decisions (Khan et al., 2021). Scenario analysis identifies potential events or circumstances that may affect a supply chain, evaluates their potential impact and likelihood, and develops appropriate response strategies (Zalitis et al., 2022). Numerical analytics involves using statistical models and data analysis techniques to generate insights and make predictions (Chen et al., 2022).

Scenario analysis and numerical analytics can be used in supply chain management to improve inventory control, demand forecasting, and production scheduling, among other areas of the supply chain (Shih & Lin, 2022).

The impact of demand shifts or supply chain disruptions, such as natural catastrophes or transportation strikes, can be modeled using scenario analysis (Al-Haidous, Al-Breiki, et al., 2022). Using numerical analytics can help supply chains run more cost-effectively by forecasting demand, spotting trends, and optimizing inventory levels (Bianco et al., 2023).

Overall, scenario analysis and numerical analytics are crucial tools in supply chain management that enable organizations to make data-driven decisions, anticipate risks, and respond proactively to market or supply chain environment changes. Organizations can leverage these tools to improve their supply chain performance, reduce costs, and enhance customer satisfaction (Al-Haidous, Al-Breiki, et al., 2022).

1.4.6 Robotics technologies for sustainable supply chain operations

Robotics technologies are increasingly developed in supply chain operations to boost productivity, reduce total supply chain costs, and promote environmentally friendly procedures (Khan et al., 2021). Robotics can help in all three of these areas. Sustainable supply chain management entails integrating environmental, social, and economic factors into supply chain operations (Macdonald et al., 2018). Robotics technology can contribute to environmental sustainability by reducing waste and carbon emissions through improved energy efficiency, reduced packaging material, and optimized transportation routes (Dolgui & Ivanov, 2020). For instance, last-mile deliveries can be handled by autonomous cars and drones, eliminating the need for diesel-powered trucks and lowering emissions associated with transportation (Khan et al., 2021). Robotics can help reduce labor expenses and enhance worker safety regarding social sustainability. Robots can complete risky or repetitive duties, freeing up human workers to concentrate on more specialized, satisfying, and decision-making employment. Additionally, since they can operate continuously, robots increase production effectiveness and lessen the need for overtime from human employees (Modgil et al., 2021). Robotics technology offers supply chain effectiveness, reduced costs, and greater competitiveness, all of which support the long-term economic viability of businesses. It automates warehouse operations, enhances inventory control, and reduces transportation expenses (Choi et al., 2021). Generally speaking, robotics technology can alter supply chain operations and encourage sustainable behaviors. By utilizing these technologies, companies can operate more efficiently on all fronts—environmental, social, and economic sustainability—creating a supply chain that is viable and more resilient.

1.5 Conclusions and research scopes

Given the disruptions, risks, and issues identified in supply chain networks, there has been a growing concern in recent years about their sustainability. Scholars and practitioners are working toward developing sustainable supply chains that balance economic, social, and environmental performance. The utilization of computational intelligence tools, such as ML, AI, and robotics, has been identified as a potential strategy for achieving sustainability goals in supply chain management. While these technologies have significantly advanced supply chains, further research is necessary.

One primary area for future research is the development of innovative computational models tailored to the intricate and dynamic nature of managing supply chains. To facilitate more efficient and sustainable supply chain designs, researchers can focus on creating advanced optimization algorithms that capture the complexity of real-world systems. Additionally, more sophisticated AI and ML models can be developed to adapt to changing supply chain conditions.

A vital research area is the development of integrated computational methods for sustainable supply chains. To develop optimization methods that can solve various supply chain problems, more research is needed to learn how ML models can detect patterns in data from supply chains. Researchers can also work on developing models that consider data from other sources, such as social media, weather reports, and news feeds, to gain more accurate and timely information about supply chain conditions. To meet customer expectations and for higher profitability, the decision-makers of supply chains should emphasize customer preferences or integrate "outside-in" supply networks.

When applying computational intelligence tools to supply chain management, governance warrants further attention. While these technologies offer numerous benefits, they also raise ethical and societal concerns. Researchers could explore the potential impacts of AI and ML models on employment and human rights within supply chain management decision-making. Additionally, studies could focus on developing frameworks for ethically and sustainably applying these technologies to supply chain management. For instance, social sustainability during large-scale disruptions, such as the COVID-19 pandemic, is critical for future research.

Finally, more empirical research is needed to evaluate the effectiveness of computational intelligence tools in achieving supply chain management sustainability objectives. Researchers could conduct case studies of organizations that have implemented these technologies to identify best practices and key success factors. Additionally, their research could focus on developing metrics and indicators to assess how these technologies influence sustainability performance in areas such as carbon emissions, waste reduction, and social responsibility.

Therefore, the application of computational intelligence methods, including robotics, AI, and ML, has extreme capabilities to increase the sustainability of supply chains in various aspects. However, substantial research must be conducted for these technologies to reach their full potential. Future research should focus on developing novel computational models, integrating various computational methods, addressing governance challenges, and conducting empirical research to evaluate the effectiveness of these technologies and strategies in solving supply chain problems and achieving sustainable development goals. Ultimately, these contributions will help develop more resilient and robust supply chains, better equipped to withstand difficulties and disruptions in the future.

References

Ade Irawan, C., Dan-Asabe Abdulrahman, M., Salhi, S., & Luis, M. (2022). An efficient matheuristic algorithm for bi-objective sustainable closed-loop supply chain networks. *IMA Journal of Management Mathematics*, *33*(4), 603−636. Available from https://doi.org/10.1093/imaman/dpac003.

Aldrighetti, R., Battini, D., Ivanov, D., & Zennaro, I. (2021). Costs of resilience and disruptions in supply chain network design models: A review and future research directions. *International Journal of Production Economics*, *235*, 108103. Available from https://doi.org/10.1016/j.ijpe.2021.108103.

Al-Haidous, S., Al-Breiki, M., Bicer, Y., & Al-Ansari, T. (2022). Evaluating lng supply chain resilience using swot analysis: The case of Qatar. *Energies*, *15*(1). Available from https://doi.org/10.3390/en15010079.

Al-Haidous, S., Govindan, R., Elomri, A., & Al-Ansari, T. (2022). An optimization approach to increasing sustainability and enhancing resilience against environmental constraints in LNG supply chains: A Qatar case study. *Energy Reports*, *8*, 9742−9756. Available from https://doi.org/10.1016/j.egyr.2022.07.120.

Alhalalmeh, M. I. (2022). The impact of supply chain 4.0 technologies on its strategic outcomes. *Uncertain Supply Chain Management*, *10*(4), 1203−1210. Available from https://doi.org/10.5267/j.uscm.2022.8.008.

Ali, S. M., Arafin, A., Moktadir, M. A., Rahman, T., & Zahan, N. (2018). Barriers to reverse logistics in the computer supply chain using interpretive structural model. *Global Journal of Flexible Systems Management*, *19*, 53−68. Available from https://doi.org/10.1007/s40171-017-0176-2.

Ang, E., Iancu, D. A., & Swinney, R. (2017). Disruption risk and optimal sourcing in multitier supply networks. *Management Science*, *63*(8), 2397−2771. Available from https://doi.org/10.1287/mnsc.2016.2471.

Aouad, A., Levi, R., & Segev, D. (2019). Approximation algorithms for dynamic assortment optimization models. *Mathematics of Operations Research*, *44*(2), 377−766. Available from https://doi.org/10.1287/moor.2018.0933.

Aurisano, N., Weber, R., & Fantke, P. (2021). Enabling a circular economy for chemicals in plastics. *Current Opinion in Green and Sustainable Chemistry*, *31*, 100513. Available from https://doi.org/10.1016/j.cogsc.2021.100513.

Bag, S. (2016). Fuzzy VIKOR approach for selection of big data analyst in procurement management. *Journal of Transport and Supply Chain Management*, *1*. Available from https://doi.org/10.4102/jtscm.v10i1.230.

Bastas, A., & Garza-Reyes, J. A. (2022). Impact of the COVID-19 pandemic on manufacturing operations and supply chain resilience: Effects and response strategies. *Journal of Manufacturing Technology Management*, *33*(5), 962−985. Available from https://doi.org/10.1108/JMTM-09-2021-0357.

Belhadi, A., Kamble, S., Fosso Wamba, S., & Queiroz, M. M. (2022). Building supply-chain resilience: An artificial intelligence-based technique and decision-making framework. *International Journal of Production Research*, *60*(14), 4487−4507. Available from https://doi.org/10.1080/00207543.2021.1950935.

Belhadi, A., Kamble, S., Jabbour, C. J. C., Gunasekaran, A., Ndubisi, N. O., & Venkatesh, M. (2021). Manufacturing and service supply chain resilience to the COVID-19 outbreak: Lessons learned from the automobile and airline industries. *Technological Forecasting and Social Change*, *163*, 120447. Available from https://doi.org/10.1016/j.techfore.2020.120447.

Bianco, D., Bueno, A., Godinho Filho, M., Latan, H., Miller Devós Ganga, G., Frank, A. G., & Chiappetta Jabbour, C. J. (2023). The role of Industry 4.0 in developing resilience for manufacturing companies during COVID-19. *International Journal of Production Economics*, *256*, 108728. Available from https://doi.org/10.1016/j.ijpe.2022.108728.

Bui, T. D., Tsai, F. M., Tseng, M. L., Tan, R. R., Yu, K. D. S., & Lim, M. K. (2021). Sustainable supply chain management towards disruption and organizational ambidexterity: A data driven analysis. *Sustainable Production and Consumption*, *26*, 373−410. Available from https://doi.org/10.1016/j.spc.2020.09.017.

Cavalcante, I. M., Frazzon, E. M., Forcellini, F. A., & Ivanov, D. (2019). A supervised machine learning approach to data-driven simulation of resilient supplier selection in digital manufacturing. *International Journal of Information Management*, *49*, 86−97. Available from https://doi.org/10.1016/j.ijinfomgt.2019.03.004.

Chen, H., Hsu, C. W., Shih, Y. Y., & Caskey, D. (2022). The reshoring decision under uncertainty in the post-COVID-19 era. *Journal of Business and Industrial Marketing*, *10*, 2064−2074. Available from https://doi.org/10.1108/JBIM-01-2021-0066.

Cheramin, M., Saha, A. K., Cheng, J., Paul, S. K., & Jin, H. (2021). Resilient NdFeB magnet recycling under the impacts of COVID-19 pandemic: Stochastic programming and Benders decomposition. *Transportation Research Part E: Logistics and Transportation Review*, *155*, 102505. Available from https://doi.org/10.1016/j.tre.2021.102505.

Choi, T. M. (2021). Risk analysis in logistics systems: A research agenda during and after the COVID-19 pandemic. *Transportation Research Part E: Logistics and Transportation Review*, *145*, 102190. Available from https://doi.org/10.1016/j.tre.2020.102190.

Choi, T., Kumar, S., Yue, X., & Chan, H. (2021). Disruptive technologies and operations management in the industry 4.0 era and beyond. *Production and Operations Management*, *31*(1), 9−31. Available from https://doi.org/10.1111/poms.13622.

Chopra, S., Sodhi, M. M., & Lücker, F. (2021). Achieving supply chain efficiency and resilience by using multi-level commons. *Decision Sciences*, *52*(4), 817−832. Available from https://doi.org/10.1111/deci.12526.

Chowdhury, P., Paul, S. K., Kaisar, S., & Moktadir, M. A. (2021). COVID-19 pandemic related supply chain studies: A systematic review. *Transportation Research Part E: Logistics and Transportation Review*, *148*, 102271. Available from https://doi.org/10.1016/j.tre.2021.102271.

de Vargas Mores, G., Finocchio, C. P. S., Barichello, R., & Pedrozo, E. A. (2018). Sustainability and innovation in the Brazilian supply chain of green plastic. *Journal of Cleaner Production*, *177*, 12−18. Available from https://doi.org/10.1016/j.jclepro.2017.12.138.

Dheeraj, N., & Vishal, N. (2012). An overview of green supply chain management in India. *Journal of Recent Sciences*, *1*(6), 77−82.

Dohale, V., Verma, P., Gunasekaran, A., & Ambilkar, P. (2021). COVID-19 and supply chain risk mitigation: A case study from India. *International Journal of Logistics Management*, *34*(2). Available from https://doi.org/10.1108/IJLM-04-2021-0197.

Dolgui, A., & Ivanov, D. (2020). Exploring supply chain structural dynamics: New disruptive technologies and disruption risks. *International Journal of Production Economics*, *229*, 107886. Available from https://doi.org/10.1016/j.ijpe.2020.107886.

Dolgui, A., & Ivanov, D. (2021). Ripple effect and supply chain disruption management: New trends and research directions. *International Journal of Production Research*, *59*(1), 102−109. Available from https://doi.org/10.1080/00207543.2021.1840148.

Dubey, R., Gunasekaran, A., Childe, S. J., Fosso Wamba, S., Roubaud, D., & Foropon, C. (2019). Empirical investigation of data analytics capability and organizational flexibility as complements to supply chain resilience. *International Journal of Production Research*, *59*(1), 110−128. Available from https://doi.org/10.1080/00207543.2019.1582820.

Dwivedi, Y. K., Hughes, L., Ismagilova, E., Aarts, G., Coombs, C., Crick, T., Duan, Y., Dwivedi, R., Edwards, J., Eirug, A., Galanos, V., Ilavarasan, P. V., Janssen, M., Jones, P., Kar, A. K., Kizgin, H., Kronemann, B., Lal, B., Lucini, B., ... Williams, M. D. (2019). Artificial Intelligence (AI): Multidisciplinary perspectives on emerging challenges, opportunities, and agenda for research, practice and policy. *International Journal of Information Management*, *57*, 101994. Available from https://doi.org/10.1016/j.ijinfomgt.2019.08.002.

Fernández-Miguel, A., Riccardi, M. P., Veglio, V., García-Muiña, F. E., Fernández del Hoyo, A. P., & Settembre-Blundo, D. (2022). Disruption in resource-intensive supply chains: Reshoring and nearshoring as strategies to enable them to become more resilient and sustainable. *Sustainability (Switzerland)*, *14*(17). Available from https://doi.org/10.3390/su141710909.

Gholami-Zanjani, S. M., Jabalameli, M. S., Klibi, W., & Pishvaee, M. S. (2021). A robust location-inventory model for food supply chains operating under disruptions with ripple effects. *International Journal of Production Research*, *59*(1), 301−324. Available from https://doi.org/10.1080/00207543.2020.1834159.

Ghufran, M., Khan, K. I. A., Ullah, F., Alaloul, W. S., & Musarat, M. A. (2022). Key enablers of resilient and sustainable construction supply chains: A systems thinking approach. *Sustainability*, *14*(19), 11815. Available from https://doi.org/10.3390/su141911815.

Golan, M. S., Jernegan, L. H., & Linkov, I. (2020). Trends and applications of resilience analytics in supply chain modeling: Systematic literature review in the context of the COVID-19 pandemic. *Environment Systems and Decisions*, *40*(2), 222−243. Available from https://doi.org/10.1007/s10669-020-09777-w.

Govindan, K., Azevedo, S. G., Carvalho, H., & Cruz-Machado, V. (2014). Impact of supply chain management practices on sustainability. *Journal of Cleaner Production*, *85*, 212−225. Available from https://doi.org/10.1016/j.jclepro.2014.05.068.

Grzybowska, K., & Tubis, A. A. (2022). Supply chain resilience in reality VUCA—An international delphi study. *Sustainability*, *14*(17). Available from https://doi.org/ 10.3390/su141710711.

Handfield, R., Sun, H., & Rothenberg, L. (2020). Assessing supply chain risk for apparel production in low cost countries using newsfeed analysis. *Supply Chain Management: An International Journal*, *25*(6), 803−821. Available from https://doi.org/10.1108/ SCM-11-2019-0423.

Heidari-Fathian, H., & Pasandideh, S. H. R. (2018). Green-blood supply chain network design: Robust optimization, bounded objective function & Lagrangian relaxation. *Computers and Industrial Engineering*, *122*, 95−105. Available from https://doi.org/ 10.1016/j.cie.2018.05.051.

Hsu, C. H., Li, M. G., Zhang, T. Y., Chang, A. Y., Shangguan, S. Z., & Liu, W. L. (2022). Deploying big data enablers to strengthen supply chain resilience to mitigate sustainable risks based on integrated HOQ-MCDM framework. *Mathematics*, *10*(8), 1−35. Available from https://doi.org/10.3390/math10081233.

Hultberg, E., & Pal, R. (2021). Lessons on business model scalability for circular economy in the fashion retail value chain: Towards a conceptual model. *Sustainable Production and Consumption*, *28*, 686−698. Available from https://doi.org/10.1016/j.spc.2021.06.033.

Iakovou, E., Vlachos, D., & Xanthopoulos, A. (2010). A stochastic inventory management model for a dual sourcing supply chain with disruptions. *International Journal of Systems Science*, *41*(3), 315−324. Available from https://doi.org/10.1080/00207720903326894.

Islam, S., Karia, N., Fauzi, F. B. A., & Soliman, M. S. M. (2017). A review on green supply chain aspects and practices. *Management and Marketing*, *12*(1), 12−36. Available from https://doi.org/10.1515/mmcks-2017-0002.

Ivanov, D. (2017). Simulation-based ripple effect modeling in the supply chain. *International Journal of Production Research*, *55*(7), 2083−2101. Available from https:// doi.org/10.1080/00207543.2016.1275873.

Ivanov, D. (2020). Predicting the impacts of epidemic outbreaks on global supply chains: A simulation-based analysis on the coronavirus outbreak (COVID-19/SARS-CoV-2) case. *Transportation Research Part E: Logistics and Transportation Review*, *136*, 101922. Available from https://doi.org/10.1016/j.tre.2020.101922.

Ivanov, D. (2021a). Exiting the COVID-19 pandemic: After-shock risks and avoidance of disruption tails in supply chains. *Annals of Operations Research*. Available from https:// doi.org/10.1007/s10479-021-04047-7.

Ivanov, D. (2021b). Lean resilience: AURA (Active Usage of Resilience Assets) framework for post-COVID-19 supply chain management. *International Journal of Logistics Management*. Available from https://doi.org/10.1108/IJLM-11-2020-0448.

Ivanov, D. (2021c). Supply chain viability and the COVID-19 pandemic: A conceptual and formal generalisation of four major adaptation strategies. *International Journal of Production Research*, *59*(12), 3535−3552. Available from https://doi.org/10.1080/ 00207543.2021.1890852.

Ivanov, D., & Dolgui, A. (2021). OR-methods for coping with the ripple effect in supply chains during COVID-19 pandemic: Managerial insights and research implications. *International Journal of Production Economics*, *232*, 107921. Available from https://doi. org/10.1016/j.ijpe.2020.107921.

Ivanov, D., & Rozhkov, M. (2020). Coordination of production and ordering policies under capacity disruption and product write-off risk: An analytical study with real-data based simulations of a fast moving consumer goods company. *Annals of Operations Research*, *291*, 384−407. Available from https://doi.org/10.1007/s10479-017-2643-8.

Kamalahmadi, M., Shekarian, M., & Mellat Parast, M. (2021). The impact of flexibility and redundancy on improving supply chain resilience to disruptions. *International Journal of Production Research*, *60*(6), 1992−2020. Available from https://doi.org/10.1080/00207543.2021.1883759.

Kamble, S. S., Gunasekaran, A., & Gawankar, S. A. (2020). Achieving sustainable performance in a data-driven agriculture supply chain: A review for research and applications. *International Journal of Production Economics*, *219*, 179−194. Available from https://doi.org/10.1016/j.ijpe.2019.05.022.

Karmaker, C. L., Ahmed, T., Ahmed, S., Ali, S. M., Moktadir, M. A., & Kabir, G. (2021). Improving supply chain sustainability in the context of COVID-19 pandemic in an emerging economy: Exploring drivers using an integrated model. *Sustainable Production and Consumption*, *26*, 411−427. Available from https://doi.org/10.1016/j.spc.2020.09.019.

Katsikopoulos, K. V., Egozcue, M., & Garcia, L. F. (2022). A simple model for mixing intuition and analysis. *European Journal of Operational Research*, *303*(2), 779−789. Available from https://doi.org/10.1016/j.ejor.2022.03.005.

Khan, S. A. R., Ponce, P., Tanveer, M., Aguirre-Padilla, N., Mahmood, H., & Shah, S. A. A. (2021). Technological innovation and circular economy practices: Business strategies to mitigate the effects of COVID-19. *Sustainability (Switzerland)*, *13*(15), 1−17. Available from https://doi.org/10.3390/su13158479.

Lee, A. W. L., Neo, E. R. K., Khoo, Z. Y., Yeo, Z., Tan, Y. S., Chng, S., Yan, W., Lok, B. K., & Low, J. S. C. (2021). Life cycle assessment of single-use surgical and embedded filtration layer (EFL) reusable face mask. *Resources, Conservation and Recycling*, *170*, 105580. Available from https://doi.org/10.1016/j.resconrec.2021.105580.

Liu, J., Chen, M., & Liu, H. (2020). The role of big data analytics in enabling green supply chain management: A literature review. *Journal of Data, Information and Management*, *2*, 75−83. Available from https://doi.org/10.1007/s42488-019-00020-z.

Lohmer, J., Bugert, N., & Lasch, R. (2020). Analysis of resilience strategies and ripple effect in blockchain-coordinated supply chains: An agent-based simulation study. *International Journal of Production Economics*, *228*, 107882. Available from https://doi.org/10.1016/j.ijpe.2020.107882.

Longo, F., Mirabelli, G., Solina, V., Alberto, U., De Paola, G., Giordano, L., & Ziparo, M. (2022). A simulation-based framework for manufacturing design and resilience assessment: A case study in the wood sector. *Applied Sciences*, *12*(15), 7614. Available from https://doi.org/10.3390/app12157614.

Macdonald, J. R., Zobel, C. W., Melnyk, S. A., & Griffis, S. E. (2018). Supply chain risk and resilience: Theory building through structured experiments and simulation. *International Journal of Production Research*, *56*(12), 4337−4355. Available from https://doi.org/10.1080/00207543.2017.1421787.

Mackay, J., Munoz, A., & Pepper, M. (2019). Conceptualising redundancy and flexibility towards supply chain robustness and resilience. *Journal of Risk Research*, *23*(12), 1541−1561. Available from https://doi.org/10.1080/13669877.2019.1694964.

Mandal, S. (2014). Supply chain resilience: A state-of-the-art review and research directions. *International Journal of Disaster Resilience in the Built Environment*, *5*(4), 427−453. Available from https://doi.org/10.1108/IJDRBE-03-2013-0003.

Marshall, D., McCarthy, L., Heavey, C., & McGrath, P. (2015). Environmental and social supply chain management sustainability practices: Construct development and measurement. *Production Planning and Control*, *26*(8), 673–690. Available from https://doi.org/10.1080/09537287.2014.963726.

Mehrjerdi, Y. Z., & Shafiee, M. (2021). A resilient and sustainable closed-loop supply chain using multiple sourcing and information sharing strategies. *Journal of Cleaner Production*, *289*, 125141. Available from https://doi.org/10.1016/j.jclepro.2020.125141.

Modgil, S., Singh, R. K., & Hannibal, C. (2021). Artificial intelligence for supply chain resilience: Learning from Covid-19. *International Journal of Logistics Management*, *33*(4), 1246–1268. Available from https://doi.org/10.1108/IJLM-02-2021-0094.

Mohamadi, A., & Yaghoubi, S. (2017). A bi-objective stochastic model for emergency medical services network design with backup services for disasters under disruptions: An earthquake case study. *International Journal of Disaster Risk Reduction*, *23*, 204–217. Available from https://doi.org/10.1016/j.ijdrr.2017.05.003.

Mohammed, A., Naghshineh, B., Spiegler, V., & Carvalho, H. (2021). Conceptualising a supply and demand resilience methodology: A hybrid DEMATEL-TOPSIS-possibilistic multi-objective optimization approach. *Computers and Industrial Engineering*, *160*, 107589. Available from https://doi.org/10.1016/j.cie.2021.107589.

Moktadir, M. A., Ali, S. M., Jabbour, C. J. C., Paul, A., Ahmed, S., Sultana, R., & Rahman, T. (2019). Key factors for energy-efficient supply chains: Implications for energy policy in emerging economies. *Energy*, *189*, 116129. Available from https://doi.org/10.1016/j.energy.2019.116129.

Moktadir, M. A., Ali, S. M., Rajesh, R., & Paul, S. K. (2018). Modeling the interrelationships among barriers to sustainable supply chain management in leather industry. *Journal of Cleaner Production*, *181*, 631–651. Available from https://doi.org/10.1016/j.jclepro.2018.01.245.

Moktadir, M. A., Mahmud, Y., Banaitis, A., Sarder, T., & Khan, M. R. (2021). Key performance indicators for adopting sustainability practices in footwear supply chains. *E a M: Ekonomie a Management*, *24*(1), 197–213. Available from https://doi.org/10.15240/TUL/001/2021-1-013.

Mostert, C., Sameer, H., Glanz, D., & Bringezu, S. (2021). Climate and resource footprint assessment and visualization of recycled concrete for circular economy. *Resources, Conservation and Recycling*, *174*, 105767. Available from https://doi.org/10.1016/j.resconrec.2021.105767.

Munny, A. A., Ali, S. M., Kabir, G., Moktadir, M. A., Rahman, T., & Mahtab, Z. (2019). Enablers of social sustainability in the supply chain: An example of footwear industry from an emerging economy. *Sustainable Production and Consumption*, *20*, 230–242. Available from https://doi.org/10.1016/j.spc.2019.07.003.

Nasir, S. B., Ahmed, T., Karmaker, C. L., Ali, S. M., Paul, S. K., & Majumdar, A. (2022). Supply chain viability in the context of COVID-19 pandemic in small and medium-sized enterprises: Implications for sustainable development goals. *Journal of Enterprise Information Management*, *35*(1), 100–124. Available from https://doi.org/10.1108/JEIM-02-2021-0091.

Njomane, L., & Telukdarie, A. (2022). Impact of COVID-19 food supply chain: Comparing the use of IoT in three South African supermarkets. *Technology in Society*, *71*, 102051. Available from https://doi.org/10.1016/j.techsoc.2022.102051.

Papadopoulos, T., Gunasekaran, A., Dubey, R., Altay, N., Childe, S. J., & Fosso-Wamba, S. (2017). The role of Big Data in explaining disaster resilience in supply chains for sustainability. *Journal of Cleaner Production*, *142*, 1108–1118. Available from https://doi.org/10.1016/j.jclepro.2016.03.059.

Paul, S. K., Chowdhury, P., Moktadir, A., & Lau, K. H. (2021). Supply chain recovery challenges in the wake of COVID-19 pandemic. *Journal of Business Research*, *136*, 316–329. Available from https://doi.org/10.1016/j.jbusres.2021.07.056.

Paul, S. K., Sarker, R., & Essam, D. (2016). Managing risk and disruption in production-inventory and supply chain systems: A review. *Journal of Industrial and Management Optimization*, *12*(3), 1009−1029. Available from https://doi.org/10.3934/jimo.2016.12.1009.

Pavlov, A., Ivanov, D., Werner, F., Dolgui, A., & Sokolov, B. (2019). Integrated detection of disruption scenarios, the ripple effect dispersal and recovery paths in supply chains. *Annals of Operations Research*, *319*, 609−631. Available from https://doi.org/10.1007/s10479-019-03454-1.

Peng, X., Ji, S., Thompson, R. G., & Zhang, L. (2021). Resilience planning for Physical Internet enabled hyperconnected production-inventory-distribution systems. *Computers and Industrial Engineering*, *158*, 107413. Available from https://doi.org/10.1016/j.cie.2021.107413.

Pettit, T. J., Croxton, K. L., & Fiksel, J. (2019). The evolution of resilience in supply chain management: A retrospective on ensuring supply chain resilience. *Journal of Business Logistics*, *40*(1), 56−65. Available from https://doi.org/10.1111/jbl.12202.

Pound, P., & Campbell, R. (2015). Exploring the feasibility of theory synthesis: A worked example in the field of health related risk-taking. *Social Science and Medicine*, *124*, 57−65. Available from https://doi.org/10.1016/j.socscimed.2014.11.029.

Prost, L., Berthet, E. T. A., Cerf, M., Jeuffroy, M. H., Labatut, J., & Meynard, J. M. (2017). Innovative design for agriculture in the move towards sustainability: Scientific challenges. *Research in Engineering Design*, *28*, 119−129. Available from https://doi.org/10.1007/s00163-016-0233-4.

Rahman, T., Paul, S. K., Shukla, N., Agarwal, R., & Taghikhah, F. (2022). Supply chain resilience initiatives and strategies: A systematic review. *Computers and Industrial Engineering*, *170*, 108317. Available from https://doi.org/10.1016/j.cie.2022.108317.

Rajesh, R. (2020). A grey-layered ANP based decision support model for analyzing strategies of resilience in electronic supply chains. *Engineering Applications of Artificial Intelligence*, *87*, 103338. Available from https://doi.org/10.1016/j.engappai.2019.103338.

Razavian, E., Alem Tabriz, A., Zandieh, M., & Hamidizadeh, M. R. (2021). An integrated material-financial risk-averse resilient supply chain model with a real-world application. *Computers and Industrial Engineering*, *161*, 107629. Available from https://doi.org/10.1016/j.cie.2021.107629.

Rehman, O. ur, & Ali, Y. (2021). Enhancing healthcare supply chain resilience: Decision-making in a fuzzy environment. *International Journal of Logistics Management*, *33*(2), 520−546. Available from https://doi.org/10.1108/IJLM-01-2021-0004.

Sabouhi, F., Jabalameli, M. S., & Jabbarzadeh, A. (2021). An optimization approach for sustainable and resilient supply chain design with regional considerations. *Computers and Industrial Engineering*, *159*, 107510. Available from https://doi.org/10.1016/j.cie.2021.107510.

Sahu, A. K., Datta, S., & Mahapatra, S. S. (2016). Evaluation and selection of resilient suppliers in fuzzy environment: Exploration of fuzzy-VIKOR. *Benchmarking*, *23*(3), 651−673. Available from https://doi.org/10.1108/BIJ-11-2014-0109.

Sarker, M. R., Ahmed, F., Deb, A. K., & Chowdhury, M. (2018). Identifying barriers for implementing Green Supply Chain Management (Gscm) In footwear industry of Bangladesh: A Delphi study approach. *Leather and Footwear Journal*. Available from https://doi.org/10.24264/lfj.18.3.1.

Sazvar, Z., Tafakkori, K., Oladzad, N., & Nayeri, S. (2021). A capacity planning approach for sustainable-resilient supply chain network design under uncertainty: A case study of vaccine supply chain. *Computers and Industrial Engineering*, *159*, 107406. Available from https://doi.org/10.1016/j.cie.2021.107406.

Schätter, F., Hansen, O., Wiens, M., & Schultmann, F. (2019). A decision support methodology for a disaster-caused business continuity management. *Decision Support Systems, 118*, 10−20. Available from https://doi.org/10.1016/j.dss.2018.12.006.

Shih, Y. Y., & Lin, C. A. (2022). Co-location with marketing value activities as manufacturing upgrading in a COVID-19 outbreak era. *Journal of Business Research, 148*, 410−419. Available from https://doi.org/10.1016/j.jbusres.2022.04.060.

Singh, N. P., & Singh, S. (2019). Building supply chain risk resilience: Role of big data analytics in supply chain disruption mitigation. *Benchmarking, 26*(7), 2318−2342. Available from https://doi.org/10.1108/BIJ-10-2018-0346.

Sonar, H., Gunasekaran, A., Agrawal, S., & Roy, M. (2022). Role of lean, agile, resilient, green, and sustainable paradigm in supplier selection. *Cleaner Logistics and Supply Chain, 4*, 100059. Available from https://doi.org/10.1016/j.clscn.2022.100059.

Taghikhah, F., Erfani, E., Bakhshayeshi, I., Tayari, S., Karatopouzis, A., & Hanna, B. (2022). *Artificial intelligence and sustainability: Solutions to social and environmental challenges. Artificial intelligence and data science in environmental sensing* (pp. 93−108). Academic Press In. Available from https://doi.org/10.1016/B978-0-323-90508-4.00006-X.

Taghikhah, F., Voinov, A., Shukla, N., & Filatova, T. (2020). Exploring consumer behavior and policy options in organic food adoption: Insights from the Australian wine sector. *Environmental Science and Policy, 109*, 116−124. Available from https://doi.org/10.1016/j.envsci.2020.04.001.

Tan, W. J., Cai, W., & Zhang, A. N. (2020). Structural-aware simulation analysis of supply chain resilience. *International Journal of Production Research, 58*(17), 5175−5195. Available from https://doi.org/10.1080/00207543.2019.1705421.

Tseng, M. L., Bui, T. D., Lim, M. K., Fujii, M., & Mishra, U. (2022). Assessing data-driven sustainable supply chain management indicators for the textile industry under industrial disruption and ambidexterity. *International Journal of Production Economics, 245*, 108401. Available from https://doi.org/10.1016/j.ijpe.2021.108401.

Um, J., & Han, N. (2021). Understanding the relationships between global supply chain risk and supply chain resilience: The role of mitigating strategies. *Supply Chain Management, 26*(2), 240−255. Available from https://doi.org/10.1108/SCM-06-2020-0248.

Vali-Siar, M. M., & Roghanian, E. (2022). Sustainable, resilient and responsive mixed supply chain network design under hybrid uncertainty with considering COVID-19 pandemic disruption. *Sustainable Production and Consumption, 30*, 278−300. Available from https://doi.org/10.1016/j.spc.2021.12.003.

Vijayan, G., Kamarulzaman, N. H., Mohamed, Z. A., & Abdullah, A. M. (2014). Sustainability in food retail industry through reverse logistics. *International Journal of Supply Chain Management, 3*(2), 11.

Wang, M., & Yao, J. (2021). Intertwined supply network design under facility and transportation disruption from the viability perspective. *International Journal of Production Research, 61*(8), 2513−2543. Available from https://doi.org/10.1080/00207543.2021.1930237.

Wang, Y., & Webster, S. (2021). Product flexibility strategy under supply and demand risk. *Manufacturing & Service Operations Management, 24*(3), 1261−1885. Available from https://doi.org/10.1287/msom.2021.1037.

Werner, M. J. E., Yamada, A. P. L., Domingos, E. G. N., Leite, L. R., & Pereira, C. R. (2021). Exploring organizational resilience through key performance indicators. *Journal of Industrial and Production Engineering, 38*(1), 51−65. Available from https://doi.org/10.1080/21681015.2020.1839582.

Wu, H. liang, Huang, J., Zhang, C. J. P., He, Z., & Ming, W. K. (2020). Facemask shortage and the novel coronavirus disease (COVID-19) outbreak: Reflections on public health measures. *EClinicalMedicine, 21*. Available from https://doi.org/10.1016/j.eclinm.2020.100329.

Yavari, M., & Ajalli, P. (2021). Suppliers' coalition strategy for green-Resilient supply chain network design. *Journal of Industrial and Production Engineering*, *38*(3), 197−212. Available from https://doi.org/10.1080/21681015.2021.1883134.

Zalitis, I., Dolgicers, A., Zemite, L., Ganter, S., Kopustinskas, V., Vamanu, B., Finger, J., Fuggini, C., Bode, I., Kozadajevs, J., & Häring, I. (2022). Mitigation of the impact of disturbances in gas transmission systems. *International Journal of Critical Infrastructure Protection*, *39*, 100569. Available from https://doi.org/10.1016/j.ijcip.2022.100569.

Zhao, K., Zuo, Z., & Blackhurst, J. V. (2019). Modeling supply chain adaptation for disruptions: An empirically grounded complex adaptive systems approach. *Journal of Operations Management*, *65*(2), 190−212. Available from https://doi.org/10.1002/joom.1009.

Zhao, R., Liu, Y., Zhang, N., & Huang, T. (2017). An optimization model for green supply chain management by using a big data analytic approach. *Journal of Cleaner Production*, *142*, 1085−1097. Available from https://doi.org/10.1016/j.jclepro.2016.03.006.

CHAPTER TWO

Big data analytics in construction: laying the groundwork for improved project outcomes

Arpit Singh[1], Ashish Dwivedi[1], Malini Mittal Bishnoi[2] and Swamynathan Ramakrishnan[3]

[1]Jindal Global Business School, O.P. Jindal Global University, Sonipat, Haryana, India
[2]School of Humanities, Arts and Applied Sciences, Amity University, Dubai, United Arab Emirates
[3]Amity Business School, Amity University, Dubai, United Arab Emirates

2.1 Introduction

Construction companies constantly grapple with unprompted circumstances that tend to disrupt productivity and efficiency challenges to construction organizations (Hwang et al., 2022). Historically, manual procedures and disjointed data management systems have been used in the construction sector, which has resulted in inefficiencies, delays, and cost overruns. The larger challenge remains mitigation and responsiveness to unpredictable occurrences with "Big Data." "Big data" refers to any huge and complex magnitude of information that must be processed and managed by advanced analytics technologies (Vassakis et al., 2018). This data can originate from various organized and unstructured sources, such as cameras, sensors, mobile devices, and log files (Wang et al., 2022). Big data analytics (BDA) can be a credible answer to the question and other productivity-related concerns of construction organizations (Regona et al., 2022). With the introduction of cutting-edge technology and the expansion of data availability, there is a high chance of adopting BDA to enhance the results of building projects.

Big data analytics is the process of gathering, storing, processing, and analyzing massive amount of data to derive insightful conclusions and support decision-making (Saggi & Jain, 2018). BDA can assist in optimizing

Computational Intelligence Techniques for Sustainable Supply Chain Management.
DOI: https://doi.org/10.1016/B978-0-443-18464-2.00003-0

resource allocation, identifying potential dangers and bottlenecks, strengthening safety protocols, and improving overall project management in the context of the construction sector. Big data in construction implies the colossal amount of data such as project, financial, employee safety, equipment, and customer data collected and stored in cloud-based systems that are accessible only by selected users, ranging from general contractors and developers to subcontractors and individual tradesman, from any location and at any time (Kaewnaknaew et al., 2022). Big data in project management may be used to monitor and track project progress, identify possible delays or difficulties, and improve resource allocation. Project managers, for example, can utilize data analytics tools to track project deadlines, identify essential path tasks, and forecast project results based on historical data. Big data may be used to examine financial performance, track spending, and find cost-cutting options. Contractors, for example, can utilize data analytics tools to examine spending trends, find cost-cutting opportunities, and enhance cash flow management. Big data may be utilized to improve construction site safety by identifying potential hazards and adopting proactive safety measures (Li et al., 2015). Site managers, for example, might utilize sensor data and video analytics to monitor worker behavior and detect safety issues, such as workers who are not wearing the appropriate protective equipment.

Data from equipment may be utilized to optimize equipment utilization, decrease downtime, and enhance maintenance schedules. Contractors, for example, can utilize sensor data to monitor equipment usage and forecast when a repair is needed, lowering the chance of equipment failure and maintenance expenses. By evaluating customer feedback, identifying areas for development, and adapting services to fit consumer wants, big data may be utilized to increase customer happiness and retention. Contractors, for example, can utilize data analytics tools to examine customer feedback and sentiment data to discover areas where customer service can be improved.

2.1.1 Leveraging big data analytics for the construction industry

BDA has the potential to reinvent the operational dynamics of organizations by encouraging innovations in products and services, and refurbishing organizational capabilities (Hire et al., 2022). The construction industry generates huge amount of new information, from blueprints to building information, cost and project completion estimates are continuously added to the repositories (Adekunle et al., 2022). Unfortunately, the data collected from varied sources are extremely unstructured and disorganized (Akinosho et al., 2020). The large data collected need to be processed and comprehended correctly to obtain meaningful insights. This makes the technology adoption essential to deal with big data generated in the construction industry. In recent times, Big data is one of the most valued commodities with a current value in the range of \$30–\$35 billion (Li, Chen, et al., 2022). The construction sector is growing more competitive, and businesses using BDA have a competitive advantage over those not. Researchers may assist firms in overcoming the problems associated with implementing big data technologies by examining these challenges. BDA may assist in enhancing construction safety and quality by providing firms with useful insights into their operations.

To attain a competitive edge in the market every industry can leverage the latest big data processing and storage tools to stay ahead of the game. Industries may obtain a competitive advantage by utilizing the most recent big data processing and storage techniques. These technologies may assist businesses in making better decisions by giving vital insights into customer behavior, market trends, and other important elements. Big data may also help businesses simplify operations, automate procedures, and save expenses, allowing them to run more effectively. Companies may provide a better customer experience by acquiring a deeper knowledge of their consumers' requirements and preferences by employing BDA. Furthermore, big data may assist businesses in staying ahead of the competition by recognizing emerging trends and possibilities, allowing them to be more agile and responsive in the marketplace. By utilizing the most recent big data techniques, firms may gain a major competitive advantage in their respective sectors. Despite its reluctance to accept new methods and technologies, the construction sector has recently shown a surge in innovation as a result of an infusion of big data technology (Regona et al., 2022). BDA offers numerous benefits and scope for improving productivity in the construction sector. Data analytics technology mitigates construction time and material-related costs by organizing and cleaning unstructured data, identifying potential errors, and fixing them before they occur (Zhao & Li, 2022). Data analytics technology is being used in construction to reduce time and material costs by organizing and cleaning unstructured data and detecting possible problems before they occur (Bilal et al., 2016). Material estimation errors, scheduling delays, ineffective resource allocation, and safety accidents are all examples of possible problems. BDA may assist in identifying potential areas for error and optimizing resource allocation to decrease expenses. BDA may predict future schedule delays and enhance project planning to mitigate these delays by evaluating past project data. Moreover, BDA may assist in identifying possible safety issues and mitigating risk. Overall, data analytics technology may assist construction organizations in increasing productivity, lowering costs, and reducing the risk of errors and delays. This enables project managers to make informed decisions and minimizes human error.

2.1.2 Perils of the construction industry

The construction and energy industries are arguably the biggest polluters in the world (Diemer et al., 2022). An estimated 39% of process-related carbon dioxide is released from the construction sector annually (Zhang et al., 2019). Building information modeling (BIM) is a digital procedure used in the construction industry to visually represent a building's physical and functional qualities (Ding et al., 2019). BIM enables construction professionals to collaborate and communicate more effectively throughout the building lifecycle, from design and construction through operation and maintenance. By utilizing BIM, stakeholders may more efficiently communicate project data and coordinate their actions, resulting in enhanced project results, decreased risk, and increased efficiency. BIM also enables the optimization of building design and construction processes, which results in cost savings and improved overall performance. Overall, BIM has become a crucial tool in the construction sector, aiding in cooperation, risk reduction, and project results. Integrating data analytics technology with BIM can assist managers in correctly estimating the materials and energy requirements for the projects. This reduces construction waste and encourages planners to find alternative energy-efficient resources (Megahed & Hassan, 2022). One of the most difficult issues that construction workers confront on the job is a lack of communication. A lack of communication can result in an information deficit, which arises when employees cannot access the information they require to conduct their tasks successfully. An information gap can occur in the construction sector when workers are not informed about changes to project requirements, timetables, or budgets. This can lead to errors, rework, and delays, all of which can influence the project's overall success. For example, if a worker is not informed of a change in the design of a building component, they may continue to utilize obsolete blueprints, resulting in errors and rework.

Similarly, if workers are not advised of a material delivery delay, they may be unable to adapt their schedules leading to further delays and cost over-runs. Fortunately, big data systems make information easily accessible and shared among crew members (Wang et al., 2022). Big data may assist construction firms in making information easier to access and distribute among crew members by building a centralized repository of project data that can be accessed and shared in real time by all stakeholders. This can assist in guaranteeing that all crew members have the latest and most accurate project information, decreasing the likelihood of errors and delays caused by lack of knowledge. Big data may help with this process by combining data from several sources, such as project management software, sensors, and other devices. By merging various data sources, construction companies may generate a holistic perspective of the project that all stakeholders, regardless of location or role on the project, can access. BDA may be used to clean and organize data in addition to integrating it, making it simpler to access and distribute. Data analytics techniques, for example, can be used to detect and repair problems in project data or to categorize data based on project features. This can assist in guaranteeing that employees have access to up-to-date project information, lowering the likelihood of mistakes and delays. Additionally, big data technologies like cloud computing and mobile devices allow workers to access project information from any location or time. This can let crew workers collaborate and communicate, even if they are working in various places or at different times. This avoids errors caused by misinterpretation, strengthens shareholder connections, and keeps everyone informed in the event of a last-minute change or interruption (Yoffe et al., 2022). Construction workers are the most vulnerable professionals to injuries and accidents than any other industry (Hasan & Kamardeen, 2022). Smart construction wearables and safety management software are two important examples that are gaining traction in the construction sector. Smart construction wearables are devices worn by construction workers that monitor and track various aspects of their work, such as mobility, location, and vital signs. On the other hand, safety management software is a digital application used to manage and monitor job-site safety standards and processes (Ungurean & Vatavu, 2022). Safety management software devices capitalize on the gathered data on health and activity information from past construction projects and work on the data to provide safety protocols to construction crews and detection alarms in the event of safety hazards (Xu et al., 2022). Big data is changing the game for construction businesses that aim to boost productivity, cooperation, and worker safety on job sites in an industry that is typically slow to adopt new technologies (Wu et al., 2022). Big data assists construction organizations in several ways, including offering a full perspective of the project, which can be used to detect possible concerns and possibilities for improvement. Construction companies may develop a centralized repository of project data that can be accessed and shared in real time by all stakeholders by combining data from numerous sources such as project management software, sensors, and other devices. This can assist in guaranteeing that all employees have access to the most up-to-date project information, lowering the risk of errors and delays caused by a lack of knowledge. Big data is also assisting construction firms by increasing communication and collaboration among crew members. By offering a single platform for exchanging information and updates, big data technology can improve cooperation and communication among crew members, even if they are working in various places or at different times. This can promote collaboration and productivity on the job site, resulting in improved project outcomes overall.

2.1.3 Magnitude of the potential of big data for the construction industry

Companies of all sizes may profit from the potential of big data as data analytics technologies evolve and become more accessible. Some real-life examples that highlight the challenges in the construction industry and the need for adopting new smart technologies, including BDA are as follows:

- Project Delays: Conflicts in scheduling, problems with resource allocation, and unforeseen events are just a few of the reasons why construction projects frequently have delays. For instance, a delay in the supply of materials might throw off the schedule for the entire project if a construction business is working on a significant infrastructure project. Construction organizations can use BDA to analyze past project data, spot possible bottlenecks, and allocate resources more efficiently to prevent delays.
- Safety risks: Worker safety is of the utmost importance because construction sites are inherently dangerous places. Accidents can still happen despite safety procedures because of things like human mistakes, defective equipment, or insufficient safety precautions. By using sensor data and video analytics to track employee behavior, identify safety

concerns in real time, and put preventative safety measures in place, BDA can significantly improve construction site safety.

- Cost overruns: Managing costs in construction projects is a difficult task. Inaccurate material estimates can cause cost overruns, ineffective resource allocation, and unforeseen charges. Construction organizations can use BDA to analyze previous project data, spot spending trends, and optimize resource allocation. This may entail finding areas for cost-cutting, streamlining the procurement procedure, and improving cash flow management.

- Sustainability and Environmental Impact: There is an increasing demand for sustainable practices due to the building industry's enormous environmental impact. For instance, minimizing building waste and implementing energy-efficient solutions can help lower the sector's carbon footprint. BIM and BDA can be combined to help construction companies better predict their material and energy needs, improve the energy efficiency of their building designs, and cut down on wasteful construction processes (Ding et al., 2019).

- Effective communication and coordination between project stakeholders are crucial for the project's success. However, construction projects entail several teams and stakeholders working at various times and locations. Miscommunication, delays, and mistakes may result from this. By offering a central repository of project data that can be accessed and shared in real-time, big data technology can help with these problems. This enhances teamwork, guarantees that all project stakeholders are aware of the most recent project information, and lowers the possibility of errors and delays brought on by poor communication.

2.1.4 Research contribution

Researchers can discover strategies to address the problems presented by construction firms in implementing big data technology by identifying these challenges and promoting the wider implementation of BDA. As a result, the construction industry's efficiency and production may improve. BDA may bring useful insights to construction businesses, allowing them to make better decisions. These researchers encompass diverse professionals, including academicians and industry experts. Their studies involve in-depth analysis of data and examination of the challenges faced during the implementation of big data technologies. By delving into these identified challenges, researchers can assist construction firms in overcoming the obstacles and realizing the full potential of big data technologies.

The research contributions come from various fields, such as computer science, information technology (IT), business management, and data science. This interdisciplinary approach ensures a comprehensive investigation of the causes behind the non-adoption of big data technology in construction organizations. By bridging these different fields of study, researchers contribute to the existing body of research on big data technology in the construction industry.

To summarize, the current research study aims to uncover the underlying causes of why construction organizations have been slow to adopt big data technology. By identifying these causes, the study provides valuable insights into the barriers and challenges hindering big data technology implementation. This knowledge contributes to a deeper understanding of the factors that influence the non-adoption of BDA in the construction sector.

This manuscript is divided into the following sections. Section 2.1 sets the context of the role of big data technology in organizations in enhancing efficiency and productivity. It is followed by a literature review in Section 2.2 citing some past studies in the field of BDA adoption in organizations. Section 2.3 illustrates the methods used to conduct the research including expert selection, questionnaire building, and finalizing the measurement instrument. Results are presented in Section 2.4 followed by discussions and practical implications in Sections 2.5 and 2.6 respectively. The paper reflects the conclusion and future scope in Section 2.7.

2.2 Literature review

BDA has the potential to reshape how the construction sector operates in multiple ways that can lead to significant additions in productivity and efficiency. Several studies were done extensively to carefully examine the adoption of BDA in the construction sector to identify the key factors that impact the wider adoption of BDA in the industry. Some of the most recent and notable ones are mentioned in the subsequent section.

In research investigating the benefits and drawbacks of implementing artificial intelligence (AI) in the construction business, a systematic literature review was conducted based on the Preferred Reporting Items for Systematic Reviews and Meta-Analyses (PRISMA) protocol. The study focused on constructing project lifecycle stages of planning, design, and construction. The findings from the study reveal that AI can be especially useful during the planning stage, as it helps to enhance event, risk, and cost predictions. The most promising use of AI in construction is reducing time spent on repetitive operations and improving work processes through BDA. However, the industry's fragmentation is the most challenging part, which causes data gathering and retention challenges. The report sheds light on the benefits and drawbacks of AI adoption for various stakeholders in the construction sector (Regona et al., 2022). The construction sector is found to lag behind its contemporaries in the adoption of smart technologies from the fourth industrial revolution. This study aims to identify the major challenges impeding the adoption of these technologies in the construction industry and devise effective strategies for confronting those challenges. The study revealed that the top challenges were data and information sharing, regulatory compliance, and data ownership, which can be dealt with effectively by training skilled construction workers, engagement from the government and incentives, and communications and change management (Hwang et al., 2022). In a survey-based study that explored the major challenges in the adoption of Industry 4.0 (I4.0) techniques such as digitalization, BIM, and IoT, it was found that resistance to change, unclear benefits, and implementation costs were the major challenges that needed immediate attention (Demirkesen & Tezel, 2022). Jordanian small and medium enterprises (SME) saw the relative advantage, complexity, security, top management support, organizational readiness, and government support as the major drivers of Big data adoption, while competition and compatibility were discounted as insignificant factors (Lutfi et al., 2022). Construction 4.0 has sparked interest in the construction sector, thanks to the application of BIM, cyber-physical systems, and digital technology (Alaloul et al., 2022). Although blockchain, distributed ledger technology (DLT) has the potential to alter the business, its adoption is hampered by obstacles. This study uses a sustainability lens to assess the challenges to DLT adoption in construction. The ordinal priority decision-making technique was used to identify the relevance of sustainability qualities and obstacles, and it was discovered that DLT deployment necessitates enhanced data management, sophisticated applications, client demand, and effective taxes and reporting. Solving these issues is critical for improving supply chain management, transparency, fair competition, and social sustainability in the construction sector (Sadeghi et al., 2022). In a study to investigate the major detrimental factors in the adoption of BDA in the development of smart cities, 13 BDA adoption barriers were identified and prioritized using the best-worst method (BWM). The results reflect that data complexity, lack of BDA adoption framework, and lack of BDA technologies are the main hindrances to incorporating BDA in smart city development (Khan, 2022). The administrative risks of smart contract adoption in the construction sector were investigated as the degree of adopting new technologies is fairly limited in the sector. To assess administrative risks, the study performed a literature survey and a pilot study, and it used a focus group discussion to consider risk reduction measures. The final framework identified five primary risk factors, with the top five being regulatory change, a lack of a driving force, works not accounted for in planning, flaws in present legal systems, and a lack of a dispute resolution process. The study advises that semi-automated smart contract drafting be improved as a more viable risk-reduction technique (Gurgun & Koc, 2022). A study conducted in the Malaysian construction industry to assess the adoption of IoT revealed that lack of safety and security, lack of standards, lack of benefit awareness, improper introduction, and lack of robust connectivity are the main detrimental factors hindering the adoption (Kar & Dwivedi, 2020). This study assesses the use of digital technologies in off-site construction (OSC), such as BIM, Internet of Things (IoT), and others, to establish existing practices, limits, and possibilities for enhancing OSC. To build a comprehensive picture of existing practice and highlight opportunities for future development, a systematic study and scientometric analysis of fifteen common digital technologies utilized in OSC projects are done (Wang et al., 2020). While the existing research provides valuable insights into the challenges of BDA adoption in the construction sector, several gaps and limitations can be identified. Firstly, the literature often focuses on identifying challenges without providing in-depth analysis or solutions to overcome them. There is a need for more research that offers practical strategies and frameworks to address the identified challenges.

Furthermore, the studies primarily focus on individual challenges in isolation, overlooking the interdependencies and interactions among different obstacles. A more holistic approach that considers the systemic nature of the challenges and explores their cumulative impact on BDA adoption would enhance the understanding of the complexities involved.

Moreover, the literature predominantly focuses on large construction organizations, neglecting the challenges small and medium-sized enterprises (SMEs) face. SMEs form a significant portion of the construction sector and have unique characteristics and resource constraints that may impact their ability to adopt BDA. Future research should pay attention to the specific challenges faced by SMEs and provide tailored solutions for their successful adoption of BDA.

While the adoption of BDA holds immense potential for the construction industry, several challenges must be addressed to facilitate its widespread implementation. This critical literature review highlights the key challenges, including data quality and standardization, limited data sharing, technological infrastructure, workforce readiness, cost concerns, and regulatory considerations. By addressing these gaps, the construction industry can overcome the hurdles and leverage the transformative power of BDA to drive innovation, efficiency, and competitiveness.

The current study aims to add to the existing literature on the adoption of big data technology in the construction sector and achieve the following research objectives (ROs):

RQ1: What major challenges impede organizations from adopting the new technology?

RQ2: How to prioritize the challenges based on the degree of severity associated with the identified challenges.

More specifically, this study poses the following two research questions (RQs):

RO1: Identification of critical challenges in adopting BDA in the construction sector.

RO2: Categorize challenges from the most severe to the least by ranking the challenges.

2.3 Methodology

2.3.1 Measuring instrument and expert selection

Based on the literature survey and the author's practical and research experience, an initial list of twelve challenges was offered to aid in the decision-making process for selecting the most severe task in terms of the most significant and impactful challenge hindering the adoption of BDA in the construction sector. The experts were provided with the list to acquire their views and ideas on the data collected, which would boost the list's efficacy in defining the primary hurdles in implementing BDA considering the construction sector. Some of the problems on the list did not seem essential to them in their company, but they advised to keep them since they thought it would be valuable for other comparable firms in their field.

Academicians, especially researchers active in construction supply chain management and computer science, were consulted in the following round of list refinement. Some obstacles were introduced and adjusted after an extensive and comprehensive brainstorming session. Table 2.1 displays the final set of challenges in the adoption of BDA considering construction organizations.

2.3.2 Experts' selection and refinement of measuring instrument

In the present study, professionals from the construction industry and IT, as well as scholars and academics from supply chain management and computer science, were identified. During the initial step of improving the measurement device, staff from the building supply chain management department were contacted. The authors held a focus group discussion with the workers about the present status of BDA implementation in their workplaces. The company was in its early stages of expansion, sitting on the verge of its operational outreach.

2.3.3 Sample selection and data collection

The final questionnaire was distributed to a panel of nine specialists. The specialists were chosen from three categories: construction organizations, IT organizations, and academicians active in the field of construction management and computer science who have specialized knowledge and expertise in their field, including in the areas of data analytics, construction, and technology adoption. The academicians selected for the research are at the forefront of research and development and can bring new ideas and perspectives to the adoption of BDA in the construction sector. BDA research is considered complicated and requires specialized knowledge and abilities, and experts in software development, IT, academia, and construction management. Software developers have the technical abilities required to design and implement systems and tools for collecting, storing,

Table 2.1 List of major challenges in the adoption of BDA (Big data analytics) in the construction industry.

	Challenges	References
C1	Lack of data storage facility due to legacy systems	Regona et al. (2022), Sreedevi et al. (2022), Mishra and Tyagi (2022)
C2	Handling huge volumes and a variety of data	Gao et al. (2020), Zeadally et al. (2020), Amanullah et al. (2020)
C3	Handling high velocity of data	Mohamed et al. (2020), Ghasemaghaei and Calic (2020), Deepa et al. (2022)
C4	Data integration with existing systems	Deepa et al. (2022), Selvaraj and Sundaravaradhan (2020)
C5	Channelizing the data toward the organizational output	Kumar (2022), Naeem et al. (2022), Li, Liu, et al. (2022)
C6	Establishing a prototype	Naeem et al. (2022), Zhang et al. (2022)
C7	Incorrect data processing and inconsistency in input and output information	Sreedevi et al. (2022), Li, Liu, et al. (2022), Chen (2022), Mailewa et al. (2022)
C8	Encryption of sensitive information to prevent cyber attacks	Mailewa et al. (2022), Gupta and Lakhwani (2022), Hamza et al. (2022)
C9	Big data system's scalability issues	Kumar (2022), Naeem et al. (2022), Fathi et al. (2022)
C10	Keeping track of what the incoming data is and where is it coming from	Naeem et al. (2022), Zhang et al. (2022), Jayagopal and Basser (2022)
C11	Heavy costs in implementation and maintenance	Li, Chen, et al. (2022), Wang et al. (2022), Raut et al. (2019)
C12	Lack of proper internet connectivity and other IT-related issues	Lu and Xu (2019), Dai et al. (2020), Aceto et al. (2020)
C13	Lack of methods for evaluating and improving the system's performance	Lee (2021), Stergiou et al. (2020), Le et al. (2020)
C14	Lack of expertise and skills	Bag et al. (2021), Urbinati et al. (2019), Sestino et al. (2020)
C15	Issues pertaining to data sharing on real real-time basis	Gupta et al. (2019), Zhang et al. (2022), Han and Trimi (2022)
C16	Unavoidable systemic analytical limitations	Oo and Thein (2022), Lu and Xu (2019), Dai et al. (2020)
C17	Unclear fault tolerance in algorithms	Saadoon et al. (2022), Myllyaho et al. (2022), Morales et al. (2022)
C18	Unclear tangible benefits	AL-Khatib & Shuhaiber (2022), Ren (2022), Atuahene et al. (2022)
C19	Hesitation	Soni et al. (2022), Ikegwu et al. (2022), Elhusseiny and Crispim (2022)

and analyzing enormous volumes of data. Academics have experience in the theoretical and methodological elements of data analysis, whereas IT specialists have an understanding of the infrastructure and technologies required to handle and analyze big data. These specialists collaborate to build and implement successful BDA solutions that take into consideration both technical and theoretical elements of the area. Furthermore, experts from the construction business contribute significantly to studies on BDA.

Throughout the various stages of a construction project, from design and planning through execution and maintenance, vast volume of data are generated. These professionals understand the specialized processes, workflows, and data types created in the construction sector, which are utilized to build and deploy construction-specific data analytics solutions. Furthermore, they understand the business and operational difficulties that construction organizations face and can assist in identifying areas where BDA can be leveraged to enhance efficiency, cut costs, and increase production. The comments gathered from experts from other fields during the decision-making provide a comprehensive knowledge of the primary hurdles preventing BDA implementation in construction firms. The opinion of experts on the impact of the challenges are listed in Table 2.1 on a scale of 1−5, where the meaning of each number is as mentioned as follows:

1: No impact. 2: Low Impact, 3: Moderate Impact, 4: High Impact, and 5: Extremely High Impact.

The survey data was subjected to a multi-criteria ranking technique using a dominance-based rough sets approach (DRSA) to rank the concerns depending on their severity. Singh and Misra (2020) present a comprehensive method for ranking using DRSA.

2.4 Results

The results of the DRSA-based ranking application based on expert survey data are detailed in the following sections.

2.4.1 Pairwise comparison table

Table 2.2 shows the pairwise comparison table (PCT) resulting from the pairwise comparisons the experts performed on the set of challenges.

Table 2.2 Pairwise comparison table (PCT) (partial) of criteria.

X,Y	$(q_1(X), q_1(Y))$	$(q_2(X), q_2(Y))$	$(q_3(X), q_3(Y))$	$(q_4(X), q_4(Y))$	$(q_5(X), q_5(Y))$	$(q_6(X), q_6(Y))$	$(q_7(X), q_7(Y))$	$(q_8(X), q_8(Y))$	$(q_9(X), q_9(Y))$	Relation
C1, C1	4, 4	4, 4	4, 4	4, 4	4, 4	4, 4	4, 4	4, 4	4, 4	S
C1, C11	4, 4	4, 4	4, 4	4, 4	4, 4	4, 4	4, 4	4, 4	4, 4	S
C1, C14	4, 4	4, 5	4, 4	4, 4	4, 4	4, 3	4, 4	4, 3	4, 3	S
C1, C15	4, 4	4, 4	4, 4	4, 4	4, 5	4, 4	4, 4	4, 5	4, 4	S
C1, C16	4, 2	4, 3	4, 3	4, 4	4, 4	4, 2	4, 4	4, 4	4, 2	S^c
C11, C1	4, 4	4, 4	4, 4	4, 4	4, 4	4, 4	4, 4	4, 4	4, 4	S
C11, C11	4, 4	4, 4	4, 4	4, 4	4, 4	4, 4	4, 4	4, 4	4, 4	S
C11, C14	4, 4	4, 5	4, 4	4, 4	4, 4	4, 3	4, 4	4, 4	4, 3	S
C11, C15	4, 4	4, 4	4, 4	4, 4	4, 5	4, 4	4, 4	4, 5	4, 4	S
C11, C16	4, 2	4, 3	4, 3	4, 4	4, 4	4, 2	4, 4	4, 4	4, 2	S
C14, C1	4, 4	5, 4	4, 4	4, 4	4, 4	3, 4	4, 4	4, 4	3, 4	S^c
C14, C11	4, 4	5, 4	4, 4	4, 4	4, 4	3, 4	4, 4	4, 4	3, 4	S^c
C14, C14	4, 4	5, 5	4, 4	4, 4	4, 4	3, 3	4, 4	4, 4	3, 3	S
C14, C15	4, 4	5, 4	4, 4	4, 4	4, 5	3, 4	4, 4	4, 5	3, 4	S
C14, C16	4, 2	5, 3	4, 3	4, 4	4, 4	3, 2	4, 4	4, 4	3, 2	S

(*Continued*)

Table 2.2 (Continued)

X,Y	$(q_1(X), q_1(Y))$	$(q_2(X), q_2(Y))$	$(q_3(X), q_3(Y))$	$(q_4(X), q_4(Y))$	$(q_5(X), q_5(Y))$	$(q_6(X), q_6(Y))$	$(q_7(X), q_7(Y))$	$(q_8(X), q_8(Y))$	$(q_9(X), q_9(Y))$	Relation
C15, C1	4, 4	4, 4	4, 4	4, 4	5, 4	4, 4	4, 4	5, 4	4, 4	S^c
C15, C11	4, 4	4, 4	4, 4	4, 4	5, 4	4, 4	4, 4	5, 4	4, 4	S^c
C15, C14	4, 4	4, 5	4, 4	4, 4	5, 4	4, 3	4, 4	5, 4	4, 3	S^c
C15, C15	4, 4	4, 4	4, 4	4, 4	5, 5	4, 4	4, 4	5, 5	4, 4	S
C15, C16	4, 2	4, 3	4, 3	4, 4	5, 4	4, 2	4, 4	5, 4	4, 2	S
C16, C1	2, 4	3, 4	3, 4	4, 4	4, 4	2, 4	4, 4	4, 4	2, 4	S^c
C16, C11	2, 4	3, 4	3, 4	4, 4	4, 4	2, 4	4, 4	4, 4	2, 4	S^c
C16, C14	2, 4	3, 5	3, 4	4, 4	4, 4	2, 3	4, 4	4, 4	2, 3	S^c
C16, C15	2, 4	3, 4	3, 4	4, 4	4, 5	2, 4	4, 4	4, 5	2, 4	S^c
C16, C16	2, 2	3, 3	3, 3	4, 4	4, 4	2, 2	4, 4	4, 4	2, 2	S

Table 2.3 Induced decision rules using VC-DomLEM algorithm.

Decision rules	Relation
1.{PAIR(D_1) D (2,2)}	S
2.{PAIR(D_2) D (5,5)}	S
3.{PAIR(D_6) D (3,3)}	S
4.{(2,4) D PAIR(D_1)}	S^c

PCT comprises the results of a pairwise comparison of all the criteria for assessing the degree of problems in BDA implementation considering the construction sector. Each object combination in the comparison is allocated to either a thorough outranking or a non-outranking connection. As a result, there are two types of decision classes: those with thorough outranking and those without. PCT is used as input for DRSA analysis, which yields lower and upper approximations of sets. The use of DRSA in PCT results in the decision rules as described in the next section.

2.4.2 Induced decision rules

The decision rules induced by the VC-DomLEM algorithm (Błaszczyński et al., 2009) are shown in Table 2.3.

The decision rules in Table 2.3 are explained:
1. Rule number 1 states that if decision maker D_1 evaluates criteria x as 2 or low impact and criteria y as 2 or low impact then xSy or in other words, criteria x is superior as compared to criteria y.
2. Rule number 2 states that if decision maker D_2 evaluates criteria x as 5 or extremely high impact and criteria y as 5 or Very high impact then xSy or in other words, criteria x is superior as compared to criteria y.
3. Rule number 3 states that if decision maker D_6 evaluates criteria x as 3 or moderate impact and criteria y as 3 or moderate impact then xSy or in other words, criteria x is superior as compared to criteria y.
4. Rule number 4 states that if decision maker D_1 evaluates criteria x as 2 or low impact and criteria y as 4 or very high impact then xS^cy or in other words, criteria x is inferior as compared to criteria y.

2.4.3 Preference graph

The preference graph resulting from the application of induced rules shown in Table 2.3 to all pairs of objects from Table 2.2 is shown in Fig. 2.1.

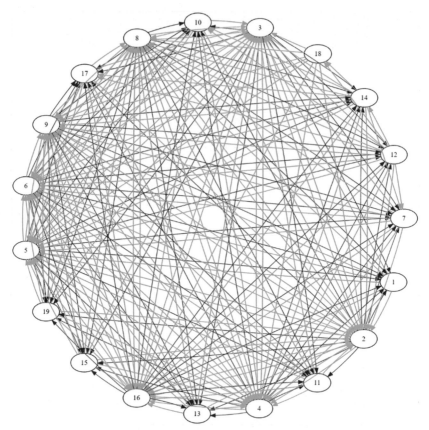

Figure 2.1 Preference graph illustrating outranking and non-outranking relations. Green arcs denote outranking relation S; red arcs denote non-outranking relation S^c.

The nodes in Fig. 2.1 indicate the challenges of implementing BDA in the construction industry. The presence of both preference relations is shown by nodes with green and red arcs. Summing the arcs with a positive weight allocated to the green arc and a negative weight assigned to the red arc yields the net preference relation. The nodes that do not have an arc linked with them indicate an overall outranking preference relation and no disagreement in the choice of preference assignment from all decision-makers.

2.4.4 Final ranking

Table 2.4 shows the final ranking of challenges determined by applying the Net Flow Score (NFS) approach to the preference graph in Fig. 2.1.

Table 2.4 Final ranking using the net flow score (NFS) method.

	Challenges	Net flow score	Rank
C 12	Lack of proper internet connectivity and other IT-related issues	21	1
C 14	Lack of expertise and skills	19	2
C 1	Lack of data storage facility due to legacy systems	16	3
C 11	Heavy costs in implementation and maintenance	16	3
C 13	Lack of methods for evaluating and improving the system's performance	16	3
C 15	Issues pertaining to data sharing on real real-time basis	16	3
C 19	Hesitation	16	3
C 7	Incorrect data processing and inconsistency in input and output information	14	4
C 10	Keeping track of what the incoming data is and where is it coming from	14	4
C 17	Unclear fault tolerance in algorithms	14	4
C 18	Unclear tangible benefits	6	5
C 2	Handling huge volumes and a variety of data	–21	6
C 3	Handling high velocity of data	–21	6
C 4	Data integration with existing systems	–21	6
C 5	Channelizing the data towards organizational output	–21	6
C 6	Establishing a prototype	–21	6
C 8	Encryption of sensitive information to prevent cyber attacks	–21	6
C 9	Big data system's scalability issues	–21	6
C 16	Unavoidable systemic analytical limitations	–21	6

The NFS of a criterion (node in the graph) is computed by subtracting the non-outranking preference relation from the outranking preference relation. It is the difference between the green and red arcs on the preference graph, as explained previously.

2.5 Discussion

The construction sector is a traditional industry that has relied mostly on manual and outdated technologies to carry out daily operations. This sector has shown a significant increase in embracing new technologies, such as BDA to improve productivity and efficiency. However, there is significant scope for the construction sector to holistically embrace BDA across geographies and departments. To extract meaningful insights from the findings, we will consider the challenges with positive NFS. A negative score indicates that the challenges are not considered important in the assessment of the reasons behind the non-adoption of BDA. Among the major challenges plaguing the adoption of BDA in the construction sector is the Lack of proper internet connectivity and other IT-related issues (C12). Construction sites are often located in remote areas where internet connectivity is poor or even non-existent. This makes it challenging to transfer large amounts of data for analysis, a critical component of BDA. It is crucial to transfer real-time data collected from sensors, drones, and other monitoring devices to the central location of analysis. However, slow internet connectivity can result in slow data transfer and incomplete data sets that can compromise the quality and accuracy of the results of BDA.

In addition to poor internet connectivity, other IT-related issues, such as lack of skilled workforce and expertise (C14) can exacerbate the issue of non-adoption of the BDA framework. The construction industry has high demands for skilled labor, making it difficult to attract and retain skilled IT professionals. The shortage of IT professionals can lead to inordinate delays in implementing and maintaining BDA projects. The lack of standard data formats is another important IT-related issue that forces the major stakeholders to reach a consensus on data standards. This lack of standardization makes it difficult to collect, store, analyze, and present results from various data sources that are universally consistent and acceptable. The lack of internet connectivity and other IT-related issues impeding the adoption of BDA in organizations have been highlighted in Raut et al. (2021).

Lack of data storage facilities due to legacy systems (C1) and heavy costs in implementation and maintenance (C11) can present insurmountable challenges in the adoption of BDA in construction organizations. Legacy systems cannot hold large volume of data. Additionally, they are not designed to handle the velocity and variety of data generated by modern devices such as sensors and drones. This can lead to a shortage of storage space, resulting in data loss and compromising the quality of data analysis. The high costs involved in implementing and maintaining BDA projects are an enormous financial burden on organizations deciding to incorporate the BDA framework. The construction sector operates with a lean budget, making it difficult for organizations to invest in hardware, software, and staffing to implement and maintain BDA. This directly affects the timely completion of the projects thereby affecting the profits. The influence of heavy investments in erecting the infrastructure for the BDA framework was discussed in the study by Kumar et al. (2022) and Wang et al. (2018).

The lack of methods for evaluating and improving the system's performance (C13) is ranked third in the challenges to the adoption of BDA in the construction sector. Continuous performance monitoring and assessment are crucial components of any BDA system, as they assist organizations in identifying and rectifying limitations in their systems. Without proper methods for evaluating and improving the system's performance, organizations may not be able to appropriately position themselves to explain the quality of the derived results from BDA implementation. Additionally, erroneous and inefficient BDA systems can lead to inferior quality outcomes with poor data quality, incorrect data analysis, and a lack of actionable insights. Organizations that cannot properly evaluate and improve system performance are likely to experience a decline in the return on investment (RoI) from BDA initiatives. Issues pertaining to data sharing on a real-time basis (C15) are the next major obstacle in the implementation of BDA in construction organizations. The construction sector is characterized by several disconnected systems handling different operations simultaneously, making it difficult to collect and share data in real time. The sensitive nature of data, such as financial and safety information, can make organizations hesitant to share it openly and in real time. The lack of standardization and interoperability between different systems can make real-time data sharing difficult. Also, the reticence and slowness of the construction organizations to adopt new technologies further worsen data sharing on a real-time basis.

Hesitation (C19) is a crucial factor in influencing the decisions of construction management to embrace the BDA framework for multiple reasons. Many construction organizations lack the technical expertise to implement and utilize BDA tools. There may be resistance from employees to change their traditional methods of operation to a completely new framework that requires to be updated with the latest technologies. Construction organizations might be hesitant to adopt the BDA framework due to concerns over the privacy and security of sensitive project and client information.

Incorrect data processing and inconsistency in input and output information (C7) and keeping track of what the incoming data is and where is it coming from (C10) affect the BDA adoption in construction organizations. This can cause the users to lose confidence in the decision-making process aided by the BDA framework. The analysis and recommendations generated from the inaccurate and inconsistent data can lead to faulty outcomes, wasting time, money, and other resources. The data source for the BDA systems should be credible and trustworthy. If the correct source of data is unknown then the results of the analysis will be futile with no tangible benefit.

Unclear fault tolerance in algorithms (C17) and unclear tangible benefits (C18) lead to uncertainty in adopting BDA systems in construction organizations. Unclear fault tolerance in algorithms implies the lack of understanding and knowledge of the working of algorithms and, thereby unclear reliability of systems. If the fault tolerance of algorithms is not well-defined, it may not be able to detect and correct errors and inaccuracies in data. Consequently, this leads to incorrect recommendations and insights, which can negatively impact an organization's decision to adopt BDA systems. To justify their investment in BDA, construction companies must first comprehend the benefits of the technology. Getting buy-in and investment from stakeholders, particularly senior management and key decision-makers, can be difficult if the advantages are not obvious or conveyed. For example, if the potential for cost savings or enhanced project performance is not communicated, firms may struggle to perceive the value of implementing BDA. Unclear tangible benefits and ambiguous fault tolerance in algorithms impeding the adoption of BDA in organizations were highlighted in several articles (Gupta & Mishra, 2022; Nti et al., 2022; Saeed & Arshed, 2022; Wang et al., 2022).

2.6 Managerial implications

Based on the output of the study, there are several managerial implications for construction organizations.

1. Lack of proper internet connectivity is the major roadblock in implementing BDA in construction organizations. Therefore, construction companies should consider upgrading their internet connectivity to ensure reliable data transfer for better outcomes.

2. A streamlined hiring drive should be carried out to employ skilled IT professionals to manage the IT systems and resolve any technical issues.

3. To ensure data consistency and standardization, organizations across the construction sector can use common data formats, metadata, and dictionaries. This will help restore order and uniformity in data collection, storage, analysis, and communication with the stakeholders.

4. Periodical training sessions should be conducted for the workforce to update them on the technical know-how of BDA systems. This should be integrated with frequent seminars, conferences, and orientations highlighting the utility of BDA systems in construction organizations.

5. Implementing cloud computing solutions can help organizations improve data transfer, storage, and analysis. Cloud-based solutions can help reduce the cost of hardware and software and offer scalability and flexibility. This can be assisted by IT vendors that can also ensure that the IT systems are up-to-date, secure, and meet their needs.

6. It is paramount that organizations should evaluate the current systems, hardware, and software to identify the areas that require upgrades or replacement.

7. Defining the BDA adoption's objectives, scope, and outcomes clearly will lead the implementation process and aid in the prioritization of investments in hardware, software, and personnel.

8. Construction organizations should create a framework for data management, data governance, and data security to guarantee that the data collected is of high quality and secure against unwanted access.

9. To overcome the hesitancy in employees' acceptance of BDA systems in construction organizations, in-house technical expertise or partnerships with third-party vendors specializing in BDA frameworks can help spread education about BDA technologies and inculcate confidence and trust in the new paradigm. Using case studies, success stories, and pilot projects to demonstrate the advantages of BDA may assist businesses in comprehending the real benefits of BDA and how it can enhance their operations.

To remove myths and misconceptions, construction organizations should provide clear and extensive descriptions of how algorithms function, including any restrictions or hazards associated. Offering proof-of-concept demonstrations or pilot projects to demonstrate the real-world benefits of BDA can help conduct rigorous risk assessments to address any privacy and security concerns and ensure that appropriate safeguards are in place to secure sensitive information.

2.7 Conclusion, limitations, and future scopes

Finally, the construction industry has demonstrated substantial growth in the adoption of BDA technology to boost productivity and efficiency. There is, nevertheless, momentous opportunity for development. The sector's major challenges in adopting BDA include a lack of proper internet connectivity and IT-related issues, high implementation and maintenance costs, lack of methods for evaluating and improving system performance, issues with real-time data sharing, reticence incorrect data processing and inconsistency, unclear fault tolerance in algorithms, and the difficulty of tracking incoming data. To overcome one of the major challenges of the lack of proper internet connectivity and IT-related issues, reliable and high-speed internet connectivity which is essential for seamless data transfer and real-time analysis must be provided by the construction organizations. Additionally, the construction industry needs to overcome IT-related challenges such as system maintenance, software compatibility, and integration of different technological platforms. Algorithms with higher fault tolerance are required to handle the complexity and variability of construction data. Developing robust algorithms that can handle incomplete or erroneous data will enhance the reliability and accuracy of BDA applications in the construction industry. Data integration and storage capabilities need to be improved to manage the vast volume of construction data generated from multiple sources. Efficient data integration and storage systems will facilitate seamless data access, retrieval, and analysis, enabling the construction industry to harness the full potential of BDA. To address these issues, the construction industry must invest in the required gear, software, and personnel and work towards standardized data formats. Attracting and retaining qualified IT workers to manage BDA initiatives is also critical. To gain the benefits of BDA, the construction industry must solve these issues and aim for a comprehensive BDA implementation.

Some of the important limitations of the study include the following. The study's potentially small sample size could limit the findings' capacity to be generalized. The results might not be indicative of the entire industry if the sample is small or limited to a certain area or type of construction organization. The study relies on surveys for the data collection, which may create bias or have restrictions on capturing the whole spectrum of difficulties. Different data collection techniques could offer various viewpoints and insights. The study participants' descriptions of their challenges in adopting BDA may contain biases or subjective interpretations. Personal experiences, perceptions, or organizational variables may impact the accuracy and dependability of their responses. The study was completed in a particular time frame, which could have limited the comprehension of the dynamic nature of BDA adoption issues. The challenges that the construction sector faces could change over time as a result of technology breakthroughs, market shifts, or legislative developments.

Future research in this area should strive to solve these obstacles and generate innovative ways to promote the wider deployment of BDA in the building industry. This can include improving systems for internet connectivity, developing methods for increasing workforce expertise and skill set, lowering costs and improving performance evaluation systems, improving data sharing methods, addressing security and privacy concerns, developing algorithms with higher fault tolerance, improving data integration, and enhancing data storage capabilities. Furthermore, future studies might concentrate on the potential benefits of BDA in the construction sector, as well as novel applications and creative uses of BDA in construction operations. These might involve research on how BDA affects project deadlines, quality control, cost management, safety, and sustainability. Research can also be directed towards the creation of BDA frameworks adapted to the construction industry's specific demands and requirements.

References

Aceto, G., Persico, V., & Pescapé, A. (2020). Industry 4.0 and health: Internet of things, big data, and cloud computing for healthcare 4.0. *Journal of Industrial Information Integration, 18*100129.

Adekunle, P., Aigbavboa, C., Akinradewo, O., Oke, A., & Aghimien, D. (2022). Construction information management: Benefits to the construction industry. *Sustainability, 14*(18), 11366.

Akinosho, T. D., Oyedele, L. O., Bilal, M., Ajayi, A. O., Delgado, M. D., Akinade, O. O., & Ahmed, A. A. (2020). Deep learning in the construction industry: A review of present status and future innovations. *Journal of Building Engineering, 32*101827.

Alaloul, W. S., Alzubi, K. M., Malkawi, A. B., Al Salaheen, M., & Musarat, M. A. (2022). Productivity monitoring in building construction projects: a systematic review. *Engineering, Construction and Architectural Management, 29*(7), 2760−2785.

AL-Khatib, A. W., & Shuhaiber, A. (2022). Green intellectual capital and green supply chain performance: Does big data analytics capabilities matter? *Sustainability, 14*(16), 10054.

Amanullah, M. A., Habeeb, R. A. A., Nasaruddin, F. H., Gani, A., Ahmed, E., Nainar, A. S. M., & Imran, M. (2020). Deep learning and big data technologies for IoT security. *Computer Communications, 151*, 495−517.

Atuahene, B.T., Kanjanabootra, S., & Gajendran, T. (2022). Preliminary benefits of big data in the construction industry: A case study. In *Proceedings of the Institution of Civil Engineers-Management, Procurement and Law, 175*(2), 67−77.

Bag, S., Pretorius, J. H. C., Gupta, S., & Dwivedi, Y. K. (2021). Role of institutional pressures and resources in the adoption of big data analytics powered artificial intelligence, sustainable manufacturing practices and circular economy capabilities. *Technological Forecasting and Social Change, 163*120420.

Błaszczyński, J., Słowinski, R., & Szelag, M. (2009). *VC-DomLEM: Rule induction algorithm for variable consistency rough set approaches.* Technical Report RA-07/09, Poznań University of Technology.

Bilal, M., Oyedele, L. O., Qadir, J., Munir, K., Ajayi, S. O., Akinade, O. O., . . . Pasha, M. (2016). Big Data in the construction industry: A review of present status, opportunities, and future trends. *Advanced Engineering Informatics, 30*(3), 500−521.

Chen, M. (2022). The influence of big data analysis of intelligent manufacturing under machine learning on start-ups enterprise. *Enterprise Information Systems, 16*(2), 347−362.

Dai, H. N., Wang, H., Xu, G., Wan, J., & Imran, M. (2020). Big data analytics for manufacturing internet of things: Opportunities, challenges and enabling technologies. *Enterprise Information Systems, 14*(9−10), 1279−1303.

Deepa, N., Pham, Q. V., Nguyen, D. C., Bhattacharya, S., Prabadevi, B., Gadekallu, T. R., & Pathirana, P. N. (2022). A survey on blockchain for big data: Approaches, opportunities, and future directions. *Future Generation Computer Systems, 131*, 209−226.

Demirkesen, S., & Tezel, A. (2022). Investigating major challenges for industry 4.0 adoption among construction companies. *Engineering, Construction and Architectural Management, 29*(3), 1470−1503.

Diemer, A., Nedelciu, C. E., Morales, M., Batisse, C., & Cantuarias-Villessuzanne, C. (2022). Waste management and circular economy in the french building and construction sector. *Frontiers in Sustainability, 38*40091.

Ding, M., Flaig, R. W., Jiang, H. L., & Yaghi, O. M. (2019). Carbon capture and conversion using metal−organic frameworks and MOF-based materials. *Chemical Society Reviews, 48*(10), 2783−2828.

Elhusseiny, H. M., & Crispim, J. (2022). SMEs, barriers and opportunities on adopting industry 4.0: A review. *Procedia Computer Science, 196*, 864−871.

Fathi, M., Haghi Kashani, M., Jameii, S. M., & Mahdipour, E. (2022). Big data analytics in weather forecasting: A systematic review. *Archives of Computational Methods in Engineering, 29*(2), 1247−1275.

Gao, R. X., Wang, L., Helu, M., & Teti, R. (2020). Big data analytics for smart factories of the future. *CIRP Annals, 69*(2), 668−692.

Ghasemaghaei, M., & Calic, G. (2020). Assessing the impact of big data on firm innovation performance: Big data is not always better data. *Journal of Business Research, 108,* 147−162.

Gupta, G., & Lakhwani, K. (2022). An enhanced approach to improve the encryption of big data using intelligent classification technique. *Multimedia Tools and Applications,* 1−34.

Gupta, S., & Mishra, R. (2022, July). The Necessity to Adopt Big Data Technologies for Efficient Performance Evaluation in the Modern Era. In *Congress on Intelligent Systems: Proceedings of CIS 2021, Volume 2* (pp. 613−623). Singapore: Springer Nature Singapore.

Gupta, S., Chen, H., Hazen, B. T., Kaur, S., & Gonzalez, E. D. S. (2019). Circular economy and big data analytics: A stakeholder perspective. *Technological Forecasting and Social Change, 144,* 466−474.

Gurgun, A. P., & Koc, K. (2022). Administrative risks challenging the adoption of smart contracts in construction projects. *Engineering, Construction and Architectural Management, 29*(2), 989−1015.

Hamza, R., Hassan, A., Ali, A., Bashir, M. B., Alqhtani, S. M., Tawfeeg, T. M., & Yousif, A. (2022). Towards secure Big Data analysis via fully homomorphic encryption algorithms. *Entropy, 24*(4), 519.

Han, H., & Trimi, S. (2022). Towards a data science platform for improving SME collaboration through Industry 4.0 technologies. *Technological Forecasting and Social Change, 174*121242.

Hasan, A., & Kamardeen, I. (2022). Occupational health and safety barriers for gender diversity in the Australian construction industry. *Journal of Construction Engineering and Management, 148*(9)04022100.

Hire, S., Sandbhor, S., & Ruikar, K. (2022). Bibliometric survey for adoption of building information modeling (BIM) in construction industry−a safety perspective. *Archives of Computational Methods in Engineering, 29*(1), 679−693.

Hwang, B. G., Ngo, J., & Teo, J. Z. K. (2022). Challenges and strategies for the adoption of smart technologies in the construction industry: The case of Singapore. *Journal of Management in Engineering, 38*(1)05021014.

Ikegwu, A. C., Nweke, H. F., Anikwe, C. V., Alo, U. R., & Okonkwo, O. R. (2022). Big data analytics for data-driven industry: A review of data sources, tools, challenges, solutions, and research directions. *Cluster Computing, 25,* 3344−3387.

Jayagopal, V., & Basser, K. K. (2022). *Data management and big data analytics: Data management in digital economy. Research anthology on big data analytics, architectures, and applications* (pp. 1614−1633). IGI Global.

Kaewnaknaew, C., Siripipatthanakul, S., Phayaphrom, B., & Limna, P. (2022). Modelling of talent management on construction companies' performance: A model of business analytics in Bangkok. *International Journal of Behavioral Analytics, 2*(1).

Kar, A. K., & Dwivedi, Y. K. (2020). Theory building with big data-driven research−Moving away from the "What" towards the "Why.". *International Journal of Information Management, 54*102205.

Khan, S. (2022). Barriers of big data analytics for smart cities development: A context of emerging economies. *International Journal of Management Science and Engineering Management, 17*(2), 123−131.

Kumar, M. (2022). Scalable malware detection system using big data and distributed machine learning approach. *Soft Computing, 26*(8), 3987−4003.

Kumar, N., Kumar, G., & Singh, R. K. (2022). Analysis of barriers intensity for investment in big data analytics for sustainable manufacturing operations in post-COVID-19 pandemic era. *Journal of Enterprise Information Management, 35*(1), 179−213.

Le, T. T., Fu, W., & Moore, J. H. (2020). Scaling tree-based automated machine learning to biomedical big data with a feature set selector. *Bioinformatics (Oxford, England)*, *36* (1), 250–256.

Lee, C. (2021). Factors influencing the credibility of performance measurement in non-profits. *International Review of Public Administration*, *26*(2), 156–174.

Li, C., Chen, Y., & Shang, Y. (2022). A review of industrial big data for decision making in intelligent manufacturing. *Engineering Science and Technology, an International Journal*, *29*101021.

Li, X., Liu, H., Wang, W., Zheng, Y., Lv, H., & Lv, Z. (2022). Big data analysis of the internet of things in the digital twins of smart city based on deep learning. *Future Generation Computer Systems*, *128*, 167–177.

Li, S., Xu, L. D., & Zhao, S. (2015). The internet of things: a survey. *Information Systems Frontiers*, *17*, 243–259.

Lu, Y., & Xu, X. (2019). Cloud-based manufacturing equipment and big data analytics to enable on-demand manufacturing services. *Robotics and Computer-Integrated Manufacturing*, *57*, 92–102.

Lutfi, A., Alsyouf, A., Almaiah, M. A., Alrawad, M., Abdo, A. A. K., Al-Khasawneh, A. L., & Saad, M. (2022). Factors influencing the adoption of big data analytics in the digital transformation era: Case study of Jordanian SMEs. *Sustainability*, *14*(3), 1802.

Mailewa, A., Mengel, S., Gittner, L., & Khan, H. (2022). Mechanisms and techniques to enhance the security of big data analytic framework with mongodb and Linux containers. *Array*, *15*100236.

Megahed, N. A., & Hassan, A. M. (2022). Evolution of BIM to DTs: A paradigm shift for the post-pandemic AECO industry. *Urban Science*, *6*(4), 67.

Mishra, S., & Tyagi, A. K. (2022). *The role of machine learning techniques in internet of things-based cloud applications. Artificial intelligence-based internet of things systems* (pp. 105–135). Springer.

Mohamed, A., Najafabadi, M. K., Wah, Y. B., Zaman, E. A. K., & Maskat, R. (2020). The state of the art and taxonomy of big data analytics: View from new big data framework. *Artificial Intelligence Review*, *53*, 989–1037.

Morales, J. Y. R., Mendoza, J. A. B., Torres, G. O., Vázquez, F. D. J. S., Rojas, A. C., & Vidal, A. F. P. (2022). Fault-tolerant control implemented to Hammerstein–Wiener model: Application to bio-ethanol dehydration. *Fuel*, *308*121836.

Myllyaho, L., Raatikainen, M., Männistö, T., Nurminen, J. K., & Mikkonen, T. (2022). On misbehaviour and fault tolerance in machine learning systems. *Journal of Systems and Software*, *183*111096.

Naeem, M., Jamal, T., Diaz-Martinez, J., Butt, S. A., Montesano, N., Tariq, M. I., ... De-La-Hoz-Valdiris, E. (2022). *Trends and future perspective challenges in big data. Advances in intelligent data analysis and applications* (pp. 309–325). Springer.

Nti, I. K., Quarcoo, J. A., Aning, J., & Fosu, G. K. (2022). A mini-review of machine learning in big data analytics: Applications, challenges, and prospects. *Big Data Mining and Analytics*, *5*(2), 81–97.

Oo, M. C. M., & Thein, T. (2022). An efficient predictive analytics system for high dimensional big data. *Journal of King Saud University-Computer and Information Sciences*, *34*(1), 1521–1532.

Raut, R. D., Mangla, S. K., Narwane, V. S., Gardas, B. B., Priyadarshinee, P., & Narkhede, B. E. (2019). Linking big data analytics and operational sustainability practices for sustainable business management. *Journal of Cleaner Production*, *224*, 10–24.

Raut, R., Narwane, V., Kumar Mangla, S., Yadav, V. S., Narkhede, B. E., & Luthra, S. (2021). Unlocking causal relations of barriers to big data analytics in manufacturing firms. *Industrial Management & Data Systems*, *121*(9), 1939–1968.

Regona, M., Yigitcanlar, T., Xia, B., & Li, R. Y. M. (2022). Opportunities and adoption challenges of AI in the construction industry: A PRISMA review. *Journal of Open Innovation: Technology, Market, and Complexity, 8*(1), 45.

Ren, S. (2022). Optimization of enterprise financial management and decision-making systems based on big data. *Journal of Mathematics, 2022.*

Saadoon, M., Hamid, S. H. A., Sofian, H., Altarturi, H. H., Azizul, Z. H., & Nasuha, N. (2022). Fault tolerance in big data storage and processing systems: A review on challenges and solutions. *Ain Shams Engineering Journal, 13*(2)101538.

Sadeghi, M., Mahmoudi, A., & Deng, X. (2022). Adopting distributed ledger technology for the sustainable construction industry: Evaluating the barriers using Ordinal Priority Approach. *Environmental Science and Pollution Research, 29*(7), 10495−10520.

Saeed, M., & Arshed, N. (2022). Revolutionizing insurance sector in India: A case of blockchain adoption challenges. *International Journal of Contemporary Economics and Administrative Sciences, 12*(1), 300−324.

Saggi, M. K., & Jain, S. (2018). A survey towards an integration of big data analytics to big insights for value-creation. *Information Processing & Management, 54*(5), 758−790.

Selvaraj, S., & Sundaravaradhan, S. (2020). Challenges and opportunities in IoT healthcare systems: A systematic review. *SN Applied Sciences, 2*(1), 139.

Sestino, A., Prete, M. I., Piper, L., & Guido, G. (2020). Internet of things and Big Data as enablers for business digitalization strategies. *Technovation, 98*102173.

Singh, A., & Misra, S. C. (2020). A Dominance based rough set analysis for investigating employee perception of safety at workplace and safety compliance. *Safety Science, 127,* 104702.

Soni, G., Kumar, S., Mahto, R. V., Mangla, S. K., Mittal, M. L., & Lim, W. M. (2022). A decision-making framework for Industry 4.0 technology implementation: The case of FinTech and sustainable supply chain finance for SMEs. *Technological Forecasting and Social Change, 180*121686.

Sreedevi, A. G., Harshitha, T. N., Sugumaran, V., & Shankar, P. (2022). Application of cognitive computing in healthcare, cybersecurity, big data and IoT: A literature review. *Information Processing & Management, 59*(2)102888.

Stergiou, C. L., Psannis, K. E., & Gupta, B. B. (2020). IoT-based big data secure management in the fog over a 6G wireless network. *The Journal IEEE Internet of Things, 8*(7), 5164−5171.

Ungurean, O. C., & Vatavu, R. D. (2022). *Users with motor impairments' preferences for smart wearables to access and interact with ambient intelligence applications and services. International symposium on ambient intelligence* (pp. 11−21). Cham: Springer.

Urbinati, A., Bogers, M., Chiesa, V., & Frattini, F. (2019). Creating and capturing value from Big Data: A multiple-case study analysis of provider companies. *Technovation, 84,* 21−36.

Vassakis, K., Petrakis, E., & Kopanakis, I. (2018). *Big data analytics: Applications, prospects and challenges.* Mobile big data: A roadmap from models to technologies (pp. 3−20). Cham: Springer.

Wang, Y., Wang, J., & Wang, X. (2020). COVID-19, supply chain disruption and China's hog market: a dynamic analysis. *China Agricultural Economic Review, 12*(3), 427−443.

Wang, J., Xu, C., Zhang, J., & Zhong, R. (2022). Big data analytics for intelligent manufacturing systems: A review. *Journal of Manufacturing Systems, 62,* 738−752.

Wang, Y., Kung, L., & Byrd, T. A. (2018). Big data analytics: Understanding its capabilities and potential benefits for healthcare organizations. *Technological Forecasting and Social Change, 126,* 3−13.

Wu, C., Li, X., Guo, Y., Wang, J., Ren, Z., Wang, M., & Yang, Z. (2022). Natural language processing for smart construction: Current status and future directions. *Automation in Construction, 134*104059.
Xu, M., Nie, X., Li, H., Cheng, J. C., & Mei, Z. (2022). Smart construction sites: A promising approach to improving on-site HSE management performance. *Journal of Building Engineering, 49*104007.
Yoffe, H., Plaut, P., & Grobman, Y. J. (2022). Towards sustainability evaluation of urban landscapes using big data: A case study of Israel's architecture, engineering and construction industry. *Landscape Research, 47*(1), 49–67.
Zeadally, S., Siddiqui, F., Baig, Z., & Ibrahim, A. (2020). Smart healthcare: Challenges and potential solutions using internet of things (IoT) and big data analytics. *PSU Research Review, 4*(2), 149–168.
Zhang, D., Pee, L. G., Pan, S. L., & Cui, L. (2022). Big data analytics, resource orchestration, and digital sustainability: A case study of smart city development. *Government Information Quarterly, 39*(1)101626.
Zhang, Y., Yan, D., Hu, S., & Guo, S. (2019). Modelling of energy consumption and carbon emission from the building construction sector in China, a process-based LCA approach. *Energy Policy, 134*110949.
Zhao, J., & Li, S. (2022). Life cycle cost assessment and multi-criteria decision analysis of environment-friendly building insulation materials-A review. *Energy and Buildings, 254*111582.

CHAPTER THREE

Role of unmanned air vehicles in sustainable supply chain: queuing theory and ant colony optimization approach

Muhammad Ikram[1], Idiano D'Adamo[2] and Charbel Jose Chiappetta Jabbour[3]

[1]School of Business Administration, Al Akhawayn University in Ifrane, Ifrane, Morocco
[2]Department of Computer, Control and Management Engineering, Sapienza University of Rome, Rome, Italy
[3]Department of Information Systems, Supply Chain Management & Decision Support, NEOMA Business School, Mont-Saint-Aignan, France

3.1 Introduction

The supply chain sector is a vital component of contemporary business activities, encompassing the administration of products and services from their genesis to the final consumer. This intricate process engages various stakeholders, such as manufacturers, suppliers, distributors, retailers, and end-users. Effectively managing the supply chain is imperative for business success, as it influences aspects like client contentment, cost-effectiveness, and environmental sustainability (Parajuli et al., 2019).

A recent publication by MarketsandMarkets has projected an increase in the UAV logistics and transportation market from USD 11.20 billion in 2022 to USD 29.06 billion by 2027, with a compound annual growth rate (CAGR) of 21.01% (MarketsandMarkets, 2020). Concurrently, an independent report by DRONEII has also forecasted a rise in the global UAV market from USD 14 billion in 2018 to above USD 43 billion in 2024, marking a CAGR of 20.5% (DRONEII, 2019). Industry leaders like Amazon, DHL, Federal Express, Google, and Facebook are actively investigating the applicability of drone technology in logistics, package delivery, and internet connection transmission (Lee & Choi, 2016; Wang et al., 2017). The envisioned role of

Computational Intelligence Techniques for Sustainable Supply Chain Management.
DOI: https://doi.org/10.1016/B978-0-443-18464-2.00013-3

drones in these companies' business models primarily revolves around last-mile delivery and direct transportation of packages from the depot to the end users (Yurek & Ozmutlu, 2018).

In the contemporary business landscape, sustainable supply chain management (SCM) has emerged as a significant concept. This approach entails the fusion of environmental, social, and economic considerations within the supply chain operations. Its primary objective is to protect the environment while also ensuring economic sustainability (Sarkis, 2020). By adopting sustainable supply chain practices, businesses can bolster their reputation, decrease expenses, and elevate stakeholder value.

A major challenge confronting the supply chain sector is the transportation of goods. As a vital component of the supply chain process, transportation accounts for a considerable portion of the overall costs. Conventional transportation methods, such as trucks and ships, are often slow and emit large amounts of carbon dioxide. Unmanned Air Vehicles (UAVs), or drones, have emerged as an innovative solution to this issue.

A UAV is a remotely piloted aircraft that does not require a human operator onboard. They have been utilized in diverse applications, such as military surveillance, photography, and video production. The adoption of UAVs within the supply chain industry has accelerated in recent years due to their ability to increase delivery speeds, decrease transportation expenses, and diminish carbon footprints (Triche et al., 2020). Capable of navigating traffic and bypassing congestion, UAVs facilitate expedited deliveries (Chiang et al., 2019). Additionally, UAVs can be deployed quickly and do not require extensive infrastructure, such as airports or ports, making them a cost-effective transportation option. Moreover, UAVs can reduce CO_2 substantially, enhancing supply chain sustainability (Chiang et al., 2019).

While UAVs hold immense promise for the supply chain sector, their incorporation is challenging. A primary challenge lies in the regulatory framework governing their use. Aviation authorities worldwide enforce varying regulations, which can impede seamless integration (Sun et al., 2019). Additionally, privacy and security concerns related to UAV deployment must be addressed. To implement UAVs in the supply chain, these challenges need to be examined and navigated thoroughly. By carefully considering regulatory constraints and addressing privacy and security issues, businesses can harness the potential of UAVs to transform their supply chain operations.

Costs associated with supply chain and logistics encompass transportation, warehousing, and order processing expenses. Companies can diminish these by integrating UAVs into their SCM and logistics procedures. From a cost perspective, UAVs are potentially more economical and environmentally friendly compared to trucks, given they are battery-operated instead of relying on gasoline (Ha et al., 2018). The incorporation of UAVs in delivery operations is expedient and cost-effective in terms of per-kilometer transportation expense (Wang et al., 2019). Shavarani (2019) predicted that aerial delivery systems could lead to annual cost savings upward of USD 50 million. Drones, in contrast to traditional delivery vehicles like trucks, incur lower maintenance costs and have the potential to decrease labor expenses by performing tasks independently (Dorling et al., 2016). Moshref-Javadi et al. (2020) propose that drones, when combined with trucks, can lead to significant cost savings, especially for deliveries in suburban regions with moderate density. UAVs facilitate prompt delivery of products to customers, thereby supporting efficient SCM and logistics operations and circumventing delay-related costs.

The maneuverability of air defense weapons has consistently been a focal point in ongoing research concerning UAV path planning (Qadir et al., 2021). As defined by Vagale et al. (2021), path planning for autonomous vehicles involves devising an optimal route from the starting point to the destination while circumventing obstacles. The authors further emphasize that path planning has been instrumental in enhancing the level of autonomy for UAVs.

A path planning technology must meet four fundamental requirements: high computation efficiency, safety, optimization, and feasibility (Puente-Castro et al., 2022). UAV performance and autonomy can be improved through innovations in path planning (Wan et al., 2022).

The purpose of this chapter is to discuss the challenges of collision avoidance in supply chain scenarios involving multiple UAV deployments at a predetermined arrival rate. This study explores three-dimensional (3D) path planning for UAVs by utilizing queuing theory, in which space is divided into three sections between two points—A and B. UAVs deployed in the supply chain system must have an exponentially distributed arrival rate and navigate from point A to point B, with the arrival information updated periodically over a fixed time interval.

This chapter aims to develop and implement a path-planning algorithm for UAVs that allows new UAVs to alter their routes when they encounter potential collisions. The UAVs' speed must also be adjustable to avoid congestion in situations where one UAV is ahead of the others. By effectively addressing these challenges, the study aims to improve the safe and efficient integration of UAVs into supply chain systems.

UAV's integration into SCM can open up new opportunities for businesses. For instance, this technology can facilitate same-day delivery services, resulting in enhanced customer satisfaction and increased revenue, even in cases involving multiple UAV collisions. Moreover, UAVs enable businesses to expand their reach into remote areas, helping them gain a greater share of the market (Colajanni et al., 2022).

Several strategies can be used to optimize the integration of UAVs into the supply chain. The queueing theory offers a mathematical framework for analyzing and optimizing waiting lines, among other things. The application of queueing theory can help pinpoint supply chain bottlenecks and optimize the flow of goods. Ant Colony Optimization (ACO) uses a metaheuristic algorithm based on ant behavior to solve optimization problems, which has proven effective. ACO can enhance efficiency and reduce costs by optimizing UAV routing and scheduling. These approaches can allow businesses to fully leverage the potential of UAVs in SCM.

The integration of UAV technology, queueing theory, and ACO results in greater efficiency and sustainability of supply chain processes. In addition to producing new opportunities for businesses, UAV integration in the supply chain can also enable same-day delivery services, leading to higher customer satisfaction and higher revenue.

This chapter explores the role of UAVs in sustainable supply chains and applies queueing theory and ACO to supply chain process optimization. Furthermore, this chapter discusses potential research avenues for UAV integration into supply chains as well as challenges and opportunities associated with this process. This chapter contributes to the existing literature on sustainable SCM by providing valuable insights into operating multiple UAVs within a supply chain system.

This chapter is organized as follows. Section 3.2 delivers a review of contemporary literature on the application of UAVs in SCM, emphasizing the use of various algorithms. Section 3.3 provides an overview of different advanced methods achieved by integrating UAVs with SCM. In Section 3.4, results are demonstrated through a novel approach that tackles the issue of collision avoidance when deploying multiple UAVs. The final section of this chapter concludes the discussion, detailing the limitations and outlining potential future directions.

3.2 Literature review

UAVs have captivated the interest of researchers and practitioners alike in recent years. This section examines various approaches employed to optimize the integration of UAVs into supply chains and the role of UAVs in SCM.

Numerous studies have investigated UAVs' potential for SCM from a transportation perspective. For example, Wang et al. (2021) put forth a drone-enabled transportation system for last-mile delivery. The authors devised a model to optimize the delivery process and demonstrated that drone usage could significantly decrease delivery time and costs. In a similar vein, Zheng et al. (2018) suggested a multi-UAV cooperative transportation system that could enhance the efficiency of last-mile delivery. Through simulation experiments, the authors illustrated that this system could markedly reduce delivery time and boost UAV utilization rates.

Additionally, UAVs have been used in the supply chain for inventory management. Kurniawan et al. (2019) suggested an inventory management system employing UAVs for stock-taking and inventory monitoring. The authors created a model to optimize inventory levels and demonstrated that UAV usage could reduce inventory holding costs and enhance inventory accuracy.

UAVs have also been used to monitor and inspect supply chains in addition to transportation and inventory management. For instance, Lee et al. (2019) introduced a UAV-based monitoring system for container yards. By using UAVs to monitor the container yard layout, the authors demonstrated that monitoring time could be significantly reduced and container yard safety could be improved significantly.

The optimization of UAV integration has been achieved through a variety of approaches, including optimization algorithms and simulation experiments. For example, Anagnostopoulos et al. (2023) presented an optimization model for integrating UAVs into supply chains. In the study, a genetic algorithm was used to optimize UAV routing and it was found that the model could significantly lower transportation costs and carbon emissions.

A similar approach is employed to optimize the integration of UAVs into SCM using queueing theory. Queuing theory offers a mathematical framework for analyzing and optimizing waiting lines. Zhou et al. (2021) proposed a queuing model for the optimal integration of UAVs in the supply chain. The authors demonstrated that utilizing queuing theory could help pinpoint bottlenecks in the supply chain process and optimize the flow of goods.

An optimization algorithm called ACO optimizes UAV integration in SCM. ACO is a metaheuristic algorithm based on ant behavior that has been successfully used to address optimization problems. Zhou et al. (2021) suggested an ACO-based algorithm for optimizing UAV routing in the supply chain. As a result of the algorithm, the authors showed that the supply chain was more efficient and transportation times were reduced.

A pioneering study on UAVs in logistics was conducted by She and Ouyang (2021). The study examined the feasibility of utilizing UAVs for last-mile delivery, concluding that UAVs could decrease delivery times and costs. In addition, She and Ouyang (2021) conducted a study on using UAVs in the pharmaceutical supply chain, observing that they could improve efficiency and reliability.

The application of UAVs in disaster relief operations has also been studied in several studies. According to Kamat et al. (2022), UAVs could provide timely and effective assistance during disaster relief operations when used in humanitarian logistics. Similarly, Amiri et al. (2018) explored the application of UAVs in disaster response operations, concluding that UAVs could significantly increase the speed and efficiency of such responses. Yan et al. (2017) suggested a multi-objective optimization model for UAV-based logistics, which was shown to optimize transportation costs and delivery times.

UAVs are excellent at transporting goods rapidly, which is one of their primary benefits in SCM. In a study conducted by Maciel-Pearson et al. (2019), UAVs reduced parcel delivery times by up to 80%. Moreover, UAVs can deliver goods to remote areas where traditional transportation methods, such as trucks and ships, are not feasible. Additionally, UAVs can lower transportation costs by removing the need for airports and ports.

UAVs can significantly contribute to supply chain sustainability by reducing CO_2 emissions. Goodchild and Toy (2018) revealed that utilizing UAVs reduced CO_2 emissions by up to 95% compared to traditional delivery methods. The diminished carbon footprint can also result in improved corporate social responsibility (CSR) for businesses.

There are a number of challenges associated with UAV use in SCM, and one of them is the regulatory framework. The use of UAVs is regulated by aviation authorities, and these regulations differ across countries. For instance, in the USA, the Federal Aviation Administration (FAA) enforces strict regulations for using UAVs for commercial purposes. These regulations mandate businesses to obtain a remote pilot certificate and restrict the altitude and distance of UAVs. In contrast, regulations in countries such as China and Japan are less restrictive, leading to faster adoption of UAVs in the supply chain process (Kim, 2019).

Path planning for UAVs constitutes a significant portion of the literature reviewed. There is an extensive use of algorithms as a significant part of UAV research involves route optimization. Nevertheless, the need for further advancement and breakthroughs in UAV path planning remains. The prospects in this field are outlined as follows.

In industrial settings, UAVs typically fly below 400 m in low-altitude conditions. This complex environment presents unique challenges for UAV navigation algorithms. Thus, these algorithms must account for varying conditions such as altitude, wind speed, temperature, and humidity, adapting to the specific flight environment.

There's a scarcity of research focused on UAV navigation in urban contexts. Potential areas for exploration include improving the detail in flight environment modeling, increasing the efficiency of navigation algorithms, and enhancing the coordinated operations between UAVs and delivery vehicles. There's also scope to merge various algorithms to boost the algorithm's real-time decision-making efficiency. Currently, most navigation strategies target individual UAVs. With the trend moving toward UAV groups and smarter tech, future studies might explore advanced navigation techniques for UAV groupings.

There have been numerous algorithms and methods developed for the path planning of UAVs by various authors, including Wahab et al. (2020). Furthermore, potential fields, cell decompositions, and potential fields are some of the general path-planning techniques. It is important to consider both static and dynamic constraints in comprehensive path planning to avoid UAV collisions with adversaries. In this regard, dynamic obstacles present more dimensions in path planning for UAV navigation in space. While path planning for static obstacles involves space configuration, dynamic obstacles introduce an additional dimension to the space configuration: time. However, Dobrokhodov et al. (2020) suggested that incorporating the time dimension in the state space of UAVs is more akin to trajectory or motion planning rather than path planning.

In a pivotal study within the realm of SCM logistics, Ben Amarat and Zong (2019) explored the complexities of routing optimization for a singular vehicular entity traversing diverse potential trajectories. The methodology underscores the incorporation of a penalty weight to the anterior route chosen by a UAV during its successive planning stages. Within a spatial framework delineated by grid structures, grid cells charted by a specific UAV are accorded elevated penalty metrics, thereby signaling auxiliary UAVs to pursue alternate navigation strategies. The algorithmic schema, in its deployment, commences with a probabilistic map-based route formulation for the primary vehicle. This is succeeded by generating a modified spatial representation, superimposing penalty metrics on grid cells affiliated with the route of the principal vehicle. In the subsequent phase, drawing from this augmented spatial schematic, a trajectory is devised for the ensuing UAV. It is worth noting that this iterative methodology is scalable and adaptable for scenarios encompassing an array of UAVs within the supply chain logistics.

It is evident from the existing literature that UAVs have significant potential with respect to SCM. Supply chain processes can be made more efficient, faster, and sustainable by using UAVs. The integration of UAVs in SCM, however, faces several challenges, including regulatory and security concerns. To the best of the author's knowledge, this study is the first of its kind to highlight the issue of UAV collisions within the supply chain network. As a result, this study fills a gap in the literature by developing a multidimensional model for making the supply chain network more efficient and avoiding collisions with UAVs.

3.3 Methodology

This study aims to determine the most efficient path between points A and B without colliding by deploying multiple UAVs in the spatial system. The UAVs are required to have an exponential distribution arrival rate, and congestion is not desired in a given cell, as that may lead to collision. The proposed technique for the solution of path planning is ACO.

This study uses mobile UAVs as artificial ants. A UAV inspired by real ants has been developed. Real ants move based on a chemical trail known as pheromone. The ants prefer the regions with the greatest concentration of this chemical trail among the available paths (Sharma et al., 2022). Pheromones, as discussed above, are left behind by the moving ants that guide the followers. In this case, the insects would tend to follow the rich area of the chemical trail, as described above. In this case, the pheromone refers to the probabilistic measure used in the ACO algorithm. The probabilistic measure was calculated using the equation below (Selvi & Umarani, 2010):

$$P_{i,j}^{n} = \frac{\left(\lambda_{i,j}\right)^{\alpha}}{\sum_{j\in\ N_{i}^{n}}^{m}\left(\lambda_{i,j}\right)^{\alpha}} \tag{3.1}$$

$P_{i,j}^{n}$ = Transition probability for then n-th ant navigates from node (i) to node (j). m = Number of iterations that represent ants. $\lambda_{i,j}$ = Pheromone trail. α = Weight value.

The traveling ants are the ones that update the pheromone trails. This corresponds to the number of iterations made for updating the paths to achieve the optimum solution according to the fitness function. The simulation was implemented in MATLAB® according to the procedures in the pseudocode below:

Algorithm: Ant Colony Optimization;
Initialization of the (ACO) algorithm parameters;
Construction of grid map;
Placement of the obstacles in random coordinates;
Locate point A (starting point); locate point B (destination of the UAVs)
If iterate = 1, 2, 3 ... N
  For ant = 1, 2, 3... n
    For step = 1, 2, 3 ... m
          Find the likelihood (probability) of n-th next ...
coordinates
          Move to next node location as determined by probability.
          If the current position is equal to destination
            Break
          End
      End
    Keep in memory the distance moved by the n-th ant.
    Calculate the pheromone amount produced by the n-th ant
  End
  Update pheromone concentration for the whole map.
End
Display results.

Here's a possible content writing based on the given pseudocode and results:

Pseudocode for the algorithm begins with the declaration of constants and the initialization of variables. There are several constants included in this model: the number of cities, the number of ants, the alpha and beta parameters for the pheromone and distance heuristics, the evaporation rate for the pheromone, and the initial pheromone level. In addition to the distance matrix and the pheromone matrix, there are probability matrices, path matrices, and probability matrices.

The main iteration of the algorithm involves looping for the number of ants, which is set to 500 in this case. Within each ant's loop, three nested loops update the probability matrix, select the next city based on the probabilities, and update the path and pheromone matrices accordingly. These loops implement the ant's decision-making process, which balances the exploration of new paths with the exploitation of known paths.

A pheromone matrix is updated by evaporation and by adding pheromone deposited by the ants after all ants have completed their tours. As determined by the length of the ants' tours, this pheromone update reinforces the paths used by the best ants.

The code that implements this algorithm was tested on a set of benchmark instances and produced results that were compared with the Genetic Algorithm (GA) approach used by Mirjalili et al. (2020) for the same instances. The results showed that the ACO approach found better solutions than the GA approach in most cases, especially for large instances. The Traveling Salesman Problem is a well-known combinatorial optimization problem that is NP-hard. This demonstrates the effectiveness of the ACO algorithm.

3.3.1 Path planning for multiple unmanned air vehicles

Drone technology is enabling a greater number of UAVs to be deployed simultaneously. As a result, separate paths may be required for the UAVs. For instance, the case may be illustrated using a simple scenario of two available paths such that the likelihood that a given UAV encounters an adversary along a path "i" is represented by π_i; if a UAV encounters an obstacle (adversary) the probability that it becomes neutralized becomes 1. Considering dual methodologies for path planning in unmanned aircraft systems, the subsequent two strategies can be employed.

The first approach centers on a unified routing system. Here, both UAVs follow an identical trajectory, culminating in a probability of reaching the desired location, symbolized as 1-p_1. Notably, p_1 encapsulates the chance that neither UAV attains the end goal, as highlighted by (Li, Ge, et al., 2020). Conversely, the secondary approach emphasizes diverse routing. Here, each UAV is designated a unique trajectory. The cumulative probability that both vehicles converge at the intended location arises from the combined probabilities associated with individual non-reachability along their respective routes. This likelihood is written as (1-p_1)(1-p_2). Since p_1 >0,p_2 >0 and p_1 < 1,p_2 < 1, then the product (p_1 p_2) which is the likelihood that no UAV reaches the target is less than 1-p_1. This pattern suggests that when UAVs embark on separate routes, their propensity to face challenges or obstacles amplifies, a perspective resonated by Li, Xiong, et al. (2020).

Based on the assessment, when consistent arrival at the destination is uncertain for UAVs, the alternative strategy proves more viable for mission success. This is underscored by the reduced probability of UAVs failing to reach their end goal. The rationale is straightforward: allocating distinct routes to the UAVs diminishes their risk of detection by potential threats in the environment.

Panda et al. (2020) proposed a path-planning approach for a single vehicle with multiple possible paths. The planning process is sequential, with penalty weights added to the previous UAV's path to avoid traversing the same cells. In a grid-based spatial layout, cells once crossed by a UAV incur a significant penalty, leading other UAVs to bypass them. A four-step algorithm is used for this approach. The first vehicle's path is planned based on a probability map. The second step is to add a penalty to the cells along the path generated for the first vehicle, resulting in a new map. The third step involves path planning for the second UAV based on the updated map. Finally, these three procedures are repeated for each subsequent UAV (Huang et al., 2023).

Yao et al. (2015) simulated this planned path in a 30 by 30 spatial system represented by square cells. Random values were uniformly produced across five simulation zones, and P(S_i) values were calculated using equation vi. The cell at coordinates (30, 30) was designated as the origin, with the target at coordinates (3, 5) marked with an asterisk. The regions with darker shading are more dangerous and have higher P(S_i) values, as shown in the diagram below. This approach provides a practical solution for optimizing path planning in multiple UAV scenarios and minimizing the risks associated with traversing hazardous regions.

Fig. 3.1A shows the results of the simulation conducted by Yao et al. (2015). On the left side of the Fig. 3.1C the results of the simulation when a weight of 6 was used, producing the safest path within the spatial system. The path deviates from the darkened grids, which represent high-risk zones, resulting in a non-straight path. This approach ensures that the UAV avoids the most dangerous regions and follows the safest possible route. On the right side of Fig. 3.1A, the same probability map was used with a different approach, utilizing a weight specified by the expression below. This approach produces a straighter path than the previous method, as seen in the diagram. However, it still avoids the high-risk regions to ensure the safety of the UAV. Both methods demonstrate the effectiveness of the path-planning approach proposed by Schwarting et al. (2018), as they successfully navigate through a complex spatial system with hazardous regions. The ability to avoid danger zones and navigate the safest path is crucial for the success of missions involving multiple UAVs, and this approach provides a practical solution to optimize path planning and minimize risk.

$$d_{ij} = \left\{ \begin{array}{l} -\log\left(P\left(S_j\right)\right) + c \\ \infty \end{array} \right. \tag{3.2}$$

The weighted link Eq. (3.2) used on the right-hand side of Fig. 3.1A includes a constant real number c, which was adjusted to $c = 0.1$ after modifying the path equation. Comparing the two diagrams, the right-hand side result shows the shortest path while still avoiding the high-risk regions. This approach provides a more efficient solution for path planning in complex spatial systems with multiple possible paths.

In Fig. 3.1B, the results obtained by Kapadia et al. (2016) depict a scenario involving dynamic obstacles, represented by a diamond sign in cell (23, 22). The UAV on the right-hand side shows a deviation from the original path after a change in the probability of cell (23, 22) was detected. To avoid the obstacle, the UAV followed the solid line instead of the dotted path. In a scenario where multiple UAVs are deployed, the first vehicle follows the solid line, the second UAV follows the dashed line, and the third vehicle follows the dotted line. These paths were produced based on the sequence of deployment and the probability map. Overall, these approaches demonstrate the importance of path planning in ensuring the success of

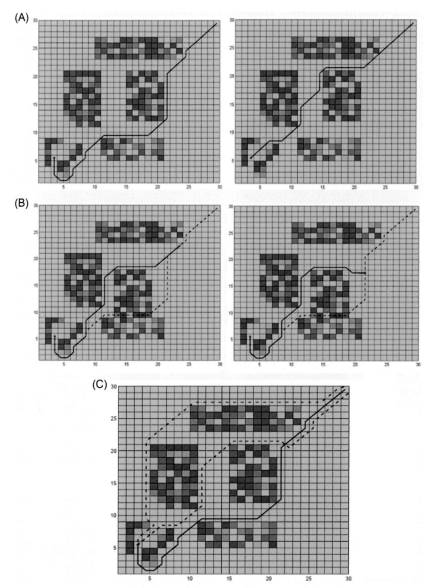

Figure 3.1 (A) Unpenalized results on the left, penalized results on the right. (B) Navigation path of UAV during alteration in probability map for both cases. (C) Paths taken by multiple UAVs during deployment. *UAVs*, unmanned air vehicles.

missions involving multiple UAVs. The ability to adapt to changes in the environment, avoid obstacles, and navigate the safest possible path is crucial for minimizing risks and achieving mission objectives.

3.3.2 Queuing theory models

Queuing theory models are a crucial analytical tool used in the optimization of air traffic problems, telephony networks, and other scenarios that involve queues or waiting lines. In air traffic control, queuing theory is used to grant permissions for take-off and landing, as well as booking-related activities. The theory is also applicable in modeling the path for multiple UAVs (Zhang et al., 2019).

To understand the application of queuing theory to UAV path planning, it is essential to review its characteristics. A queuing system consists of three main attributes: the arrival characteristics, waiting line, and service facility. The arrival characteristics refer to the input sources that generate arrivals to the service system. It has three primary attributes: the behavior of arrivals, the size of the calling population, and the pattern of arrivals (Zhang et al., 2023).

The behavior of arrivals describes the time interval between customers or arrivals according to a probability distribution. The size of the calling population represents the number of customers that are waiting in the queue, which could be finite or infinite. The pattern of arrivals refers to the randomness or predictability of the arrival process.

In the context of UAV path planning, the arrival characteristics could represent the probability distribution of the UAVs' arrival time and location. The waiting line could represent the queue of UAVs waiting to reach their destination, while the service facility could represent the path planning algorithm used to guide the UAVs toward their target. Overall, queuing theory provides a valuable framework for optimizing UAV path planning, enabling efficient resource allocation, and reducing waiting times for multiple UAVs. Researchers and practitioners can develop effective path-planning strategies for UAVs by modeling arrival characteristics, waiting lines, and service facilities in a variety of scenarios.

The behavior of arrivals in a queuing system refers to the properties of the input to the system that describe the events of the input in relation to the queue. In an ideal case, the inputs or subjects wait in the queue until they are served. However, in practical situations, such as serving arriving passengers at an airport, some behaviors, such as jumping the queue, may occur. Some subjects may fail to join the queue, which is known as balking, while others may join the waiting line but eventually leave before being served, referred to as reneging.

In the system, the size of the calling population determines how many subjects arrive each time. The arrival rate may be unlimited or limited. Finite calling refers to the arrival of subjects that involve a countable number, unlike unlimited arrival. The pattern of arrival describes the schedule of the arriving subjects. For example, a given number of drones may be arriving in a controlled space at a given time. A random arrival is also possible if the arrivals of the subjects are independent of each other. In queuing models, the arrival is approximated using probability distributions. Poisson or exponential distribution is commonly used in describing the arrival rate.

By considering the behavior of arrivals, the size of the calling population, and the pattern of arrivals, queuing theory can be used to optimize path planning for multiple UAVs. This framework can help researchers and practitioners develop strategies to manage the arrival of UAVs, reduce waiting times, and ensure efficient resource allocation.

$$P\left(n;t\right) = \frac{(\lambda t)^{n}}{n!} \, e^{-\lambda t} \qquad (3.3)$$

'n' queuing theory, the variable n is set for integers such that $n = 0$, 1, 2... which represents the number of arrivals per unit time. $P(n;t)$ is the probability of n arrivals, where lambda is the average arrival rate.

In the case of an unlimited population, models like exponential arriving rate or Poisson's arriving rate can be used in modeling exercises. Waiting lines are parts of a queue system, and their length can be either unlimited or limited. A limited queue line cannot expand to become infinite due to physical law or restrictions, while an unlimited waiting line has no restrictions on size.

In planning UAV paths, queue discipline is another vital attribute of a waiting line, which describes the rules by which subjects receive their service. For instance, First-In-First-Out (FIFO) is a common rule in most queue systems. However, the FIFO system is not applicable in emergency cases (Jayaweera & Hanoun, 2020).

The service facility has two attributes: configuration and pattern of the service system. The classification based on the queuing configurations of the system is in terms of the number of servers, channels, and the number of service stops, among others. In the spatial system of UAV path planning, the cubes for a 3D system represent service points. This is because the probability of an adversary encountering a vehicle in space occurs at the cubes.

In summary, queuing theory provides a framework for optimizing path planning in the spatial system of UAVs. By considering attributes such as arrival characteristics, waiting lines, and service facilities, researchers and practitioners can develop effective strategies to manage the arrival of UAVs, reduce waiting times, and ensure efficient resource allocation.

In queuing theory, models of queuing systems are identified using Kendall notation. The Kendall rules define the arrival patterns and distribution of service time (Voznak et al., 2011). The notation comprises three common letters of the alphabet used to describe the arrival pattern. The letter A denotes inter-arrival time distribution, while the letter B denotes service time distribution. Other symbols that fall under A and B are M, D, Ek, and G, which denote exponential, deterministic, Erlang of order k, and general, respectively.

The exponential distribution is commonly used to model inter-arrival times, while the service time can be modeled using a variety of distributions, such as exponential, deterministic, or Erlang of order k. The general distribution, denoted by the letter G, is used when the service time cannot be modeled using any of the other distributions. By using Kendall notation, researchers and practitioners can identify the appropriate queuing model for a given scenario and optimize the path planning for multiple UAVs. This framework can help manage the arrival of UAVs, reduce waiting times, and ensure efficient resource allocation in a spatial system.

3.3.3 Three-dimensional path-planning algorithms

In the previous section, we discussed how queuing theory can be used to describe the arrival of items into a system, queue behaviors, and service time. Implementing queuing theory can be done using various algorithms that are commonly used in various applications. The evolutionary computational strategy, commonly known as numerical optimization techniques, is often used in path planning for multiple robots (Song et al., 2020).

The common algorithms applied to path planning for autonomous vehicles include the Genetic Algorithm (GA), Particle Swarm Optimization (PSO) algorithm, and ACO. The paradigm employed in these algorithms in path planning is called the multiple Traveling Salesman Problem (mTSP). The principle operation of the TSP is that more than one salesman is permitted to form part of the solution. Multiple single UAV path planning problems are solved by determining the sequence of visiting the desired regions. Genetic algorithms and Ant Colony algorithms are discussed below.

GA is a typical evolutionary algorithm applied in numerous optimization problems such as UAV path planning and so on. The algorithm works like a natural biological system that involves information transfer in genes, hence the name. The Genetic Algorithm involves three main parts in its execution time, namely natural selection, crossover, and mutation. GA works in three stages (Okwu & Tartibu, 2021):

1. Initialization: The initial population is generated randomly, usually with a specified number of chromosomes.
2. Selection: The fittest chromosomes are selected based on fitness evaluation criteria, such as the highest fitness score or lowest cost.
3. Crossover and Mutation: The selected chromosomes are used to create new offspring by combining the genetic information using crossover and mutation operators. The offspring then form the next generation of the population, and the process is repeated until a satisfactory solution is found.

ACO is another algorithm that is widely used in path planning for multiple UAVs. It is inspired by the behavior of ants in their search for food. ACO involves three main components, namely the ants, the pheromone, and the environment (Rokbani et al., 2021). The ants leave a trail of pheromone as they move, and the pheromone is used by other ants to follow the path to the food source. The algorithm works by simulating this behavior in the search for the optimal path (Zhou et al., 2022).

By utilizing evolutionary algorithms such as GA and ACO in path planning for multiple UAVs, resource allocation can be optimized, waiting times can be reduced, and arrivals can be managed efficiently. These algorithms offer a promising approach to solving complex optimization problems in spatial systems. In the optimization process, the algorithm generates solutions, also known as chromosomes, through an iteration process. Each chromosome is updated according to the fitness function, and the best-fit chromosome is selected to appear in the next generation.

The mutation is not random but rather stepwise, at a rate of one percent or a mutation rate of 0.01. This means that, in the crossover population, one percent undergoes mutation. For example, if thirty chromosomes are produced, each with six bits, then the bits mutated are determined by $30 \times 6 \times 0.01 = 1.8$, which is roughly 2 bits (Brand et al., 2010).

As a result of the GA process, the generated solutions are updated according to the fitness function, and the best-fit chromosome is selected to appear in the next generation. Crossover and mutation are used to achieve diversity and optimize the solution space. The application of the Genetic Algorithm in UAV path planning can help optimize resource allocation and reduce waiting times, ensuring efficient and safe operation of multiple UAVs.

3.3.4 Ant colony optimization

ACO is another optimization algorithm that imitates the traveling salesperson paradigm of optimization. Similar to evolutionary algorithms, it mimics the social attributes of an ant system, where traveling ants leave a trail of substance that guides them in their path. The substance, the pheromone trail, acts as a feedback system that guides future ant movements toward the most efficient path (Dai et al., 2019).

ACO is useful in finding an optimal path and is commonly applied in telecommunication routing path finding, among other numerical optimization problems. Its application is similar to PSO and is, therefore, chosen for application in this study. In the context of UAV path planning, ACO can be used to optimize the path selection process, considering factors such as avoiding high-risk areas and minimizing travel time. The algorithm can help improve the efficiency and safety of multiple UAV operations in a spatial system, ensuring that resources are optimally utilized while minimizing waiting times.

3.3.5 Proposed model

The problem at hand involves 3D path planning for UAVs arriving in a space governed by exponential distribution rules. Each UAV enters location A, defined by coordinates, and aims to reach point B, also specified by coordinates. This particular issue involves the presence of obstacles, which may cause collisions if an inadequate path is chosen between the two points. As a result, optimization rules are appropriate for addressing this problem. It's important to note that airflow speed is not considered in this scenario, and the simulation will be conducted for a predetermined duration. Additionally, each cube serves as a service center according to queuing theory. Probability functions are employed to determine the optimal path in this context.

Let's consider the space below.

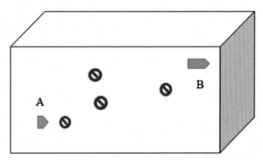

Figure 3.2 Sketch for showing the starting point and destination of the UAVs. *UAVs, unmanned air vehicles.*

The Fig. 3.2 presents a regulated space featuring an arrival point, denoted as A, for the UAVs and a destination, designated as point B. Scattered between these two points are red hazard markers representing threats that the UAVs must circumvent. As discussed in the literature review, these markers are randomly generated to ensure a high probability of collision, which the UAVs must avoid. Additionally, the speed of the UAVs is controlled to prevent collisions among the vehicles themselves.

To solve this problem, this study proposes to use the ACO algorithm implemented in MATLAB® software. A model representing the proposed solution can be observed in Fig. 3.3.

3.4 Results and discussion

This study implemented the ACO algorithm to determine the paths taken by UAVs between their origin and destination multiple supply chain points. The starting point A (0, 0, 0) signifies the entry cubicle where all UAVs enter the system, and the endpoint B (20, 20, 0) marks the destination. Between points A and B, circular obstacles were generated using the ezplot function, featuring a unit-length radius. In the controlled space, a total of fourteen obstacles were created. These obstacles were assigned threat values based on equation iv from graph theory. Fig. 3.4 provides a graphical representation of the generated obstacles between supply chain routes.

The fourteen obstacles, produced using the ezplot function, pose risks to the UAVs, which must avoid traversing through their coordinates

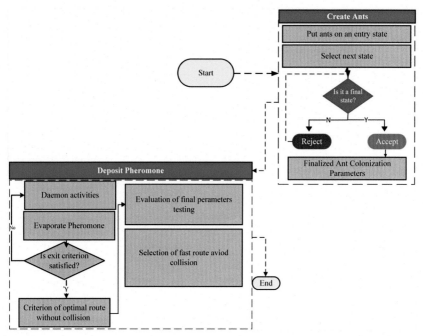

Figure 3.3 Ant colony algorithm flow chart for supply chain optimization.

within the supply chain network. In the metaheuristic optimization process for the UAV path, these points introduce penalties. Consequently, while converging toward the supply chain destination points, the UAVs circumvent these coordinates and opt for the nearest path. Fig. 3.5 illustrates the path taken by UAVs from the origin to the destination of the supply chain network represented by the blue line.

Despite deploying numerous UAVs resembling ants, a singular route was taken between supply chain points A and B, influenced by the obstacle layout. This route was determined by a safety factor, bypassing coordinates with heightened penalties. Pheromone trails served as indicators for secure delivery routes, highlighting areas with dense pheromone presence. For instance, Fig. 3.6 showcases these trails, with lines marking feasible routes. The color scale indicates that a more yellowish tint relates to denser pheromone levels, indicating safer navigation zones. Such trails present various alternate routes for other UAVs, minimizing the risks of congestion or collisions.

In the queuing framework, spatial cuboids act as service hubs, and UAVs are attracted to those exhibiting elevated pheromone levels. It is

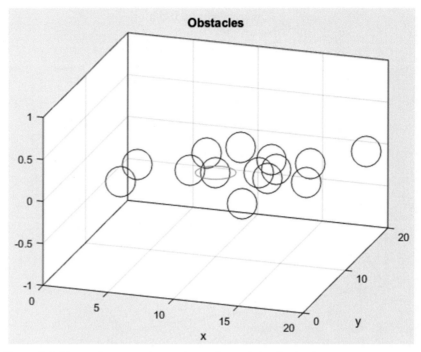

Figure 3.4 Threats between the starting point and the destination.

interesting to note that the maximum concentration reached two units as a result of the pheromone decay factor, which represents the best solution for pathway planning. The three-dimensional distribution of pheromone concentration is depicted below (Figs. 3.7 and 3.8).

The outcomes from our study were compared with findings from other experts in the field. A study by Peng and colleagues in 2015 integrated the Lyapunov Guidance Vector Field (LGVF) with Interfered Fluid Dynamical Systems (IFDS) to facilitate real-time UAV navigation, ensuring obstacle evasion in threat-laden terrains. This blended strategy outperformed the standalone techniques. Notably, the metrics in their research varied, encompassing a singular unit sampling time, a set cruise speed of 0.15, specific repulsion range parameters, and a planning step spectrum from 1 to 10, among other unique factors.

The combination of LGVF and IFDS for path planning yields better outcomes than genetic algorithms. This is clear as the created routes are unique, ensuring safe delivery from point A to B without any collisions.

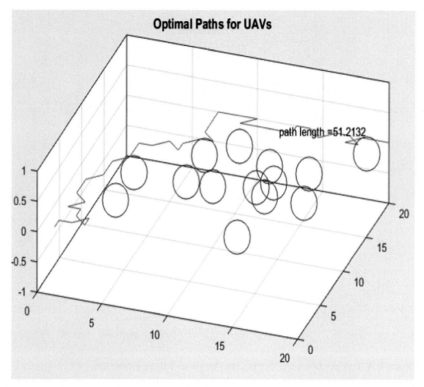

Figure 3.5 UAVs path between supply chain point A and point B. *UAVs*, unmanned air vehicles.

3.4.1 Theoretical implications

This study leads to significant theoretical advancements in multiple domains like SCM, robotics, optimization algorithms, and cybersecurity. The interdisciplinary nature of this approach may enable more cross-domain research, which could result in innovative solutions to complex problems. The ACO and IFDS-LGVF algorithms' implementation and their comparison could lead to new developments or modifications in path planning algorithms, increasing their efficiency or suitability for different scenarios.

UAVs could revolutionize how we perceive their role in global supply chains as they are used as part of global supply chains. If successful, this could expand the theoretical understanding of UAV applications and drive further research in this area.

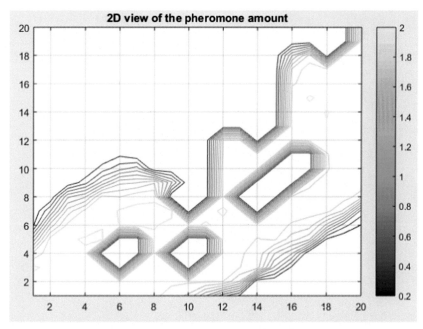

Figure 3.6 Two dimensional view of pheromone trails of supply chain route.

Security and privacy concerns in UAV usage could be addressed by developing a framework in the area of cybersecurity and information systems. This could guide future practices and policies around UAV usage and data management.

3.4.2 Managerial implications

This study provides managerial applications to industries and managers. Managers could use the findings from this study to formulate strategies for implementing UAVs in their supply chain operations. The comparison of different path planning algorithms would provide an understanding of what could work best for their specific scenarios, thus aiding strategic decisions.

The study provides insights into potential cost savings from using UAVs, which would help managers make informed budgeting decisions. The study's findings on security and privacy considerations could help managers understand the associated risks and develop risk mitigation strategies, including data protection policies and physical security measures. Managers could leverage the findings to improve process efficiency, such

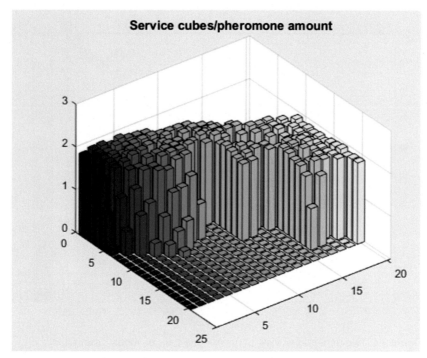

Figure 3.7 3D supply route based on pheromone trail.

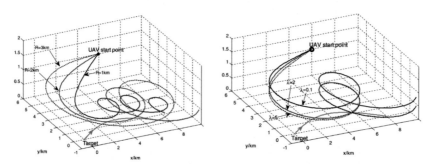

Figure 3.8 Different expansion parameters and detection radius values for LGVF + IFDS.

as reducing delays in supply chain operations by adopting UAVs, thus leading to improved service quality and customer satisfaction. Understanding the technology and its implications would allow managers to train the workforce better and manage changes effectively.

3.4.2.1 Specific action plans

- Managing multiple UAVs effectively in a supply chain requires managers to develop strategies to ensure collision-free functioning. This includes the selection of appropriate path-planning algorithms, infrastructure development, and staff training.
- Managers have to plan and budget for investments in UAVs and related technologies, including data management systems and security infrastructures.
- Managers must establish and implement risk mitigation measures. These should include data protection policies, physical security procedures, and emergency response strategies to handle potential issues effectively.
- Managers should devise comprehensive policies related to UAV usage, encapsulating operational guidelines, safety protocols, and data management practices to ensure smooth operations.
- Finally, managers should actively engage with regulatory bodies to advocate for policies that enable the use of UAVs in supply chain activities, whilst also prioritizing privacy and security concerns.

3.5 Conclusion and recommendations

This study developed a 3D path planning method using queueing theory to address the importance of UAVs in sustainable supply chains. This research aims to enhance UAV navigation between two locations in a spatial setup, where their entry follows a growth rate. The primary focus was guiding UAVs from point A (0, 0, 0) to point B (20, 20, 0) using a Cartesian framework, updating regularly at set periods. To achieve our goal, we incorporated queuing theory and compared the performance of two evolutionary algorithms. The first algorithm we employed was the ACO, which simulates the movement of ants navigating around obstacles to find the optimal solution corresponding to the destination coordinate. The second algorithm was a hybrid of LGVF and IFDS algorithms. This hybrid approach utilizes a fluid motion analogy to solve optimization problems.

Our results showed that both ACO and the LGVF-IFDS hybrid approach were effective in optimizing the 3D path planning for UAVs traveling between the two points. However, the LGVF-IFDS hybrid approach outperformed ACO in terms of optimization efficiency and computational time.

This research contributes to the advancement of sustainable supply chains by offering a comprehensive and effective 3D path-planning method for UAVs. This study provides valuable insight for future researchers and practitioners in UAVs and logistics by incorporating queueing theory and comparing two evolutionary algorithms.

The current study may rely on certain simplifications in the queuing theory and ACO models, which may not fully capture real-world complexities. The model may assume a static environment with fixed obstacle positions, while actual conditions might involve dynamic obstacles or changing conditions that affect UAV navigation. The study may focus on a specific type of UAVs or a narrow range of supply chain applications, limiting the generalizability of the results to other UAVs or broader supply chain scenarios. The ACO algorithm can be computationally intensive, especially when scaling up to larger problem sizes or more complex scenarios. Privacy and security concerns: Future research may address privacy and security implications associated with the use of UAVs in supply chains.

Developing models that account for dynamic environments and real-time changes in obstacle positions, weather conditions, or airspace restrictions could improve the applicability of the results. Investigating other optimization techniques, such as Particle Swarm Optimization, Genetic Algorithms, or Reinforcement Learning, to compare their effectiveness in UAV path planning and supply chain optimization. Exploring the applicability of the proposed approach to a broader range of UAV types and supply chain scenarios could enhance its practical relevance and increase its impact. Integration of privacy and security measures: Future research could focus on incorporating privacy and security considerations into the queuing theory and ACO models, developing algorithms that ensure the safe and responsible use of UAVs in supply chains. Policy and regulatory frameworks: Examining the effects of existing or proposed policy and regulatory frameworks on UAVs adoption and operation in supply chains, and proposing new policies to support the sustainable integration of UAVs into SCM.

References

Amiri, A., Burkart, V., Yu, A., Webster, D., & Ulven, C. (2018). *The potential of natural composite materials in structural design. Sustainable composites for aerospace applications* (pp. 269−291). Elsevier. Available from https://doi.org/10.1016/B978-0-08-102131-6.00013-X.

Anagnostopoulos, T., Komisopoulos, F., Salmon, I., & Ntalianis, K. (2023). *UAV-enabled supply chain architecture for flood recovery in smart cities. Lecture notes in networks and systems* (Vol. 528, pp. 483−496). Springer. Available from https://doi.org/10.1007/978-981-19-5845-8_34.

Ben Amarat, S., & Zong, P. (2019). 3D path planning, routing algorithms and routing protocols for unmanned air vehicles: A review. *Aircraft Engineering and Aerospace Technology*, *91*(9), 1245−1255. Available from https://doi.org/10.1108/AEAT-01-2019-0023.

Brand, M., Masuda, M., Wehner, N., & Yu, X. H. (2010). Ant colony optimization algorithm for robot path planning. In *2010 international conference on computer design and applications* (Vol. 3, pp. V3−436). IEEE.

Chiang, W. C., Li, Y., Shang, J., & Urban, T. L. (2019). Impact of drone delivery on sustainability and cost: Realizing the UAV potential through vehicle routing optimization. *Applied Energy*, *242*, 1164−1175. Available from https://doi.org/10.1016/j.apenergy.2019.03.117.

Colajanni, G., Daniele, P., & Sciacca, D. (2022). Reagents and swab tests during the COVID-19 Pandemic: An optimized supply chain management with UAVs. *Operations Research Perspectives*, *9*, 100257. Available from https://doi.org/10.1016/j.orp.2022.100257.

Dai, X., Long, S., Zhang, Z., & Gong, D. (2019). Mobile robot path planning based on ant colony algorithm with a* heuristic method. *Frontiers in Neurorobotics*, *13*, 15. Available from https://doi.org/10.3389/fnbot.2019.00015.

Dobrokhodov, V. N., Walton, C., Kaminer, I. I., & Jones, K. D. (2020). Energy-optimal guidance of hybrid ultra-long endurance UAV. *IFAC-PapersOnLine*, *53*(2), 15639−15646. Available from https://doi.org/10.1016/j.ifacol.2020.12.2500.

DRONEII. (2019). "Drone Market Report 2019−2024." Drone Industry Insights. Available at: https://www.droneii.com/project/drone-market-report.

Dorling, K., Heinrichs, J., Messier, G. G., & Magierowski, S. (2016). Vehicle routing problems for drone delivery. *IEEE Transactions on Systems, Man, and Cybern.*

Goodchild, A., & Toy, J. (2018). Delivery by drone: An evaluation of unmanned aerial vehicle technology in reducing CO2 emissions in the delivery service industry. *Transportation Research Part D: Transport and Environment*, *61*, 58−67. Available from https://doi.org/10.1016/j.trd.2017.02.017.

Ha, D., Ha, Q. M., Deville, Y., Pham, Q. D., & Hà, M. H. (2018). On the min-cost traveling salesman problem with drone. *Transportation Research Part C: Emerging Technologies*, *86*, 597−621. Available from https://doi.org/10.1016/j.trc.2017.11.015.

Huang, C., Zhou, X., Ran, X., Wang, J., Chen, H., & Deng, W. (2023). Adaptive cylinder vector particle swarm optimization with differential evolution for UAV path planning. *Engineering Applications of Artificial Intelligence*, *121*, 105942. Available from https://doi.org/10.1016/j.engappai.2023.105942.

Jayaweera, H. M., & Hanoun, S. (2020). A Dynamic Artificial Potential Field (D-APF) UAV path planning technique for following ground moving targets. *IEEE Access*, *8*, 192760−192776. Available from https://doi.org/10.1109/ACCESS.2020.3032929.

Kamat, A., Shanker, S., Barve, A., Muduli, K., Mangla, S. K., & Luthra, S. (2022). Uncovering interrelationships between barriers to unmanned aerial vehicles in humanitarian logistics. *Operations Management Research*, *15*(3−4), 1134−1160. Available from https://doi.org/10.1007/s12063-021-00235-7.

Kapadia, M., Badler, N. I., Beacco, A., Pelechano, N., Garcia, F., & Reddy, V. (2016). Multi-domain Planning in Dynamic Environments. Proceedings of the 12th ACM

SIGGRAPH/Eurographics Symposium on Computer Animation, 75—88. Available from https://doi.org/10.1007/978-3-031-02586-0-9.

Kim, D. H. (2019). *Regulations and laws pertaining to the use of unmanned aircraft systems (UAS) by ICAO, USA, China, Japan, Australia, India, and Korea. Unmanned aerial vehicles in civilian logistics and supply chain management* (pp. 169—207). IGI Global. Available from https://doi.org/10.4018/978-1-5225-7900-7.ch007.

Kurniawan, B., Vamplew, P., Papasimeon, M., Dazeley, R., & Foale, C. (2019). *An empirical study of reward structures for actor-critic reinforcement learning in air combat manoeuvring simulation. Lecture Notes in Computer Science (Including Subseries Lecture Notes in Artificial Intelligence and Lecture Notes in Bioinformatics),* 11919 LNAI (pp. 54—65). Springer. Available from https://doi.org/10.1007/978-3-030-35288-2_5.

Lee, J., Shim, D. H., Cho, S., Shin, H., Jung, S., Lee, D., & Kang, J. (2019). A mission management system for complex aerial logistics by multiple unmanned aerial vehicles in MBZIRC 2017. *Journal of Field Robotics, 36*(5), 919—939. Available from https://doi.org/10.1002/rob.21860.

Lee, S., & Choi, Y. (2016). Reviews of unmanned aerial vehicle (drone) technology trends and its applications in the mining industry. *Geosystem Engineering, 19*(4), 197—204. Available from https://doi.org/10.1080/12269328.2016.1162115.

Li, J., Xiong, Y., She, J., & Wu, M. (2020). A path planning method for sweep coverage with multiple UAVs. *IEEE Internet of Things Journal, 7*(9), 8967—8978. Available from https://doi.org/10.1109/JIOT.2020.2999083.

Li, K., Ge, F., Han, Y., Wang, Y., & Xu, W. (2020). Path planning of multiple UAVs with online changing tasks by an ORPFOA algorithm. *Engineering Applications of Artificial Intelligence, 94,* 103807. Available from https://doi.org/10.1016/j.engappai.2020.103807.

Maciel-Pearson, B. G., Akcay, S., Atapour-Abarghouei, A., Holder, C., & Breckon, T. P. (2019). Multi-task regression-based learning for autonomous unmanned aerial vehicle flight control within unstructured outdoor environments. *IEEE Robotics and Automation Letters, 4*(4), 4116—4123. Available from https://doi.org/10.1109/lra.2019.2930496.

MarketsandMarkets. (2020). Drone Logistics and Transportation Market by Solution (Warehousing, Shipping, Infrastructure, Software), Sector (Commercial, Military), Drone (Freight Drones, Passenger Drones, Ambulance Drones), and Region—Global Forecast to 2027. Available at: https://secure.livechatinc.com/.

Mirjalili, S., Song Dong, J., Sadiq, A. S., & Faris, H. (2020). *Genetic algorithm: Theory, literature review, and application in image reconstruction. Studies in* computational intelligence (811, pp. 69—85). Springer. Available from https://doi.org/10.1007/978-3-030-12127-3_5.

Moshref-Javadi, M., Lee, S., & Winkenbach, M. (2020). Design and evaluation of a multi-trip delivery model with truck and drones. *Transportation Research Part E: Logistics and Transportation Review, 136,* 101887. Available from https://doi.org/10.1016/j.tre.2020.101887.

Okwu, M. O., & Tartibu, L. K. (2021). *Genetic Algorithm. Studies in computational intelligence* (927, pp. 125—132). Springer. Available from https://doi.org/10.1007/978-3-030-61111-8_13.

Panda, M., Das, B., Subudhi, B., & Pati, B. B. (2020). A comprehensive review of path planning algorithms for autonomous underwater vehicles. *International Journal of Automation and Computing, 17*(3), 321—352. Available from https://doi.org/10.1007/s11633-019-1204-9.

Parajuli, R., Thoma, G., & Matlock, M. D. (2019). Environmental sustainability of fruit and vegetable production supply chains in the face of climate change: A review.

Science of the Total Environment, *650*, 2863−2879. Available from https://doi.org/10.1016/j.scitotenv.2018.10.019.

Puente-Castro, A., Rivero, D., Pazos, A., & Fernandez-Blanco, E. (2022). A review of artificial intelligence applied to path planning in UAV swarms. *Neural Computing and Applications*, *34*(1), 153−170. Available from https://doi.org/10.1007/s00521-021-06569-4.

Qadir, Z., Ullah, F., Munawar, H. S., & Al-Turjman, F. (2021). Addressing disasters in smart cities through UAVs path planning and 5G communications: A systematic review. *Computer Communications*, *168*, 114−135. Available from https://doi.org/10.1016/j.comcom.2021.01.003.

Rokbani, N., Kumar, R., Abraham, A., Alimi, A. M., Long, H. V., Priyadarshini, I., & Son, L. H. (2021). Bi-heuristic ant colony optimization-based approaches for traveling salesman problem. *Soft Computing*, *25*(5), 3775−3794. Available from https://doi.org/10.1007/s00500-020-05406-5.

Sarkis, J. (2020). Supply chain sustainability: Learning from the COVID-19 pandemic. *International Journal of Operations & Production Management*, *41*(1), 63−73.

Schwarting, W., Alonso-Mora, J., & Rus, D. (2018). Planning and decision-making for autonomous vehicles. *Annual Review of Control, Robotics, and Autonomous Systems*, *1*, 187−210. Available from https://doi.org/10.1146/annurev-control-060117-105157.

Selvi, V., & Umarani, D. R. (2010). Comparative analysis of ant colony and particle swarm optimization techniques. *International Journal of Computer Applications*, *5*(4), 1−6. Available from https://doi.org/10.5120/908-1286.

Sharma, A., Shoval, S., Sharma, A., & Pandey, J. K. (2022). Path planning for multiple targets interception by the Swarm of UAVs based on swarm intelligence algorithms: A review. *IETE Technical Review (Institution of Electronics and Telecommunication Engineers, India)*, *39*(3), 675−697. Available from https://doi.org/10.1080/02564602.2021.1894250.

Shavarani, S. M. (2019). Multi-level facility location-allocation problem for post-disaster humanitarian relief distribution: A case study. *Journal of Humanitarian Logistics and Supply Chain Management*, *9*(1), 70−81.

She, R., & Ouyang, Y. (2021). Efficiency of UAV-based last-mile delivery under congestion in low-altitude air. *Transportation Research Part C: Emerging Technologies*, *122*, 102878. Available from https://doi.org/10.1016/j.trc.2020.102878.

Song, P. C., Pan, J. S., & Chu, S. C. (2020). A parallel compact cuckoo search algorithm for three-dimensional path planning. *Applied Soft Computing*, *94*106443. Available from https://doi.org/10.1016/j.asoc.2020.106443.

Sun, X., Ng, D. W. K., Ding, Z., Xu, Y., & Zhong, Z. (2019). Physical layer security in UAV systems: Challenges and opportunities. *IEEE Wireless Communications*, *26*(5), 40−47. Available from https://doi.org/10.1109/MWC.001.1900028.

Triche, R. M., Greve, A. E., & Dubin, S. J. (2020). UAVs and Their Role in the Health Supply Chain: A Case Study from Malawi. 2020 International Conference on Unmanned Aircraft Systems, ICUAS 2020, 1241−1248. Available at: https://doi.org/10.1109/ICUAS48674.2020.9214064.

Vagale, A., Bye, R. T., Oucheikh, R., Osen, O. L., & Fossen, T. I. (2021). Path planning and collision avoidance for autonomous surface vehicles II: A comparative study of algorithms. *Journal of Marine Science and Technology (Japan)*, *26*(4), 1307−1323. Available from https://doi.org/10.1007/s00773-020-00790-x.

Voznak, M., Halas, M., Borowik, B., & Kocur, Z. (2011). Delay model of RTP flows in accordance with M/D/1 and M/D/2 Kendall's notation. *International Journal of Mathematics and Computers in Simulation*, *5*(3), 242−249.

Wahab, M. N. A., Nefti-Meziani, S., & Atyabi, A. (2020). A comparative review on mobile robot path planning: Classical or meta-heuristic methods. *Annual Reviews in Control*, *50*, 233−252. Available from https://doi.org/10.1016/j.arcontrol.2020.10.001.

Wan, Y., Zhong, Y., Ma, A., & Zhang, L. (2022). An accurate UAV 3-D path planning method for disaster emergency response based on an improved multiobjective swarm intelligence algorithm. *IEEE Transactions on Cybernetics*. Available from https://doi.org/10.1109/TCYB.2022.3170580.

Wang, C., Lan, H., Saldanha-Da-gama, F., & Chen, Y. (2021). On optimizing a multi-mode last-mile parcel delivery system with vans, truck and drone. *Electronics (Switzerland)*, *10*(20), 2510. Available from https://doi.org/10.3390/electronics10202510.

Wang, X., Poikonen, S., & Golden, B. (2017). The vehicle routing problem with drones: Several worst-case results. *Optimization Letters*, *11*, 679−697. Available from https://doi.org/10.1007/s11590-016-1035-3.

Wang, C., Wang, J., Shen, Y., & Zhang, X. (2019). Autonomous navigation of UAVs in large-scale complex environments: A deep reinforcement learning approach. *IEEE Transactions on Vehicular Technology*, *68*(3), 2124−2136.

Yan, M.-d, Zhu, X., Zhang, X.-x., & Qu, Y.-h (2017). Consensus-based three-dimensionalmulti-UAV formation control strategy with high precision. *Frontiers of Information Technology and Electronic Engineering*, *18*(7), 968−977. Available from https://doi.org/10.1631/FITEE.1600004.

Yao, P., Wang, H., & Su, Z. (2015). UAV feasible path planning based on disturbed fluid and trajectory propagation. *Chinese Journal of Aeronautics*, *28*(4), 1163−1177. Available from https://doi.org/10.1016/j.cja.2015.06.014.

Yurek, E. E., & Ozmutlu, H. C. (2018). A decomposition-based iterative optimization algorithm for traveling salesman problem with drone. *Transportation Research Part C: Emerging Technologies*, *91*, 249−262. Available from https://doi.org/10.1016/j.trc.2018.04.009.

Zhang, H., Fei, Y., Li, J., Li, B., & Liu, H. (2023). Method of vertiport capacity assessment based on queuing theory of unmanned aerial vehicles. *Sustainability (Switzerland)*, *15*(1), 709. Available from https://doi.org/10.3390/su15010709.

Zhang, Z., Wu, D., Xu, W., Shang, J., Feng, Z., & Zhang, P. (2019). UAV-enabled multiple traffic backhaul based on multiple RANs: A batch-arrival-queuing-inspired approach. *IEEE Access*, *7*, 161437−161448. Available from https://doi.org/10.1109/ACCESS.2019.2951603.

Zheng, X., Wang, F., & Li, Z. (2018). A multi-UAV cooperative route planning methodology for 3D fine-resolution building model reconstruction. *ISPRS Journal of Photogrammetry and Remote Sensing*, *146*, 483−494. Available from https://doi.org/10.1016/j.isprsjprs.2018.11.004.

Zhou, J., Jin, L., Wang, X., & Sun, D. (2021). Resilient UAV traffic congestion control using fluid queuing models. *IEEE Transactions on Intelligent Transportation Systems*, *22*(12), 7561−7572. Available from https://doi.org/10.1109/TITS.2020.3004406.

Zhou, X., Ma, H., Gu, J., Chen, H., & Deng, W. (2022). Parameter adaptation-based ant colony optimization with dynamic hybrid mechanism. *Engineering Applications of Artificial Intelligence*, *114*, 105139. Available from https://doi.org/10.1016/j.engappai.2022.105139.

CHAPTER FOUR

Exploring the enablers of sustainable supply chain management: A simplified computational intelligence perspective

Muhammad Shujaat Mubarik[1] and Sharfuddin Ahmed Khan[2]
[1]Department of Marketing and Operations, Edinburgh Business School, Heriot-Watt University, Edinburgh, United Kingdom
[2]Industrial Systems Engineering, Faculty of Engineering and Applied Science, University of Regina, Canada

4.1 Introduction

From last two decades businesses have found themselves under escalating pressure from regulators, investors, and governments to play a vital role in addressing the crucial issue of climate change. Simultaneously, consumer preferences are also undergoing a seismic shift, with an exponential rise in those preferring brands committed to sustainability (Mubarik, 2023; Teng et al., 2023). In response to this mounting pressure, various firms are demonstrating their dedication to reducing carbon footprints by transforming their supply chains into bastions of sustainability. Importantly, supply chain management (SCM) assumes a key role in the pursuit of sustainability. Majority of the key environmental, social, and governance (ESG) initiatives are either directly or indirectly linked with SCM, and operations. Hence, requiring firms to adopt sustainable supply chain management (SSCM). The adoption of SSCM can be dissected into three distinct scopes:

Scope 1: This forms the bedrock of supply chain sustainability and is linked to an organization's foundational impact. It includes the emissions directly generated by the organization and the resources it consumes. Quantifying these emissions, and subsequently reducing them, falls under the ambit of Scope 1 (Khan, Mubarik, & Paul, 2022; Kusi-Sarpong et al., 2022).

Computational Intelligence Techniques for Sustainable Supply Chain Management.
DOI: https://doi.org/10.1016/B978-0-443-18464-2.00006-6

Scope 2: Within Scope 2, emissions are linked with the procurement of electricity, heat, cooling, and steam. These procurements come with specific emissions.

Scope 3: Scope 3 has a wider net, entrenching an organization's customers and suppliers. While many organizations are actively engaged in the adoption of Scope 1, fewer have embarked on Scope 2, and only a handful have ventured on Scope 3.

It is important to note that the transformation of a supply chain(SC) along the lines of sustainability, with a focus on these scopes(1, 2, and 3), requires the development of robust sustainability-related key performance indicators (KPIs). This demand for change extends beyond the organization's boundaries, encompassing both upstream (suppliers) and downstream (customers) stakeholders. In this holistic shift toward sustainability, SCM emerges as a linchpin, facilitating the transition to a more environmentally conscious and socially responsible process.

Efficient and effective transformation of a supply chain, with a keen focus on these three scopes, demands the adoption of SSCM Practices. To effectively steer SC toward sustainability, it is important to capitalize on the requisite drivers while concurrently overcoming major obstacles. A plethora of research could be found identifying the key elements that facilitate or impede the smooth transition from a traditional supply chain to a sustainable one. This body of research underscores critical enablers and barriers that must be addressed for a seamless transformation (Baah & Zhihong, 2019; Mann & Harmeet, 2020; Oyedijo, et al., 2023). Key enablers include factors like supplier relationships, the establishment of a consistent supply chain web, strategic sourcing, sustainable procurement, information sharing, eco-conscious distribution, efficient reverse logistics, and sustainable packaging (Mubarik et al., 2023a).

Conversely, a battery of barriers ranging from organizational inertia to entrenched corporate culture has been identified. Despite a significant increase in the studies in recent years that have approached SSCM from various angles, a degree of confusion persists regarding the relative significance of these enablers and barriers. The majority of the attempts in this domain have relied on conventional methodologies, fraught with inherent flaws, and path dependencies resulting in inconclusive and sometimes contradictory findings (Mubarik et al., 2023b, 2023c). Clarifying and synthesizing the complex web of enablers and barriers is important to steer the complex landscape of SSCM effectively. To drive meaningful change in this realm, it is imperative to distinguish which factors yield the greatest influence and why, transcending the contradictions and vagueness that have thus far characterized this important field of study.

Recently, machine learning (ML) has emerged as a powerful tool for conducting a wide range of analyses. ML is a subset of artificial intelligence (AI) that includes training algorithms to learn from data and make decisions or predictions without being explicitly programmed. Besides numerous other potentials, one of its unique capabilities is to precisely distinguish the significance of various factors and quantify their importance. What differentiates ML is its capacity to learn from data, allowing it to generate robust and practical insights. In this context, the objective of this chapter is to harness the potential of ML tools to identify, select, and prioritize the key enablers and barriers of SSCM. By employing ML algorithms, we aim to unlock a deeper understanding of the factors that drive or hinder the adoption of sustainable practices within supply chains. In the context of prioritizing barriers and enablers of SSCM, ML is applied by analyzing historical data, survey responses, or various relevant datasets. By feeding this data into ML models, algorithms can identify patterns, priorities, and correlations among different factors that create hindrances or contribute to the success of SSCM. These models can then allot weightage or scores to each factor, efficiently prioritizing them based on their significance. The strength of ML lies in its ability to handle large and complex datasets, adapt to changing conditions, uncover hidden insights, and make it a valuable tool for optimizing and advancing sustainability efforts within SCs.

The study focuses on the textile industry by taking the data from the experts of five major textile-producing countries (Parsad et al., 2019). Sustainability is a serious problem for textile firms, which are focusing on applying sustainability standards throughout their supply chains (Oelze, 2017; Oelze, 2017). Textile companies provide items that have a major social and environmental impact throughout their existence. Simultaneously, there is a rising understanding of the need for sustainability. The study adopts a threefold approach. In the first phase, relevant barriers and enablers have been identified by reviewing the literature. In the second phase, relevant barriers and enablers were selected based on the data collected from the 27 experts from seven countries (China, India, Pakistan, Indonesia, Thailand, South Korea, and Bangladesh). In the third phase, the selected enablers and barriers have been prioritized using the ML approach.

4.1.1 Sustainable supply management: definitions and dimensions

A supply chain with fully integrated environmentally responsible, social, and ethical practices into all of the supply chain processes and transactions is called SSCM (Khan, Mubarik, & Paul, 2022). Put differently, SSCM integrates environmentally, socially, and financially viable practices into all pillars of the supply chain, plan, source, make, and deliver. All of the definitions of SSCM highlight its three important dimensions namely environmentally responsible, socially responsible, and financially responsible practices. Socially responsible practices represent a firm's adherence to the labor laws, its contribution to the community in which it's operating through CSR and outreach programs, etc. Socially responsible practices ensure the conformance to human rights inside the organization and the firm's contribution to society.

Environmentally responsible practices include the adaptation of all the processes, techniques, and approaches that improve the environmental quality either by reducing emissions or waste. Carbon emission reduction techniques are used to limit or eliminate the release of pollutants into the atmosphere that are environmentally destructive (Nathanson, 2019). Whereas waste reduction practices can be divided into two. First water waste reduction aimed to improve the quality of the water that is dirty or has had its quality lowered because of human activity. Wastewater comes from a variety of sources, including home, commercial, industrial, and agricultural activity. The techniques used to synthesize wastewater into a byproduct that will either be sent to the water cycle having minimal environmental effect or reused are referred to as water waste reduction (TL, 2021). The second type of waste management practices are aimed at solid waste reduction. The collection, processing, and disposal of solid waste that has become unusable. Unhygienic conditions can arise as a result of inappropriate MSW disposal, resulting in environmental pollution and the development of vector-borne illnesses (Nathanson, 2020). Sustainable supply chain practices also aim to decrease waste in terms of excessive inventory levels. Inventory reduction is the process of lowering inventory levels to the point where they are capable of meeting customer requirements. Inventory management is necessary to get rid of excess inventory, free storage space, save money, and increase profits (Viloria & Paula, 2016). Likewise, these practices also focus on the elevation of the quality of product/service eco-design. Product quality may be improved by integrating environmental considerations into the product creation process and balancing environmental and socioeconomic objectives. Eco-design considers environmental concerns at all phases of the system development process, intending to produce items that have the minimal level of environmental impact possible across their entire lifetime (Schäfer & Manuel, 2021, Mubarik et al., 2021b, 2021a, 2023c)

Another aspect of SSCM is ensuring economic Performance. SSCM equally focuses on the firm's financial performance and emphasizes the reduction of at least the following costs.

1. *Decrease in cost for materials purchasing:* The term cost reduction generally refers to cost reductions gained through the procurement of materials. Material acquisition costs often comprise base costs (i.e., manufacturing costs), closing costs (i.e., commission, legal fees, and brokerage fees), transportation costs, taxes and duties, and exploration and negotiation (Melanie, 2017).

2. *Decrease in cost for energy consumption:* Energy consumption refers to the amount of energy used to execute specific tasks, manufacture something, or just live in a place. Calculating energy usage can help you figure out how much to spend in an energy performance system which will help you save money in the long run (Teba, 2018).

3. *Increase in revenue from green products/services:* Green solutions allow you to expand your consumer base while also allowing you to demand higher pricing. Customers are willing to pay 10% extra for green items (e.g., organic, or made from recycled content) (Fischhoff, 2013).

Putting together, SSCM focuses on the social, environmental, and economic aspects of a firm and tends to mend the supply chain processes in conformance with these three aspects.

4.1.2 Barriers and drivers of sustainable supply chain management

4.1.2.1 Sustainable supply chain management drivers

Internal drivers are forces produced inside a company (Caniato et al., 2012) that forecast proactive sustainability action (Gonzalez-Benito & Gonzalez-Benito, 2009). These drivers are divided into four categories. (1) Corporate strategy; (2) Organizational culture; (3) organizational resources; and (4) organizational features.

4.1.2.2 Corporate strategy

Corporate strategy reflects an organizational long-term plan in the pursuit of sustainability (Ahmed et al., 2021). A comprehensive corporate strategy, outlining sustainability goals, objectives, and timelines, plays a pivotal role in the attainment of SSCM (Khan, Mubarik, & Paul, 2022; Mubarik, Naghavi, et al., 2021). Top leadership environmental awareness is critical for comprehending the profitability demands from stakeholders, which prompts the development of sustainable practices inside the company and throughout the logistics system (Kersten, 2019). A well-crafted corporate strategy also reflects the demonstrated sponsorship of the top management. Given the instrumental role of corporate strategy in the literature it can be deemed as one of the significant drivers of the SSCM in textile firms.

4.1.2.3 Organizational culture

A lot of research has demonstrated the role of organizational culture in strategy and execution. A conducive, responsible, and flexible organizational culture is considered suitable for the adoption of sustainability. Likewise, an organizational culture focused on inclusivity and openness can play a significant role in the pursuit of sustainability in general as well as in SSCM. As noted by Mubarik et al. (2018) and Kersten (2019), a conducive organizational culture serves as an internal motivator for executing sustainability activities in compliance with legal provisions and rules to fulfill the needs of stakeholders (Kersten, 2019).

4.1.2.3.1 Relational capital

A sustainable supply chain requires the firm and the members of the supply chain to build a relationship of trust and transparency. In this perspective, long-term collaboration with all the stakeholders, especially with suppliers and customers, is essential for the drive toward sustainability (Mubarik, Naghavi, et al., 2021). A closer relationship with suppliers helps an organization to better look into its upstream supply chain processes and make them sustainable. Likewise, close coordination and cooperation with customers enable a firm to make the downstream supply chain more sustainable and adopt sustainable measures like closed-loop or circular SCM (Kusi-Sarpong et al., 2022).

4.1.2.3.2 Human capital

The knowledge, skills, abilities, health, and motivation of employees of an organization are viewed as the primary drivers of sustainability

(Mubarik, Naghavi, et al., 2021; Mubarik et al., 2018; Kersten, 2019). Employees' awareness and understanding of sustainable business practices not only help adopt SSCM but also play a key role in maintaining sustainable performance. There is ample amount of evidence that could be found confirming the effective role of employees' human capital in the success of sustainability adoption. It reveals that human capital is an essential driver of SSCM irrespective of any sector/industry.

4.1.2.3.3 Structural capital
Structural capital or process capital illustrates organizational business routines, processes, databases, modus operandi, and ways to make decisions (short, medium, and long term) and execute various business transactions (Mubarik et al., 2018). Primarily it reflects the tacit infrastructure of a business and encapsulates all types of business processes, including the supply chain. A firm with intact, flexible, and efficient structural capital can have a higher potential to adopt SSCM amicably. The flexible price capital not only allows for the transformation of the SCM on the lines of sustainability but also helps to uplift SSCM performance (Khan, Mubarik, & Paul, 2022).

4.1.2.3.4 ERP/IT system
Integration with internal business departments/functions and with external stakeholders is crucial for the organization to survive and thrive. The higher level of integration allows firms to have improved cross-functional and cross-organizational information sharing. An effective and well-integrated enterprise resource planning (ERP) system allows an organization to seamlessly integrate all of its departments and also develop a closer link with external organizations (suppliers and customers). Further, this system also helps identify the various loopholes, and wastes in the processes, thus enabling the adoption of SSCM (Mobashar & Mubarak, 2020).

4.1.2.3.5 Well-mapped supply chain (supply chain mapping)
Supply chain mapping demonstrates an organizational capability to know about its upstream and downstream supply chain entities. It allows an organization to have a real-time track of information about key stakeholders. A well-mapped supply chain also helps to identify the bottlenecks, wastes, and issues in a supply chain; hence, enabling a firm to adopt SSCM (Mubarik, Naghavi, et al., 2021).

4.1.2.3.6 Size of the firm
The scale of a company has a direct impact on sustainability considerations. The number of full-time employees employed by the company may be used to determine its size (Kersten, 2019). To attain environmental targets, central organizations must deal with exterior production constraints and cultivate partnerships with suppliers.

4.1.2.3.7 Global presence
As a result, the placement of an organization in the distribution networks is regarded as a driving factor of SSCM, as it has a considerable influence on firms' sustainability-related behavior. When compared to firms with little worldwide presence, organizations with broad global operations face substantially higher demand to embrace sustainable policies. (Bai et al., 2015; González-Benito, 2009; Kersten, 2019).

4.1.2.3.8 Suppliers integration
With the sharing of knowledge on sustainability issues, we can build better SSCM at our end by resolving those issues through their experiential knowledge. Close collaboration for an extended period helps us build a better sustainable supply chain. Supplier Organizational Culture constitutes an external enabler for SSCM (Ciliberti, F.; Pontrandolfo, P.; Scozzi, B.). In addition to that, the integration, both technological and logistical, of the members of the supply chain is required for making the implementation successful (Vachon, S.; Klassen, R.).

4.1.2.3.9 ERP system
The distribution of corporate assets for the adoption of sustainable practices is critical. An effective information system should offer appropriate resources to promote the company's sustainability objectives (Haverkamp et al., 2010; Schrettle et al., 2014; Giunipero et al., 2012; Kersten, 2019).

4.1.2.3.10 Exaptive capabilities
Exaptive capabilities are a firm's characteristics and strengths, which it develops to encounter certain challenges and has been reused for some other purpose. For example, a firm develops its capabilities to fight against COVID-19. The same capabilities are deployed by a firm to counter any sudden situation that has arisen due to some other circumstances. In that case, the reuse of the capabilities attained while fighting COVID-19 is

called the exaptive capabilities (Mubarik, Naghavi, et al., 2021). We argue that such capabilities can also help an organization adopt SSCM.

4.1.2.3.11 Organizational ambidexterity
Organizational ambidexterity is the ability of a firm to strike a balance between its exploitative and explorative activities. In another way, it reflects an organization's ability to simultaneously focus on existing markets and explore new markets (Mahmood & Mubarik, 2020). An ambidextrous firm has a better ability to transform its conventional SCM to SSCM due to its better flexibility.

4.1.2.3.12 Absorptive capacity
An organization can identify, assimilate, and execute the knowledge outside an organization. The extent to which an organization learns from the external environment predominantly depends upon its absorptive capacity (Mahmood & Mubarik 2020; Ahmed et al., 2020). Since SSCM is an outside knowledge that an organization intends to adopt or internalize, the absorptive capacity can play a significant role. There is anecdotal evidence that firms with higher absorptive capacity tend to perform better than their counterparts in terms of SSCM adoption.

4.1.2.3.13 Internationalization
It is the extent to which a firm is connected with the international market through equity-based or non-equity-based international entry modes. Linkage with international markets provides a broader exposure to the organization and its ability to adopt the latest practices like SSCM gets increased. Organizations that export to a diverse range of international markets, especially developed countries markets, tend to have higher pressure from their clients to make their supply chain processes more sustainable (Khan, Mubarik, & Paul, 2022). It implies that a firm's internationalization experience could be a key enabler or driver of SSCM (Table 4.1).

4.1.2.4 Barriers to the implementation of sustainable supply chain management
There are certain organizational barriers, which hinder the adoption of SSCM. We have reviewed the literature and developed a list of such barriers that have frequently appeared in the literature. The below paragraph explains each of the identified barriers.

Table 4.1 Drivers of SSCM (sustainable supply chain management).

Drivers	Sources
Corporate strategy	Kersten (2019), Ahmed et al. (2021), Mubarik, Kusi-Sarpong, et al. (2021)
Organization Culture	Kersten (2019)
Size of the firm	Mubarik, Kusi-Sarpong, et al. (2021), Kusi-Sarpong et al. (2022)
Global presence	Bai et al. (2015), Kersten (2019), Mubarik and Bontis (2022)
Supply Chain mapping	Mubarik, Kusi-Sarpong, et al. (2021), Khan, Mubarik, and Kusi-Sarpong, Gupta, et al. (2022)
Suppliers Integration	Mubarik, Kusi-Sarpong, et al. (2021)
ERP/IT system	Mobashar and Mubarak (2020)
Human Capital	Mubarik, Kusi-Sarpong, et al. (2021)
Relational capital	Mubarik, Kusi-Sarpong, et al. (2021)
Structural Capital	Mubarik, Kusi-Sarpong, et al. (2021)
Exaptive capabilities	
Organizational Ambidexterity	
Absorptive Capacity	
Internationalization	Reza et al. (2021), Mubarik, Kusi-Sarpong, et al. (2021)

4.1.2.4.1 Lack of top management

Top management support and demonstrated sponsorship are critical for the implementation of SSCM. There is ample amount of evidence showing the organization's failure due to the lack of support on the part of top and middle management. It may limit the organization's ability to participate in sustainability activities (Kusi-Sarpong et al., 2022).

4.1.2.4.2 Lack of leadership

The leadership abilities of the managers initiating and implementing the SSCM initiative are also crucial for its successful adoption. Lack of leadership abilities and dedication of concerned managers badly hamper the transformation of SCM to SSCM (Mubarik & Bontis, 2022).

4.1.2.4.3 Cost constraints

To make a few of the SSCM processes more sustainable, an organization may be required to incur costs in terms of training, process re-engineering, equipment replacement, etc. There are few studies mentioning cost as a barrier to SSCM (Khan, Mubarik, & Paul, 2022).

4.1.2.4.4 Production constraints

Organizations have sometimes built-in bottlenecks, and constraints in the production processes, which significantly hamper the adoption of SSCM. Above and beyond, these constraints or barriers in production negatively influence the sustainability drive. These constraints could be physical in terms of production equipment's and machines' capacities, or tacit like process-driven bottlenecks and wastes (Kusi-Sarpong et al., 2022)

4.1.2.4.5 Lack of clear guidance

The main challenges to SSCM implementation include a lack of clear guidance from organizational leaders, incentive structures that fail to reward sustainability activities, contradictory information, and competing aims from the central government (Zaabi et al., 2013). The lack of clarity of the guidelines not only creates confusion for the departments adopting SSCM but also affects routine activities.

4.1.2.4.6 Vague regulations

One of the major reasons that a firm shows reluctance to take up the SSCM initiative is unclear or vague government. Vague and unclear regulations create barriers to sustainability initiatives. (Baig et al., 2020).

4.1.2.4.7 Lack of knowledge

Lack of awareness or knowledge about the SSCM practices and initiatives is a big hurdle. Firms sometimes start adopting SSCM without having adequate understanding and knowledge of these initiatives. The low level of understating, a lack of knowledge and tools, reluctance due to perceived time costs and resources required, and skepticism about the commercial advantages all create big hindrances in the way of SSCM (Zaabi et al., 2013).

4.1.2.4.8 Lack of governmental support

Although SSCM is primarily driven by the organization, external pressure and support from the government play a pivotal role in it. Studies (Narayanan et al., 2019) consider the lack of government initiatives, as one of the key reasons why organizations fail to prioritize sustainability initiatives

4.1.2.4.9 Lack of intrinsic motivation

Various Researchers (Majumdar & Sinha, 2019) denote that a lack of intrinsic motivation among employees can be one of the hindrances in maintaining a sustainable supply chain in the textile industry.

4.1.2.4.10 Lack of employees' commitment

Employees' commitment plays a key role in adopting any development in the organizations. Poorly employees' commitment often significantly hamper the adoption of SSCM (Mubarik, Naghavi, et al., 2021).

4.1.2.4.11 Lack of integration with suppliers

Firms' weak integration with the upstream supply chain, especially with primary suppliers has a detrimental effect on the adoption of SSCM. The upstream supply chain provides all the raw materials, equipment, and MRO supplies; hence, a closer integration with upstream is essential for SSCM (Mubarik et al., 2023a).

4.1.2.4.12 Process inefficiencies

Inefficient processes act as a major impediment in the way of SSCM. The adoption of SSCM cannot succeed without overcoming the processes' efficiencies (Mubarik, Bontis, et al., 2021).

Further researchers claim that strict regulations (Khan et al., 2023), high implementation costs, and unavailability of human resources (Mubarik, Naghavi, et al., 2021) are the major hurdles in the way of SSCM.

Table 4.2 exhibits all the identified barriers of SSCM with their references.

4.2 Methodology

We have adopted a three-stage approach to identify, select, and prioritize the barriers and drivers of SSCM. The following paragraphs explain each stage.

4.2.1 Identification

In the first stage, a thorough review of the literature has been conducted to identify the possible drivers and barriers of SSCM that have been discussed in the literature.

Table 4.2 Barriers of SSCM (sustainable supply chain management).

Barriers	Sources
Lack of top management support	Kusi-Sarpong et al. (2022)
Lack of leadership	Mubarik and Bontis (2022)
Cost constraints	Khan, Mubarik, and Kusi-Sarpong, Gupta, et al. (2022)
Production constraints	Kusi-Sarpong et al. (2022)
Lack of clear guidance	
Vague regulations	Baig et al. (2020)
Lack of knowledge/awareness	
Lack of governmental support	Narayanan et al. (2019)
Lack of intrinsic motivation	Majumdar and Sinha (2019)
Poor employees' commitment	Mubarik, Kusi-Sarpong, et al. (2021)
Lack of integration with suppliers	Mubarik et al. (2023a)
Process inefficiencies	Mubarik, Kusi-Sarpong, et al. (2021)
Strict regulations	Khan et al. (2023)
High implementation cost	
Unavailability of human resources	Mubarik, Kusi-Sarpong, et al. (2021)

4.2.2 Selection survey

In the second stage, following Mubarik, Kusi-Sarpong, et al. (2021) approach, a preliminary survey is conducted to select the barriers and drivers of SSCM in the context of Textile industries. For the preliminary survey, all the barriers and drivers are put into a uni-polar questionnaire with a 3-point scale, where 1 reflected not important, 2 somewhat important, and 3 important. The questionnaire is then sent to the 27 experts from 07 Asian countries: China, India, Pakistan, Indonesia, Thailand, South Korea, and Bangladesh. *An expert has been defined as an individual having experience of 10 years or more working in the supply chain or related department in any textile firm at any P&L level position or as an assistant manager or above.* The experts are selected using expert-cum-snowball sampling. The questionnaires are sent through electronic means. Data obtained from experts is analyzed using the following process.

First, the preference value of each driver and barrier is computed using the following equation, programmed in the ML algorithm. The PV reflects the cumulative importance/preference of a driver or barrier given by all the experts.

$$PV = \alpha NI + \beta SWI + £Imp \tag{4.1}$$

Whereas, values of α, β, and £ are 1, 2, and 3, respectively. *PV represents Preference Value. NI for Not Important. SWI for Somewhat Important. Imp for Important.*

After computing the PV, the cutoff value was computed using Eq. (4.2):

$$\text{Cut} - \text{off value} = (\text{HPV} + \text{LPV})/2 \tag{4.2}$$

Whereas HPV is the highest preference value of a driver or barrier. *LPV is the lowest preference value of a driver or barrier.*

The barriers or drivers above the cutoff value are considered relevant.

4.2.3 Prioritization

In the third stage, all the drivers and barriers selected in stage 2 are converted into a bipolar questionnaire, comparing each of the drivers (barriers) with the rest of the drivers (barriers). The questionnaires are then sent to the same 27 experts to rate the relative importance of each driver and barrier on a scale of 1−9, where 1 is equally important to 9 very extremely important. The collected data so then processed by applying a combination of ML and MCDM approaches to compute the relative priorities of each driver and barrier. It is important to note that ML's two major algorithms, the red deer algorithm (RDA) and genetic algorithm (GA), used for optimization are used for dealing with such optimization issues.

The participants after completing the questionnaire were interviewed in which a detailed insight into the reasons for selecting the rating for each enabler/barrier was discussed. Each participant was interviewed and during the interview, we tried to find out the reason or the explanation as to why they consider a particular item on the questionnaire as not important, important, or somewhat important. To understand how each driver and barrier can impact SSCM, we also conducted interviews with 05 experts.

4.3 Findings

4.3.1 Identification

After comprehensively reviewing the literature we have identified 14 drivers and 15 barriers to SSCM. These drivers and barriers are not textile industry-specific rather these could prevail in any industry. Hence to confirm their relevance and further select the most relevant barriers and drivers, we conducted a preference value survey as discussed in Section 3. The proceeding section briefly discusses it.

4.3.2 Selection

A total of 21 experts responded to the survey, conducted online. The summary of the respondents' opinions is illustrated below in Table 4.3 and Table 4.4. Both tables have three distinct parts: Response, WPR, and PV. The response part consists of three columns (NI, SWI, IMP), summarizing the preference of each expert for a particular driver or barrier. For example in Table 4.3, driver organization culture has been considered NI (not important) by 02 experts, SWI (somewhat important) by 14 experts, and IMP (important) by 5 experts; a total of experts 21. Part 2, WPR (weighted preference rating) has been computed for each group of responses using Eq. 4.1. It shows the preference score of experts for each driver and barrier. Part 3, CPV (cumulative preference value) shows the

Table 4.3 Summary of the response (drivers).

Drivers	NI	SWI	IMP	NI	SWI	IMP	CPV (cumulative preference value)
	Response			**WPR (weighted preference rating)^**			
Organization culture	2	14	5	0.10	0.67	0.24	2.14
Corporate strategy	3	13	5	0.14	0.62	0.24	2.10
Human capital	2	16	3	0.10	0.76	0.14	2.05
Relational capital	3	15	3	0.14	0.71	0.14	2.00
ERP/IT system	5	13	3	0.24	0.62	0.14	1.90
Structural capital	4	15	2	0.19	0.71	0.10	1.90
Internationalization	7	9	5	0.33	0.43	0.24	1.90
Suppliers integration	4	16	1	0.19	0.76	0.05	1.86
Supply chain mapping	5	15	1	0.24	0.71	0.05	1.81
Organizational ambidexterity	9	9	3	0.43	0.43	0.14	1.71
Size of the firm	10	9	2	0.48	0.43	0.10	1.62
Exaptive capabilities	9	11	1	0.43	0.52	0.05	1.62
Absorptive capacity	10	9	2	0.48	0.43	0.10	1.62
Global presence	13	7	1	0.62	0.33	0.05	1.43

Cutoff value: 1.79

Five drivers namely organizational ambidexterity, size of the firm, exaptive capabilities, absorptive capacity, and global presence have PV lower than the cutoff value of 1.79.
^WPR stands for weighted preference rating obtained by dividing the number of responses in each category by total response.
NI for Not Important; SWI for Somewhat Important; Imp for Important.

Table 4.4 Summary of the response (barriers).

Drivers	Response			WPR^			CPV
	NI	SWI	IMP	NI	SWI	IMP	
Lack of top management support	4	11	6	0.19	0.52	0.29	2.10
Lack of integration with suppliers	5	12	4	0.24	0.57	0.19	1.95
Process inefficiencies	3	16	2	0.14	0.76	0.10	1.95
Cost Constraints	6	11	4	0.29	0.52	0.19	1.90
Lack of clear guidance	4	15	2	0.19	0.71	0.10	1.90
Lack of governmental pressure	4	15	2	0.19	0.71	0.10	1.90
Lack of intrinsic motivation	4	16	1	0.19	0.76	0.05	1.86
Poor employees' commitment	6	12	3	0.29	0.57	0.14	1.86
Production constraints	7	10	4	0.33	0.48	0.19	1.86
Lack of knowledge/awareness	7	12	2	0.33	0.57	0.10	1.76
Lack of leadership	13	3	5	0.62	0.14	0.24	1.62
Vague regulations	10	10	1	0.48	0.48	0.05	1.57
Strict regulations	11	9	1	0.52	0.43	0.05	1.52
High implementation cost	13	8	0	0.62	0.38	0.00	1.38

Cutoff value: 1.74

Note: Four barriers namely lack of leadership, vague regulations, strict regulations, and high implementation cost have PV lower than the cutoff value of 1.74.
^WPR stands for weighted preference rating obtained by dividing the number of responses in each category by total response.
NI for Not Important; SWI for Somewhat Important; Imp for Important.

cumulative preference value of each driver and barrier. These PV have been further used to determine the cutoff value by taking an average of minimum and maximum CPV.

The cutoff value for drivers is 1.79, computed as the average of 2.14 (Highest CPV), and 1.43 (Lowest CPV). Likewise, the cutoff value for barriers is 1.74, computed as the average of 2.10 (Highest CPV), and 1.38 (Lowest CPV). Using these two cutoff values, 09 barriers and 09 drivers to SSCM have been selected in Table 4.5.

4.3.3 Prioritization

The selected drivers and barriers were to be prioritized according to their role in SSCM adoption in the textile industry. We employed DEMATEL approach using the ML approach for this prioritization. The benefit of ML is that the precision of the approach can be improved by providing

Table 4.5 List of selected barriers and drivers.

Drivers	Barriers
Organization culture	Lack of top management support
Corporate strategy	Lack of integration with suppliers
Human capital	Process inefficiencies
Relational capital	Cost constraints
ERP/IT system	Lack of clear guidance
Structural capital	Lack of governmental pressure
Internationalization	Lack of intrinsic motivation
Suppliers integration	Production constraints
Supply chain mapping	Lack of knowledge/awareness

Table 4.6 Priority weights derived through ML (machine learning).

Drivers	PW	Barriers	PW
Human capital	0.17	Lack of top management support	0.18
Suppliers integration	0.15	Lack of intrinsic motivation	0.16
Relational capital	0.15	Production constraints	0.13
Structural capital	0.13	Process inefficiencies	0.13
Supply chain mapping	0.11	Lack of integration with suppliers	0.12
Organization culture	0.10	Cost constraints	0.11
ERP/IT system	0.08	Lack of knowledge/awareness	0.09
Internationalization	0.06	Lack of clear guidance	0.05
Corporate strategy	0.05	Lack of governmental pressure	0.03

input and with the help of data training. The results of prioritization have been depicted in Table 4.6.

The results in Table 4.6 exhibit the highest significance of human capital (0.17) of the firms in the adoption of SSCM followed by supplier integration (0.15), relational capital (0.15), and structural capital (0.13). The apex role of human capital is not surprising as several studies (e.g., Khan, Mubarik, & Paul, 2022; Mubarik, Naghavi, et al., 2021) mention human capital as the lynchpin for the adoption of any development especially related to sustainability. Hence, the highest value of HC echoes the findings of the previous studies. Likewise, the findings on supplier integration (0.15) also concur with previous studies. Supplier integration is crucial to attaining supply chain sustainability goals (Haverkamp et al., 2010). Supplier integration in sustainability initiatives, as well as their relationships with focal organizations, were seen as critical factors in the implementation of SSCM (Kersten, 2019). Structural capital appears as a fourth essential driver of SSCM with 0.13 weightage. It confirms the significant role of structural capital in the adoption of SSCM. It is followed by supply chain mapping (0.11). The findings on SC mapping show that organizations with a well-mapped supply chain can have higher and better chances of SSCM adoption. The findings on corporate strategy show that efficient SSCM necessitates synchronization with major categories of activities, which may be implemented in different departments and disciplines, as well as anchored in corporate strategy. We noticed that once the strategy or purpose is established, it not only allows SSCM but also effectively inspires staff to engage in a generally shared and supported sustainability mission (Oelze, 2017).

Findings on the barriers reveal the most important role of top management commitment. The findings reflect that lack of top management support (0.18) stands at the top of the list followed by lack of intrinsic motivation (0.16). Studies reflect that pursuance of sustainability requires a sense of motivation from inside the human being that drives him/her toward sustainability. Motivation either directly or indirectly has an impact on the outcomes/direction of the Organization. We know for a fact that every conscious action has a motive behind it that helps us drive it toward success so basically intrinsic motivation is fuel for our thoughts and actions. Intrinsic motivation is more important and valuable as it does not have any reward or any pressure that drives the action rather it is a sense of personal satisfaction and so the intrinsic motivation to make a product more sustainable or to work toward sustainability in an organization is to satisfy the personal desire of a sustainable environment to make the earth greener and livelier for the future generations to come. Motivation from inside to drive sustainability would result in faster outcomes compared to any other case where the companies or the people are forced to work for this cause.

Production constraints (tacit and explicit) stand third with weightage (0.13). Throughout the literature, an important challenge in the implementation of sustainability practices in SCM has been the production constraints. To achieve sustainability, it is essential to overcome the various tacit or explicit hurdles in production (Gupta et al., 2020). Production constraints can partly be attributed to poor integration and lack of currency in the production process. Together with the production constraint, process inefficiencies (0.13) appear equally important. It shows that inefficient processes act as a hurdle in the way of SSCM.

Lack of integration with supplier (0.12) possesses almost equal significance. According to the findings, poor integration with suppliers can be detrimental to SSCM adoption. These findings are in line with many studies. For example, Alzawawi (2014) argues that a lack of supplier integration can influence a firm's performance. Suppliers can constitute an enabling factor for SSCM. The long-term supplier relationship is one of the important drivers as increased cooperation over time is advantageous since it considerably improves the effective implementation of sustainable supply chain strategies. (Oelze, 2017). According to our findings, cost constraint (0.11) is also a significant hurdle in the way of SSCM. Firms perceive that SSCM is an expensive proposition because of the auditing and other expenses (Oelze, 2017). Likewise, the lack of knowledge/awareness (0.09) of the workers working on the production lines significantly hinders the sustainability drive. The participants of this study expressed that many stakeholders lack the enthusiasm to make an effort toward developing and maintaining sustainable practices in their firms. The major reason behind this is that they lack knowledge regarding the long-term benefits that both the environment and the company can gain due to these practices. The initial concern of the stakeholders is usually limited to the initial capital investments that may be required to develop the business processes toward sustainability and overlooking the significant impact that it can generate in terms of the triple bottom line, that is, people, planet, and profit. Finally, lack of clear guidance (0.05) and lack of government pressure (0.03) appear as the least yet relevant barriers to SSCM. The government's poor legislation, in a few developing countries, is considered to be one of the barriers to SSCM. Government legislation is linked to the requirements of both present and upcoming regulations. Legislation promotes engagement levels and encourages organizations to adopt sustainable practices (Kersten, 2019). In other words, increasing government pressure can be an enabler in the implementation of SSCM. Government policies play a major role in the sustainability of any business.

In short, our results concur with the existing literature but provide some deeper insights, especially on the role of intellectual capital (the present study has taken its three dimensions separately: human capital, relational capital, and structural capital).

4.4 Conclusion, implications, and limitations

The overarching focus of the present study was to highlight the essential drivers and barriers to SSCM. We adopted a threefold approach whereby in the first stage a comprehensive review of the literature was conducted to identify the drivers and barriers to SSCM. In the second stage through a preference value survey, the relevant drivers and barriers were selected. In the third stage, by employing a support vector machine, and a robust application of ML, the selected barriers and drivers were prioritized. Our findings 09 drivers and 09 barriers of SSCM, which were further prioritized according to their relative importance.

The findings reveal that human capital, structural capital, and relational capital—these three dimensions are collectively known as intellectual capital—are apex drivers of SSCM in the Textile sector. A plethora of the literature has confirmed these three factors' role in the firm performance; however, their role in sustainability has not been investigated in depth. Our findings provide a starting point in this regard, suggesting future studies to explore how three dimensions of intellectual capital influence the drive toward sustainability. We have already provided expert views highlighting their significant role. Further findings highlight a lack of top management commitment, lack of intrinsic motivation, and cost constraints as the major barriers to SSCM. Although these three factors have been discussed frequently in the literature, our findings not only reveal their comparative importance but also highlight their major role in SSCM. Our findings also bring forward supplier integration and relationship management as the factors for adopting SSCM. Based on these findings we suggest firms, especially larger firms, look up to the Green Supplier Developmental Program (GSDP). The GSDP is an initiative to enhance environmental performance by developing green practices at the suppliers' level. Results also reveal if an organization can somehow minimize or avoid such costs, it can have a higher tendency to adopt SSCM. Another perspective in this regard is to consider cost as an investment. Some of the authors argue that expenses incurred to make SCM sustainable are investments for the firm, which can pay off in the future. Taking cost as an investment could also provide a different angle to deal with this burden.

Our study has some profound implications for the firms. A firm may have many reasons and pressures to adopt SSCM, as mentioned by McKinsey report, "*Companies have many reasons to focus on environmental, social, and governance (ESG) issues. They may want to satisfy their consumers, who are increasingly choosing brands with strong ESG credentials, even if the prices are higher. Or they may be seeking to stay ahead of ever more stringent regulations. Others react to pressure from banks and investors, want to improve employee engagement, or feel a need to better attract and retain talent. For most organizations, the*

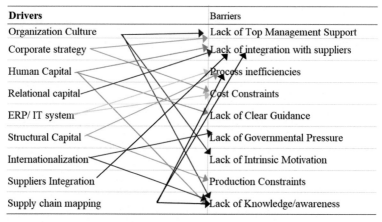

Figure 4.1 Linking drivers to overcome barriers.

answer will be a combination of these factors, which together add up to a need to understand and manage environmental impact through every part of the business—in real time." We argue that whatever the reasons are working on the identified barriers and drivers can significantly help a firm to adopt SSCM.

We suggest that organizations capitalize on their inherent drivers as potent resources to control the obstacles within the realm of SSCM. Consider, for example, how a company can strategically harness its organizational culture to effectively address major barriers such as a dearth of top management support (TMS), intrinsic motivation, and awareness. By cultivating an environment featuring progressiveness, openness, and inclusivity, an organization can foster awareness among both employees and senior leadership. This heightened awareness serves as a catalyst, inspiring the workforce and management to embrace SSCM practices with renewed vigor. Furthermore, the organization's intellectual capital (human, relational, and processes capital) can be adeptly deployed to tackle challenges ranging from awareness deficits and process inefficiencies to financial constraints and a lack of intrinsic motivation. Fig. 4.1 illustrates how these drivers could be used to tackle the barriers. It provides a comprehensive guiding framework for organizations seeking to navigate the complex terrain of SSCM optimization.

References

Alzawawi, M. (2014). Drivers and Obstacles for Creating Sustainable Supply Chain Management and Operations. *Zone I Student Papers Proceedings Archive; American Society for Engineering Education* (ASEE): Washington, DC, USA, 1−8.

Ahmed, S. S., Guozhu, J., Mubarik, S., Khan, M., & Khan, E. (2020). Intellectual capital and business performance: the role of dimensions of absorptive capacity. *Journal of Intellectual Capital*, *21*(1), 23−39.

Ahmed, M., Mubarik, M. S., & Shahbaz, M. (2021). Factors affecting the outcome of corporate sustainability policy: A review paper. *Environmental Science and Pollution Research*, *28*, 10335−10356.

Baah, C., & Zhihong, J. (2019). Sustainable supply chain management and organizational performance: The intermediary role of competitive advantage. *Journal of Management and Sustainability*, *9*(1), 119−131.

Bai, C., Sarkis, J., & Dou, Y. (2015). Corporate sustainability development in China: Review and analysis. *Industrial Management & Data Systems*, 5−40.

Baig, S. A., Abrar, M., Batool, A., Hashim, M., & Shabbir, R. (2020). Barriers to the adoption of sustainable supply chain management practices: Moderating role of firm size. *Cogent Business & Management*.

Caniato, F., Caridi, M., Crippa, L., & Moretto, A. E. (2012). Environmental sustainability in fashion supply chains: An exploratory case based research. *International Journal of Production Economics*, 659−670.

Fischhoff, M. (2013, September 5). *How going green can help your business*. Retrieved from Network for business sustainability: https://www.nbs.net/articles/how-going-green-can-help-your-business.

Giunipero, L. C., Hooker, R. E., & Denslow, D. (2012). Purchasing and supply management sustainability: Drivers and barriers. *Journal of Purchasing and Supply Management*, 258−269.

Gonzalez-Benito., & Gonzalez-Benito, O. (2009). A study of determinant factors of stakeholder environmental pressure perceived by industrial companies. *Business Strategy and the Environment*, 164−181.

González-Benito, J. G.-B. (2009). A study of determinant factors of stakeholder environmental pressure perceived by industrial companies. *Business Strategy and the Environment*, 164−181.

Gupta, H., Kusi-Sarpong, S., & Rezaei, J. (2020). Barriers and overcoming strategies to supply chain sustainability innovation. *Resources, Conservation & Recycling*.

Haverkamp, D., J Bremmers, H., & Omta, O. (2010). Stimulating environmental management performance. *British Food Journal*, 1237−1251.

Kersten, M. A. (2019). Drivers of sustainable supply chain management: Identification and classification. *Sustainability*, 23.

Khan, S. A., Gupta, H., Gunasekaran, A., Mubarik, M. S., & Lawal, J. (2023). A hybrid multi-criteria decision-making approach to evaluate interrelationships and impacts of supply chain performance factors on pharmaceutical industry. *Journal of Multi-Criteria Decision Analysis*, *30*(1−2), 62−90.

Khan, S. A., Mubarik, M. S., Kusi-Sarpong, S., Zaman, S. I., & Kazmi, S. H. A. (2021). Social sustainable supply chains in the food industry: A perspective of an emerging economy. *Corporate Social Responsibility and Environmental Management*, *28*(1), 404−418.

Khan, S. A., Mubarik, M. S., & Paul, S. K. (2022). Analyzing cause and effect relationships among drivers and barriers to circular economy implementation in the context of an emerging economy. *Journal of Cleaner Production*132618.

Khan, S. A., Mubarik, M. S., Kusi-Sarpong, S., Gupta, H., Zaman, S. I., & Mubarik, M. (2022). Blockchain technologies as enablers of supply chain mapping for sustainable supply chains. *Business Strategy and the Environment*.

Kusi-Sarpong, S., Mubarik, M. S., Khan, S. A., Brown, S., & Mubarak, M. F. (2022). Intellectual capital, blockchain-driven supply chain and sustainable production: Role of supply chain mapping. *Technological Forecasting and Social Change*, *175*121331.

Mahmood, T., & Mubarik, M. S. (2020). Balancing innovation and exploitation in the fourth industrial revolution: Role of intellectual capital and technology absorptive capacity. *Technological Forecasting and Social Change, 160,* 120248.

Majumdar, A., & Sinha, S. K. (2019). Analyzing the barriers of green textile supply chain management in Southeast Asia using interpretive structural modeling. *Sustainable Production and Consumption,* 176−187.

Mann, B. J., & Harmeet, K. (2020). Sustainable supply chain activities. *Vision,* 60−69.

Melanie. (2017). *A handle on procurement: The cost of purchasing inventory.* Available from https://www.unleashedsoftware.com/blog/procurement-cost-purchasing-inventory. (Accessed on 12 May 2017).

Mobashar, M., & Mubarak, M. F. (2020). Fostering supply chain integration through blockchain technology: A study of Malaysian manufacturing sector. *International Journal of Management and Sustainability, 9*(3), 135−147.

Mubarik, M. S. (2023). *A firm-based perspective of the notion of "Carbon neutrality": The role of supply chain mapping. Recent developments in green finance, green growth and carbon neutrality* (pp. 63−84). Elsevier.

Mubarik, M. S., & Bontis, N. (2022). Intellectual capital, leadership and competitive advantage: a study of the Malaysian electrical and electronics industry. *International Journal of Learning and Intellectual Capital, 19*(6), 562−583.

Mubarik, M. S., Bontis, N., Mubarik, M., & Mahmood, T. (2021). Intellectual capital and supply chain resilience. *Journal of Intellectual Capital.*

Mubarik, M. S., Naghavi, N., Mubarik, M., Kusi-Sarpong, S., Khan, S. A., Zaman, S. I., & Kazmi, S. H. A. (2021). Resilience and cleaner production in industry 4.0: Role of supply chain mapping and visibility. *Journal of Cleaner Production, 292*126058.

Mubarik, M. S., Kusi-Sarpong, S., Govindan, K., Khan, S. A., & Oyedijo, A. (2021). Supply chain mapping: A proposed construct. *International Journal of Production Research,* 1−17.

Mubarik, M. S., Khan, S. A., Kusi-Sarpong, S., Brown, S., & Zaman, S. I. (2023a). *Supply Chain Mapping. Sustainability, and Industry 4.0.* Taylor & Francis.

Mubarik, M. S., Khan, S. A., Kusi-Sarpong, S., & Mubarik, M. (2023b). Supply chain sustainability in VUCA: Role of BCT-driven SC mapping and 'Visiceability'. *International Journal of Logistics Research and Applications,* 1−19.

Mubarik, M. S., Khan, S. A., Acquaye, A., & Mubarik, M. (2023c). Supply chain mapping for improving "visilience": A hybrid multi-criteria decision making based methodology. *Journal of Multi-Criteria Decision Analysis.*

Mubarik, M. S., Chandran, V. G. R., & Devadason, E. S. (2018). Measuring human capital in small and medium manufacturing enterprises: what matters? *Social Indicators Research, 137,* 605−623.

Narayanan, A. E., Sridharan, R., & Kumar, P. R. (2019). Analyzing the interactions among barriers of sustainable supply chain management practices: A case study. *Journal of Manufacturing Technology Management.*

Nathanson, J.A. (2019, Dec 24). *Air pollution control.* From Encyclopedia Britannica. Available from: https://www.britannica.com/technology/air-pollution-control. (Accessed on 14 November 2021).

Nathanson, J.A. (2020, November 10). *Solid-waste management.* From Encyclopedia Britannica. Available from: https://www.britannica.com/technology/solid-waste-management. (Accessed on 14 November 2021).

Oelze, N. (2017). Sustainable supply chain management implementation−enablers and barriers in the textile industry. *Sustainability,* 15.

Oyedijo, A., Kusi-Sarpong, S., Mubarik, M. S., Khan, S. A., & Utulu, K. (2023). Multi-tier sustainable supply chain management: A case study of a global food retailer. *Supply Chain Management: An International Journal.*

Parsad, D. S., R, P., K, G., & A, K. (2019). *Critical success factors of sustainable supply chain management and organizational performance: An exploratory study. Transportation research procedia 48 (2020)* (pp. 327–344). Mumbai: Elsevier B.V.

Schrettle, S. H.-R. (2014). Turning sustainability into action: Explaining firms sustainability efforts and their impact on firm performance. *International Journal of Production Economics*, 73–84.

Reza, S., Mubarik, M. S., Naghavi, N., & Nawaz, R. R. (2021). Internationalisation challenges of SMEs: role of intellectual capital. *International Journal of Learning and Intellectual Capital*, *18*(3), 252–277.

Schäfer, M., & Manuel, L. (2021). Ecodesign—A review of reviews. *Sustainability*, 1–33.

TL, P. (2021). *Environmental ecology: Fundamental concepts of environmental ecology*. Nestfame Creations Pvt Ltd.

Teba, C. (2018, Feb 12). *Energy consumtion what doest it mean*. From Dexma Energy Intelligence By Spacewell. Availablefrom: https://www.dexma.com/blog-en/energy-consumption-definition/.(Accessed 16.03.21).

Teng, Z. L., Guo, C., Zhao, Q., & Mubarik, M. S. (2023). Antecedents of green process innovation adoption: An AHP analysis of China's gas sector. *Resources Policy, 85*103959.

Viloria, A., & Paula, V. (2016). Inventory reduction in the supply chain of finished products for multinational companies. *Indian Journal of Science and Technology*, *9*, 1–5.

Zaabi, S. A., Dhaheri, N. A., & Diabat, A. (2013). Analysis of interaction between the barriers for the implementation of sustainable supply chain management. *The International Journal of Advanced Manufacturing Technology*, 895–905.

Further reading

A, R., D, S., & H, C. (2010). Small businesses and the environment: Turning over a new leaf? *Business Strategy and the Environment*, 273–288.

Ageron, B., Gunasekaran, A., & Spalanzani, A. (2011). Sustainable supply management: An empirical study. *International Journal of Production Economics*, 168–182.

Ahi, P., & Searcy, C. (2013). A comparative literature analysis of definitions for green and sustainable supply chain management. *Journal of Cleaner Production*, 329–341.

Akkermans, H. A., Bogerd, P., Yücesan, E., & Van Wassenhove, L. N. (2003). The impact of ERP on supply chain management: Exploratory findings from a European Delphi study. *European Journal of Operational Research*, 284–301.

Alblas, A. A., Peters, K., & Wortmann, J. C. (2014). Fuzzy sustainability incentives in new product development. *International Journal of Operations & Production Management*, 513–545.

Allen., et al. (1999). The ecology of leadership: Adapting to the challenges of the latest world. *Journal of Leadership Studies*, *5*(2), 62–82.

Ardakani, D. A., & S, A. (2018). Investigating and analysing the factors affecting the development of sustainable supply chain model in the industrial sectors. *Corporate Social Responsibility And Environmental Management*, 199–212.

Awaysheh, A., & Klassen, R.D. (n.d.). *The impact of supply chain structure on the use of supplier socially responsible practices.*

Aymen Sajjad, G. E. (2015). Sustainable supply chain management: Motivators and barriers. *Business Strategy and the Environment*, 1–13.

Azmat, M., Ahmed, S., & Mubarik, M. S. (2022). *Supply chain resilience in the fourth industrial revolution. Supply chain resilience* (pp. 149–163). Cham: Springer.

Babakus, E. (2003). The effect of management commitment to service quality on employees. *Affective and Performance Outcomes.*

Bai, C., & Satir, A. (2020). Barriers for green supplier development programs in manufacturing industry. *Resources, Conservation & Recycling.*

Ball, A., & Craig, R. (2010). Using neo-institutionalism to advance social and environmental accounting. *Critical Perspectives on Accounting*, 283–293.

Bates, O., & Hazas, M. (2013). Exploring the hidden impacts of HomeSys: Energy and emissions of home sensing and automation. In: *Proceedings of the 2013 ACM Conference on Pervasive and Ubiquitous Computing Adjunct Publication*, 809–814.

Bernard, H. R. (2014). Forward. In S. Dominguez, & B. Hollstein (Eds.), *Mixed methods social networks research*. Cambridge University Press.

Boulomytis, V., Zuffo, A., & Imteaz, M. (2019). Detection of flood influence criteria in ungauged basins on a combined Delphi-AHP approach. *Operations Research Perspectives*, 6, 1–12.

Bowen, F. E., Cousins, P. D., Lamming, R. C., & A, A. (2001). The role of supply management capabilities in green supply. *Production and Operations Management*, 174–189.

Bråten, A. V., & Siurys, D. (2021). *Drivers and barriers to sustainable supply chain management* (Master's thesis, University of South-Eastern Norway). Available at: https://openarchive.usn.no/usn-xmlui/handle/11250/2783844.

Carter., & Rogers. (2008). A framework of sustainable supply chain management: Moving toward new theory. *International Journal of Physical Distribution & Logistics Management*, 360–387.

Carter, C. R., & Dresner, M. (2001). Purchasing's role in environmental management: Cross functional development of grounded theory. *The Journal of Supply Chain Management*, 12–27.

Chiou, T., Chan, H., Lettice, F., & Chung, S. (2011). The influence of greening the suppliers and green innovation on environmental performance. *Transportation Research Part E: Logistics and Transportation Review*, 822–836.

Collins, C., Steg, L., & Koning, M. (2007). Customers' values, beliefs on sustainable corporate performance, and buying behavior. *Psychology and Marketing*, 555–577.

Cooper, R. W., Frank, G. L., & Kemp, R. A. (2000). A multinational comparison of key ethical issues, helps and challenges in the purchasing and supply management profession: The key implications for business and the professions. *Journal of Business Ethics*, 83–100.

Creswell, J.W. (2013). *Steps in conducting a scholarly mixed methods study. DBER Speaker series*. University of Nebraska Discipline-Based Education Research Group (Online) Retrieved.

Creswell, J. W., Plano Clark, V. L., Gutmann, M. L., & Hanson, W. E. (2003). Advanced mixed methods research designs. In A. Tashakkori, & C. Teddlie (Eds.), *Handbook of mixed methods in social and behavioral research* (pp. 209–240). Thousand Oaks, CA: Sage.

Cardenas-Barrón, L. E. (2009). Economic production quantity with rework process at a single-stage manufacturing system with planned backorders. *Computers & Industrial Engineering*, 57, 1105–1113.

Ciliberti, F., Pontrandolfo, P., & Scozzi, B. (n.d.). *Small business social responsibility in the supply chain*.

David, L. M. (2014). *Integrating qualitative and quantitative methods: A pragmatic approach*. Sage.

Dalkey, N., & Helmer, O. (1963). An experimental application of the DELPHI method. *Management Science*, 458–467.

De Brito, M. P., Carbone, V., & Blanquart, C. M. (2008). Towards a sustainable fashion retail supply chain in Europe Organisation and performance. *International Journal of Production Economics*, 534–553.

Dubey, R., Gunasekaran, A., & Ali, S. S. (2015). Exploring the relationship between leadership, operational practices, institutional pressures and environmental performance. A framework for green supply chain. *International Journal of Production Economics*, 120–132.

Dzokoto, S. D. Dadzie, J. (2013). Barriers to sustainable construction in the Ghanaian construction industry: Consultants perspectives, Journal of Sustainable Development, 7.

Ehrgott, M., Reimann, F., Kaufmann, L., & Carter, C. R. (2013). Environmental development of emerging economy suppliers: Antecedents and outcomes. *Journal of Business Logistics*, 131−147.

Emamisaleh, K., & Rahmani, K. (2017). Sustainable supply chain in food industries: Drivers and strategic sustainability orientation. *Cogent Business & Management*.

Epstein, M. J., & Roy, M.-J. (2003). Making the business case for sustainability: Linking social and environmental actions to financial leadership. *The Journal of Corporate Citizenship*, 9, 79−96.

Esfahbodi, A., Y, Z., G, W., & T, Z. (2017). Governance pressures and performance outcomes of sustainable supply chain management − An empirical analysis of UK manufacturing industry. *Journal of Cleaner Production*, 66−78.

Esfahbodi, A., Zhang, Y., Watson, G., & Zhang, T. (n.d.). *Governance pressures and performance outcomes of sustainable supply chain management - An empirical analysis of UK manufacturing industry*. 1−36.

Fan, X., & S, Z. (2016). *Performance evaluation for the sustainable supply chain management*. Croatia: Intechopen.

Gam, H. J., & Banning, J. (2011). Addressing sustainable apparel design challenges with problem-based learning. *Textile Research Journal*, 202−215.

Glass et al. (n.d.). *Do women leaders promote sustainability? Analysing the effect of corporate governance on environmental performance*.

González, P., Sarkis, J., & Adenso-Díaz, B. (2008). Environmental management system certification and its influence on corporate practices. *International Journal of Operations & Production Management*, 1021−1041.

Govindan, K., Muduli, K., Devika, K., & Barve, A. (2016). Investigation of the influential strength of factors on adoption of green supply chain management practices: An Indian mining scenario. *Resources, Conservation and Recycling*, 185−194.

Govindan, K. M. (2016). Investigation of the influential strength of factors on adoption of green supply chain management practices: An Indian mining scenario. *Resources, Conservation and Recycling*, 185−194.

Green Kenneth, W., Pamela., Zelbst, J., Jeramy, M., & Bhadauria, V. S. (2012). Green supply chain management practices: Impact on performance. *Supply Chain Management*, 290−305.

Gröschl, S., Gabaldón, P., & Hahn, T. (2019). The co-evolution of leaders' cognitive complexity and corporate sustainability: The case of the CEO of Puma. *Journal of Business Ethics*, 155, 741−762.

Gualandris, J., & Kalchschmidt, M. (2014). Customer pressure and innovativeness: Their role in sustainable supply chain management. *Journal of Purchasing and Supply Management*, 92−103.

Guba, E. G., & Lincoln, Y. S. (1994). Competing paradigms in qualitative research. In N. K. Denzin, & Y. S. Lincoln (Eds.), *Handbook of qualitative research* (pp. 105−117). Thousand Oaks, CA: Sage Publications.

Hamdy, O. M., K, K., & B, E. (2018). Impact of sustainable supply chain management practices on Egyptian companies' performance. *European Journal of Sustainable Development*, 119−130.

Holt, D., & Ghobadian, A. (2009). An empirical study of green supply chain management practices amongst UK manufacturers. *Journal of Manufacturing Technology Management*, 933−956.

Hsu, C.-C., Tan, C. T., Zailani., Jayaram, S. H. M., & V. (2013). Supply chain drivers that foster the development of green initiatives in an emerging economy. *International Journal of Operations & Production Management*, 656−658.

Hsu, C.-C. T., Zailani, C., Jayaraman, S. H. M., & V. (2013). Supply chain drivers that foster the development of green initiatives in an emerging economy. *International Journal of Operations & Production Management*, 656−688.

Huang, C.-L. K.-H. (2010). Drivers of environmental disclosure and stakeholder expectation: Evidence from Taiwan. *Journal of Business Ethics*, 435−451.

Häkkinen, T., & Belloni, K. (2011). Barriers and drivers for sustainable building. *Building Research and Information*, *39*(3), 239−255.

Jaber, M. Y., & Khan, M. (2010). Managing yield by lot splitting in a serial production line with learning, rework and scrap. *International Journal of Production Economics*, *124*(12), 32−39.

Jia, F., Zuluaga, L., Bailey, A., & Rueda, X. (2018). Sustainable supply chain management in developing countries: An analysis of the literature. *Journal of Cleaner Production*, 46.

Johnson, R. B., & Onwuegbuzie, A. J. (2004). A research paradigm whose time has come. *Educational Researcher*, *7*(33), 14−26.

Kant, R. (2012). Textile dyeing industry an environmental hazard. *Natural Science*, *4*(1), 5. Available from https://doi.org/10.4236/ns.2012.41004.

Kasim, A., & Ismail, A. (2012). Environmentally friendly practices among restaurants: Drivers and barriers to change. *Journal of Sustainable Tourism*, 551−570.

Kazancoglu, Y., Kazancoglu, I., & Sagnak, M. (2017). Industrial management & data systems. *Industrial management & data systems performance: Application in cement industry*, 26.

Khodakarami, M., Shabani, A., Saen, R. F., & Azadi, M. (2015). Developing distinctive two-stage data envelopment analysis models: An application in evaluating the sustainability of supply chain management. *Measurement*, 62−74.

Kot, S. (2018). Sustainable supply chain management in small and medium enterprises. *Sustainability*, 1−10.

Krause, D., Vachon, S., & Klassen, R. (2009). Special topic forum on sustainable supply chain management: Introduction and reflections on the role of purchasing management. *Journal of Supply Chain Management*, 18−25.

Kuei, C., Madu, C., Chow, W., & Chen, Y. (2015). Determinants and associated performance improvement of green supply chain management in China. *Journal of Cleaner Production*, 163−173.

Louise, M., & Benn, S. (2013). Leadership for sustainability: An evolution of leadership abilities. *Journal of Business Ethics*, *112*, 369−384.

Mason, J. (2006). Mixing methods in a qualitatively driven way. *Qualitative Research*, *6*(1), 9−25.

Mathivathanan, D., K, D., & H, A. (2018). Sustainable supply chain management practices in Indian automotive industry: A multistakeholder view. *Resources, Conservation and Recycling*, 284−305.

Meixell, M. J., & Luoma, P. (2015). Stakeholder pressure in sustainable supply chain management. *Int. J. Phys. Distrib Logist. Manag.*, 69−89.

Miao, C., Humphrey, R. H., & Qian, S. (2016). Leader emotional intelligence and subordinate job satisfaction: A meta-analysis of main, mediator, and moderator effects. *Personality and Individual Differences*, *102*, 13−24.

Min, H., W., & Galle, P. (2001). Green purchasing practices of US firms. *International Journal of Operations & Production Management*, 1222−1238.

Mubarak, M. F., Shaikh, F. A., Mubarik, M., Samo, K. A., & Mastoi, S. (2019). The impact of digital transformation on business performance: A study of Pakistani SMEs. *Engineering Technology & Applied Science Research*, *9*(6), 5056−5061.

Negrete, J. D., & López, V. N. (2020). A sustainability overview of the supply chain management in textile industry. *International Journal of Trade, Economics and Finance*.

Niederberger, M., & J, S. (2020). Delphi technique in health sciences: A map. *Frontiers in Public Health*, 1−10.

Niinimäki, K., & Hassi, L. (2011). Emerging design strategies in sustainable production and consumption of textiles and textile. *Journal of Cleaner Production*, *19*, 1876−1883.

Oelze, N. (2017). Sustainable supply chain management implementation−Enablers and barriers in the Textile Industry. *Sustainability*, 17.

Osaily, N. Z. (2010). *The key barriers to implementing sustainable construction in West Bank.* Prifysgol Cymru University of Wales.

Ouyang, L.-Y., Chen, C.-K., & Chang, H.-C. (2002). Quality improvement, setup cost and lead-time reductions in lot size reorder point models with an imperfect production process. *Computers & Operations Research*, *29*(12), 1701−1717.

Pagell, M., & Wu, Z. (2009). Building a more complete theory of sustainable supply chain management using case studies of 10 exemplars. *Journal of Supply Chain Management*, 37−56.

Parkin, S. (2000). Sustainable development: the concept and the practical challenge. In Proceedings of the Institution of Civil Engineers-Civil Engineering, (Vol. 138, No. 6, pp. 3−8). Thomas Telford Ltd.

Paulraj, A., Chen, I., & Blome, C. (2011). Motives and performance outcomes of sustainable supply chain management practices: A multitheoretical perspective. *Journal of Business Ethics*, 239−258.

Paulraj, A., Chen, I. J., & Blome, C. (2015). Motives and performance outcomes of sustainable supply chain management practices: A multi-theoretical perspective. *Journal of Business Ethics JBE*, 25−299.

Piprani, A. Z., Jaafar, N. I., Ali, S. M., Mubarik, M. S., & Shahbaz, M. (2022). Multi-dimensional supply chain flexibility and supply chain resilience: The role of supply chain risks exposure. *Operations Management Research*, 1−19.

Prasad, D. S., Pradhan, R. P., Gaurav, K., & Sabat, A. K. (2019). Critical success factor os sustainable supply chain management and organizational performance: An exploratory study. *Transportation Research Procedia*, 1−18.

Preuss, L. (2009). Addressing sustainable development through public procurement: The case of local government. *An International Journal of Supply Chain Management*, 213−223.

Qorri, A., Gashi, S., & Kraslawski, A. (2020). Performance outcomes of supply chain practices for sustainable development: A meta-analysis of moderators. *Sustainable Development*, 194−216.

Reefke, H., & S, D. (2017). Key themes and research opportunities in sustainable supply chain management—Identification and evaluation. *Omega*, 195−211.

Revell, A., & Blackburn, R. (2007). The business case for sustainability? An examination of small firms in the UK's construction and restaurant sectors. *Business Strategy and the Environmental*, 404−420.

Ringdal, K. (2013). *Enhet og mangfold samfunnsvitenskapelig forskning og kvantitativ metode utg* (3. ed.). Bergen: Fagbokforl.

Rivera, J. (2004). Institutional pressures and voluntary environmental behavior in developing countries: Evidence from the Costa Rican hotel industry. *Society & Natural Resources*, 779−797.

Rose, R., & Bowen, P. (2017). Mixed methods-theory and practice. Sequential, explanatory approach. *International Journal of Quantitative and Qualitative Research Methods*, *5*, 10−27.

Rydin, Y., Amjad, U., Moore, S., Nye, M., & Withaker, M. (2006). *Sustainable construction and planning*. The Academic Report. Centre for Environemental Poilcy and Governance.

Saeed, M. A., & Kersten, W. (2019). Drivers of sustainable supply chain management: Identification and classification. *Sustainability*, 1137.

Saeed, M. A., & Kersten, W. (2019). Drivers of sustainable supply chain management: Identification and Classification. *Sustainability*, 23.

Saeed, M. A., & W, K. (2019). Drivers of sustainable supply chain management. *Sustainability*, 1−23.

Saeed, M. A., Waseek, I., & Kersten, W. (2017). Literature review of drivers of sustainable supply chain management. *Digitalization in Maritime and Sustainable Logistics*.

Sajjad, A., Eweje, G., & Tappin, D. (2015). Sustainable supply chain management: Motivators and barriers. *Business Strategy and the Environment*, 643−655.

Sarkis, J., Gongzalez-Torre., & A.-D, B. (2010). Stakeholder pressure and the adoption of environmental practices: The mediating effect of training. *Journal of Operations Management*, 163−176.

Sarkis, J., Z, Q., & L, K. (2011). An organizational theoretic review of green supply chain management literature. *International Journal of Production Economics*, 1−15.

Savin-Baden, M., & Major, C. (2023). Qualitative research: The essential guide to theory and practice. London: Routledge.

Schrettle, S., Hinz, A., Scherrer-Rathje, M., & Friedli, T. (2014). Turning sustainability into action: Explaining firms sustainability efforts and their impact on firm performance. *International Journal of Production Economics*, 73−84.

Seuring, S., & M, M. (2008). From a literature review to a conceptual framework for sustainable supply chain management. *Journal of cleaner Production*, 166−1710.

Shafique, M. N., Asghar, M. S., & Rahman, H. (2017). The impact of Green supply chain management practices on performance: Moderating role of institutional pressure with mediating effect of Green innovation. *Business Management and Education*, 91−108.

Silva, W. H., Patricia, G., José, M., Josivania, S., & Silvia, A. (2019). Sustainable supply chain management: Analyzing the past to determine a research agenda. *Logistics*, 1−15.

Smith, P. A., & Sharicz, C. (2011). The shift needed for sustainability. *The learning organization*, *18*(1), 73−86.

Soda, S. A., & Rajiv Kumar, G. (2015). GSCM: Practices, trends and prospects in Indian context. *Journal of Manufacturing Technology Management*, 889−910.

Somsuk, N., & Laosirihongthong, T. (2017). Prioritization of applicable drivers for green supply chain management implementation toward sustainability in Thailand. *International journal of Sustainable Development and World Ecology*, 175−191.

Sroufe, R., & M, S. (2017). Developing sustainable supply chains: Management insights, issues, and tools: Volume I foundations. *Business Expert Press*, 7−8.

Stange, K. C. (2006). Publishing multimethod research. *Annals of Family Medicine*, *4*(4), 292−294.

Svensson, G. (2007). Aspects of sustainable supply chain management (SSCM): Conceptual framework and empirical example. *Supply Chain Management: An International Journal*, 262−266.

Sybertz, J. (2017). *Sustainability and effective supply chain management: A literature review of sustainable supply chain*. New York: Center of Sustainable Business.

Sánchez-Flores, R. B., S, E.-S., S, O.-B., & M, E.-B. (2020). Sustainable supply chain management—A literature review on emerging economies. *Sustainability*, 1−27.

Tashakkori, A., & Teddlie, C. (2003). Handbook of mixed methods in social and behavioral research. Thousand Oaks: Sage.

Tate, W. E. (2010). Corporate social responsibility reports: A thematic analysis related to supply chain management. *Journal of Supply Chain Management*, 19−44.

Tate, W. L., Ellram, L. M., & Kirchoff, J. F. (2010). Corporate social responsibility reports: A thematic analysis related to supply chain management. *Journal of Supply Chain Management*, 19−44.

Tay, M. Y., Rahman, A., & Abdul A, Y. (2015). A review on drivers and barriers towards sustainable supply chain practices. *International Journal of Social Science and Humanity*, 6.

Taylor, G.R., & Trumbull, M. (2005). Developing a multi faced research design/para-digm. *Integrating quantitative and qualitative methods in research* (2nd ed). University Press of America.

Thaba, S. (2017). Drivers of Sustainable Supply Chain Management in South Africa a Total Interpretive Structural Method (TISM) Based Review. In: *Paper presented at the Proceedings of the World Congress on Engineering and Computer Science.*

Tummala, V., Phillips, C., & Johnson, M. (2006). Assessing supply chain management success factors: A case study. *Supply Chain Management An International Journal*, 179−192.

Turker, D., & Altuntas, C. (2014). Sustainable supply chain management in the fast fashion industry An analysis of corporate reports. *European Management Journal*, 837−849.

Vachon, S., & Klassen, R. (2006). Extending green practices across the supply chain: The impact of upstream and downstream. *International Journal of Operations & Production Management*, 26(7), 795−821.

Walker, H., Di Sisto, L., & McBain, D. (2008). Drivers and barriers to environmental supply chain management practices: Lessons from the public and private sectors. *Journal of Purchasing and Supply Management*, 69−85.

Walker, H. D. (2008). Drivers and barriers to environmental supply chain management practices: Lessons from the public and private sectors. *Journal of Purchasing and Supply Management*, 69−85.

Warasthe, R., Schulz, F., Enneking, R., & Brandenburg, M. (2020). Sustainability prerequisites and practices in textile and apparel supply chains. *Sustainability*, 12(23), 9960.

Williams, K., & Dair, C. (2007). What is stopping sustainable building in England? Barriers experienced by stakeholders in delivering. *Sustainable Developments.*

Wittstruck, D., & Teuteberg, F. (2011). Understanding the success factors of sustainable supply chain management: Empirical evidence from the electrics and electronics industry. *Corporate Social Responsibility and Environmental Management*, 141−158.

Xu, L., Mathiyazhagan, K., Govindan, K., Noorul Haq, A., Ramachandran, N. V., & Ashokkumar, A. (2013). Multiple comparative studies of green supply chain management: Pressures analysis. *Resources, Conservation and Recycling*, 26−35.

Yildiz Çankaya, S., & Sezen, B. (2018). Effects of green supply chain management practices on sustainability performance. *Journal of Manufacturing Technology Management*, 26.

Yukl, G., Mahsud, R., Prussia, G., & Hassan, S. (2019). Effectiveness of broad and specific leadership behaviors. *Personnel Review*, 48(3), 774−783.

Zailani, S., Jeyaraman, K., Vengadasan, G., & Premkumar, R. (2012). Sustainable supply chain management (SSCM) in Malaysia: A survey. *International Journal of Production Economics*, 330−340.

Zhu, Q., & Sarkis, J. (2007). The moderating effects of institutional pressures on emergent green supply chain practices and performance. *International Journal of Production Research*, 4333−4355.

Zhu, Q., & Sarkis, J. L. (2012). Examining the effects of green supply chain management practices and their mediations on performance improvements. *International Journal of Production Research*, 1377−1394.

Zhu, Q., Sarkis, J., & Lai, K. (2013). Institutional-based antecedents and performance outcomes of internal and external green supply chain management practices. *Journal of Purchasing & Supply Management*, 106−117.

Zimon, D., J, T., & R, S. (2019). Drivers of sustainable supply chain management: Practices to alignment with un sustainable development goals. *International Journal for Quality Research*, 219−236.

Scenario analysis for long-term planning: stratified decision-making models in sustainable supply chains

Amin Vafadarnikjoo

Sheffield University Management School, The University of Sheffield, Sheffield, United Kingdom

5.1 Introduction

5.1.1 Background

The circular economy (CE) aims to offer a suitable substitute for the traditional economic model of "take-make-dispose" to avoid the downsides of the traditional model and strike a balance between the triple bottom line (3BL) of planet, people, and profit (3Ps) by optimizing the value of resources and reducing waste as much as possible (Arsova et al., 2022; Calzolari et al., 2022; Ghisellini et al., 2016; MahmoumGonbadi et al., 2021). An important question yet to be answered in the CE agenda is whether a CE is compatible with economic growth (Bauwens, 2021). Bauwens (2021) argues that selecting a post-growth approach to the CE is likely to be a viable choice for making this compatibility come true. Currently, green growth or decoupling global GDP growth from economic activities and resource consumption seems far from reality (Hickel & Kallis, 2020; Krausmann et al., 2009). Some authors argue that this reconciliation between GDP growth and ecological impacts of economic activities (i.e., green growth) can still be a possible option considering large-scale circular strategies implementation or technological advancements (Bauwens, 2021; Hickel & Kallis, 2020). For example, among various proposed CE strategies (Geissdoerfer et al., 2018), narrowing loops or resource efficiency, which is about utilizing fewer resources in

Computational Intelligence Techniques for Sustainable Supply Chain Management.
DOI: https://doi.org/10.1016/B978-0-443-18464-2.00012-1

the process of manufacturing products or offering services, can still yield profitability. However, these changes do not seem to be leading to an overall reduction in materials and perhaps would result in a rebound effect (Castro et al., 2022). The rebound effect causes a complete or partial offset of environmental gains obtained from higher efficiency by a surge in the total number of produced or consumed goods (Bauwens, 2021; Hickel & Kallis, 2020). The rebound effect is still unknown to many stakeholders in developed economies, which necessitates awareness raising, for example, the Dutch textile industry (Siderius & Poldner, 2021). Based on Merton's (1936) definition that unintended consequences are "unforeseen results that fall outside of the intentions of a purposive action," rebound effects can be labeled as one unintended consequence of the CE strategy implementation of resource efficiency. Carter et al. (2020) proposed a conceptual model within sustainable supply chain management by using paradox theory to theorize the unintended consequences.

5.1.2 Gaps

Analyzing these phenomena and dealing with all challenges in circular supply chains would become more complicated accounting for various uncertain decision dimensions such as climate change, socio-economic situations, unintended consequences, rebound effects, supply chain resilience, reverse and forward integration capabilities, long-term planning period, organizational analysis levels and so on. Thus, it would be very practical for the CE agenda in general and sustainable supply chains to incorporate these uncertain dimensions to predict more reliable future circular scenarios. Thus, a decision framework modeling to select a viable potential future circular strategy is very important, hugely complex, and context-specific, which needs to integrate various criteria and deal with a high level of uncertainty. Taking advantage of computational techniques and decision models can be deemed as a useful approach for guiding policy-makers and decision-makers to realize the best potential strategy for a greener and circular supply chain. Furthermore, the CE could be seen as a potential solution for achieving sustainable development goals (SDGs) (Hák et al., 2016; Rodriguez-Anton et al., 2019; Schroeder et al., 2019). Taking into account both the CE and relevant SDGs within the analysis can result in both economic growth and less environmental damage.

5.1.3 Contributions

A summary of the contributions in this chapter are:

1. The stratified decision-making model (SDMM) (Vafadarnikjoo et al., 2023) is described as a suitable tool to explore the potential future scenarios for CE implementation within sustainable supply chain management under two dimensions of reverse integration and decent work and economic growth (SDG8).

2. The applicability of the SDMM will be shown in an uncertain decision-making context of sustainable supply chains using circularity scenario analysis and numerical simulation. The SDMM examines the changes from different potential states over a longer period (5 + years) (Webb, 2019) by considering the transitions between the reverse integration status and various levels for the SDGs (i.e., decent work and economic growth-SDG8).

3. This method is useful in discovering efficient strategies to identify the best pathway for future CE in the context of sustainable supply chain management.

The structure of the rest of this chapter is as follows: In Section 5.5.2, key variables and definitions are discussed. In Section 5.3, the SDMM is explained in detail. The previous applications of the model and prior research are discussed in Section 5.4. The application of the SDMM in sustainable supply chain management by circularity scenario analysis is discussed in Section 5.5. Future research opportunities and conclusions are presented in Section 5.6 and Section 5.7, respectively.

5.2 Key variables and definitions

5.2.1 Games of chance

One type of game against nature is the games of chance (Luce et al., 1989). They are classified as either involving risk or uncertainty (Table 5.1). In the game of chance with risk, nature's moves are not clear to the player, but he knows the significant probability of nature's actions and thus can predict the success probability of those actions. In this type of game, the expected monetary/utility value (EMV) is suitable for making a decision (Colman, 1982). In the other type of games in which

Table 5.1 General game of chance model (Vafadarnikjoo et al., 2023).

P 1	Nature			
	ACTION 1	**ACTION 2**	. . .	**ACTION** M
Alternative 1	$payoff_{11}$	$payoff_{12}$	. . .	$payoff_{1M}$
Alternative 2	$payoff_{21}$	$payoff_{22}$	. . .	$payoff_{2M}$
. . .	. . .	. . .	. . .	. . .
Alternative N	$payoff_{N1}$	$payoff_{N2}$	. . .	$payoff_{NM}$

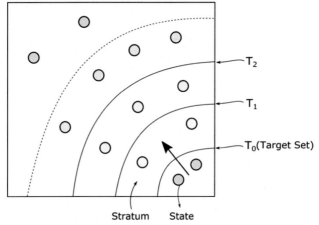

Figure 5.1 Target set, stratum, and state in CST. *CST*, Concept of stratification. Source: *Adapted from Vafadarnikjoo (2020).*

uncertainty is involved rather than risk, the probabilities of nature's states are unclear. In the literature, various rules are used for decision-making in such circumstances (Ulansky & Raza, 2021; Vafadarnikjoo et al., 2023).

5.2.2 Concept of stratification

The CST (concept of stratification) was developed to advance stratification in which a system should pass through several states to arrive at the target set. There might be a target set that comprises multiple target states. The inputs and outputs of each state are stratified in an incremental manner based on their distance from the target set (Asadabadi, 2018; Zadeh, 2016). The CST has many definitions (Vafadarnikjoo et al., 2023). The CST would be highly dynamic through an incremental enlargement process. This process aims to identify potential routes to the target (Asadabadi et al., 2022; Vafadarnikjoo et al., 2023) (Fig. 5.1).

5.3 The stratified decision-making model

As a result of a combination of CST and games of chance involving risk, the SDMM is generated (Vafadarnikjoo et al., 2023). The SDMM has N status (SS) and M outcomes (OC). There are n_i strategies under each SS_i which lead to different payoff (pf) values under various outcomes of nature. Table 5.2 illustrates the payoff matrix of the proposed model.

5.3.1 Status transition probability matrix

The matrix P is presented in Eq. (5.1), in which p_{ij} represents the probability of a transition between SS_i and SS_j. The model comprises N status. In Fig. 5.2, the status transitions are depicted.

$$P = \left[p_{ij} \right]_{N \times N} \tag{5.1}$$

5.3.2 Outcome transition probability matrix

The matrix Q is presented in Eq. (5.2), in which q_{ij} represents the probability of transition from OC_i to OC_j. The model comprises M outcomes. The outcome transitions are illustrated in Fig. 5.3.

$$Q = \left[q_{ij} \right]_{M \times M} \tag{5.2}$$

Table 5.2 The stratified game table (Vafadarnikjoo et al., 2023).

Player 1		Nature			
		OUTCOME 1	OUTCOME 2	$\ldots$	OUTCOME M
Status 1	Strategy 1	pf_{111}	pf_{112}	$\ldots$	pf_{11M}
	Strategy 2	pf_{121}	pf_{122}	$\ldots$	pf_{12M}
	$\ldots$	$\ldots$	$\ldots$	$\ldots$	$\ldots$
	Strategy n_1	pf_{1n_11}	pf_{1n_12}	$\ldots$	pf_{1n_1M}
Status 2	Strategy 1	pf_{211}	pf_{212}	$\ldots$	pf_{21M}
	Strategy 2	pf_{221}	pf_{222}	$\ldots$	pf_{22M}
	$\ldots$	$\ldots$	$\ldots$	$\ldots$	$\ldots$
	Strategy n_2	pf_{2n_21}	pf_{2n_22}	$\ldots$	pf_{2n_2M}
$\ldots$	$\ldots$	$\ldots$	$\ldots$	$\ldots$	$\ldots$
Status N	Strategy 1	pf_{N11}	pf_{N12}	$\ldots$	pf_{N1M}
	Strategy 2	pf_{N21}	pf_{N22}	$\ldots$	pf_{N2M}
	$\ldots$	$\ldots$	$\ldots$	$\ldots$	$\ldots$
	Strategy n_N	pf_{Nn_N1}	pf_{Nn_N2}	$\ldots$	pf_{Nn_NM}

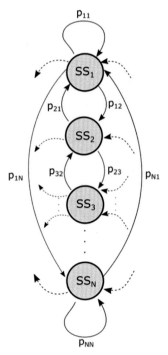

Figure 5.2 Status transitions and corresponding probabilities. Source: *Adapted from Vafadarnikjoo et al. (2023).*

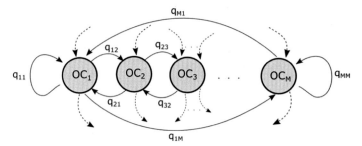

Figure 5.3 Outcome transitions and corresponding probabilities. *Source: Adapted from Vafadarnikjoo et al. (2023).*

5.3.3 State transition probability matrix

The matrix S is shown in Eq. (5.3), in which s_{ij} is the transition probability from state i (SE_i) to state j (SE_j). It comprises $N \times M$ states, which are illustrated in Fig. 5.4.

$$S = \left[s_{ij}\right]_{N \times M} \qquad (5.3)$$

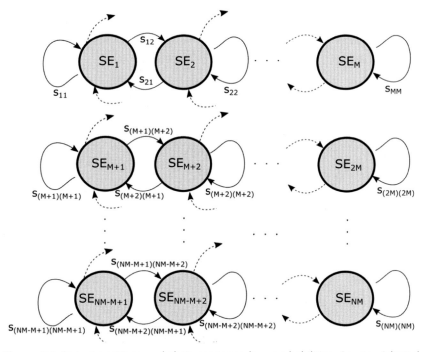

Figure 5.4 State transitions and their corresponding probabilities. Source: *Adapted from Vafadarnikjoo et al. (2023).*

5.3.4 Model assumptions

1. Similar strategies under different statuses in the model ($n_1 = n_2 = \ldots = n_N = B$).
2. It is aimed at maximizing the payoff value ($Z = Maxpf_{ijk}$) or utility value ($Z = Maxu_{ijk}$) in which $u_{ijk} \in [0, 1]$.
3. The payoff/utility values will not change because of the change in state.
4. Probabilities of status transition add up to 1, as shown in Eq. (5.4), also all the outcomes' transition probabilities add up to 1, as presented in Eq. (5.5).

$$\sum_{j=1}^{N} p_{ij} = 1 \quad \forall i = 1, \ldots, N \tag{5.4}$$

$$\sum_{j=1}^{M} q_{ij} = 1 \quad \forall i = 1, \ldots, M \tag{5.5}$$

5. The minimum number of status, outcomes, and strategies are two (i.e., $N \geq 2$, $M \geq 2$, $B \geq 2$).6

5.3.5 Solution approach

Given x (i.e., the current state of the system), the value of strategy b (v_b^x) can be calculated using Eq. (5.7) given that $b = 1, \ldots, B$ ($NM = N \times M$) noting Eq. (5.6).

$$j = \begin{cases} \{1, M+1, 2M+1, \ldots, NM-M+1\} & k=1 \\ \{2, M+2, 2M+2, \ldots, NM-M+2\} & k=2 \\ \quad\vdots & \vdots \\ \{M, 2M, 3M, \ldots, NM\} & k=M \end{cases} \tag{5.6}$$

In the case of utility values, then Eq. (5.8) is used.

$$v_b^x = \sum_{i=1}^{N} \sum_{j=iM-M+1}^{iM} s_{xj} pf_{ibk}$$

$$\forall b = 1, \ldots, B, \forall x = 1, \ldots, NM, k = \{1, 2, \ldots, M\} \tag{5.7}$$

$$v_b^x = \sum_{i=1}^{N} \sum_{j=iM-M+1}^{iM} s_{xj} u_{ibk}$$

$$\forall b = 1, \ldots, B, \forall x = 1, \ldots, NM, 0 \leq u_{ibk} \leq 1, k = \{1, 2, \ldots, M\} \tag{5.8}$$

Table 5.3 illustrates the payoff/utility decision matrix (after transition). The final strategy can be obtained by calculating the EMV value for each strategy. The implementation steps for the SDMM are depicted in Fig. 5.5.

For instance, the EMV for strategy b (EMV^b) given equal probabilities can be obtained using Eq. (5.9).

$$EMV^b = \frac{\sum_{i=1}^{NM} v_b^i}{NM} \forall b = 1, \ldots, B \tag{5.9}$$

Table 5.3 The payoff/utility decision matrix (after transition) (Vafadarnikjoo et al., 2023).

STRATEGY	STATE			
	STATE 1	STATE 2	...	STATE NM
Strategy 1	v_1^1	v_1^2	...	v_1^{NM}
strategy 2	v_2^1	v_2^2	...	v_2^{NM}
...	...	...	...	...
strategy B	v_B^1	v_B^2	...	v_B^{NM}

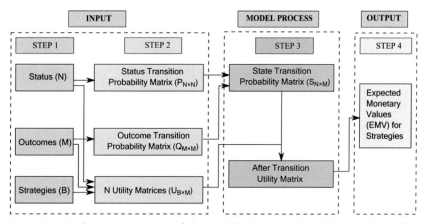

Figure 5.5 Implementation steps of the model. Source: *Adapted from Vafadarnikjoo et al. (2023).*

5.4 How has the model been used?

Vafadarnikjoo et al. (2023) applied SDMM to assess flood risk impacts in Scotland. Using the shared socio-economic pathway (SSP) and the flooding risk's impact, the UK socio-economic status impacting the adaptation choices is studied. However, the CST has been explored in a few studies. The summary of the literature is provided in Table 5.4.

5.5 Application in sustainable supply chain management

The model's applicability is shown in a case example to identify and prioritize the most suitable scenario for the circular future in a sustainable supply chain management setting (Table 5.6) while accounting for the dynamic nature of dimensions of reverse integration (Table 5.5) and SDGs (i.e., decent work and economic growth-SDG8), which is explained in the following section.

5.5.1 Decent work and economic growth (SDG8)

International Labor Organization (ILO) (ILO, 2023) listed sustainable development goals (SDG) relevant to global supply chains as follows: SDG8 (Decent Work and Economic Growth), SDG9 (Industry Innovation and Infrastructure), SDG16 (Peace, Justice, and Strong Institutions), and SDG17

Table 5.4 Stratification and decision making.

Authors	Method	Context	Country
Vafadarnikjoo et al. (2023)	Stratified decision-making model (SDMM)	Flooding risk strategy evaluation	Scotland-UK
Selvaraj and JeongHwan (2022)	A decision-making technique to achieve stratified target performance (DEMTASTAP)	Innovation policy investment	South Korea
Ecer and Torkayesh (2022)	A stratified fuzzy decision-making approach	Sustainable circular supplier selection problem in the textile industry	Turkey
Torkayesh and Simic (2022)	the hierarchical stratified best-worst method (H-SBWM)	Recycling location selection problem	Turkey
Asadabadi et al. (2022)	stratified best-worst method (SBWM) and the technique for order of preference by similarity to ideal solution (TOPSIS)	Innovation for environmental sustainability in supplier selection	Iran
Asadabadi and Zwikael (2021)	an extended version of the stratified multiple criteria decision-making (MCDM)	In time and cost estimations in project management in the construction industry	N/A
Torkayesh et al. (2021)	stratified best-worst method (SBWM)	Sustainable waste disposal technology selection problem	Iran
Asadabadi et al. (2017)	use of a fuzzy inference system (FIS) and the concept of stratification	Logistic informatics modeling	N/A

(Partnerships for the Goals). Küfeoğlu (2022) explains how economic growth and decent work are interrelated by reviewing 37 business models where emergent technologies were employed and value was created in SDG-8. Schroeder et al. (2019) identified that the most powerful relationships exist between CE practices and SDG6 (Clean Water and Sanitation), SDG7 (Affordable and Clean Energy), SDG8 (Decent Work and Economic Growth), SDG12 (Responsible Consumption and Production), and SDG15 (Life on Land). As a result, SDG8 appeared to be contributing to the global supply chains as well as CE practices. In the SDMM model in this chapter,

Table 5.5 Supply chain reverse integration measurement items and levels of integration (Bimpizas-Pinis et al., 2022).

Measurement items	Characteristics	Reverse integration (supplier)	Reverse integration (customer)
Sharing information	Sales information sharing; sharing overall plans with producers; sharing sales forecast; delivery status and stock level sharing	Low, medium, high	Low, medium, high
Developing collaborative approaches	Participating in the production process; risk sharing	Low, medium, high	Low, medium, high
Joint decision-making	Product modification; process modification; quality and cost control	Low, medium, high	Low, medium, high
System coupling	Supporting key suppliers by sharing logistical infrastructure; just-in-time; vendor-managed inventory	Low, medium, high	Low, medium, high

Table 5.6 Circularity scenarios (Bauwens et al., 2020).

No.	Circularity scenario	Characteristics
1	Planned circularity (PCI)	Strong coercive measures; command-and-control regulations
2	Circular modernism (CIM)	Setting standards for eco-efficient design for recycling by the government; R&D investment by the government; eco-modernist viewpoint; production-focused transformation
3	Bottom-up sufficiency (BUS)	Shorter supply chains; localized production
4	Peer-to-peer circularity (PPC)	Associated with the sharing economy, the gig economy, the collaborative consumption, or the access economy.

SDG8 has been considered as one dimension to evaluate scenarios for circular futures. The levels of the SDG8 are measured at three levels, that is, Good, Moderate, and Poor.

5.5.2 Supply chain reverse integration

To promote collaboration between partners in supply chains, the concept of supply chain integration (SCI) has been understood as a pivotal approach (Bimpizas-Pinis et al., 2022; Flynn et al., 2016). Bimpizas-Pinis et al. (2022) argued that with the emergence of the CE, SCI needs to be adopted in the traditional and linear supply chains to overcome the applicability challenges of SCI within the circular supply chains. For measuring supply chain integration, four measurement items, including joint decision-making, developing collaborative approaches, system coupling, and sharing information, were used to evaluate the level of SCI on both supplier and customer sides, resulting in low, medium, or high SCI (Bimpizas-Pinis et al., 2022) (see Table 5.5).

5.5.3 Circularity scenarios

Bauwens et al. (2020) indicated that the transition toward a CE will take place in socio-technical systems (STS). However, it is often seen that utilized approaches for transition to a CE take a micro perspective, such as adopting viewpoints of a single company or sector. This necessitates a systems view to conceptualize CE. Bauwens et al. (2020) explained that technology and institutions are two central change agents in STS as the dynamic interrelationships between technologies and institutions can play a pivotal role in their evolution over time. Bauwens et al. (2020) defined four scenarios of future circularity scenarios using two axes of the essence of technologies applied and institutional configurations (See Fig. 5.6).

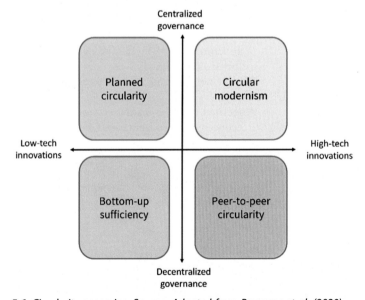

Figure 5.6 Circularity scenarios. Source: *Adapted from Bauwens et al. (2020).*

The essence or nature of technology can be categorized into low and high-tech innovations. The first category, low-tech innovations, offers more simple and less advanced features, whereas, on the other hand, high-tech innovations are complex and advanced. High-tech innovations require intensive R&D and extensive learning and experience to be familiarized with them (Bauwens et al., 2020). In Fig. 5.6, the horizontal axis defines the extent of technology innovations. Institutional configurations can be classified as centralized and decentralized governance of CE. In centralized governance, economic and political power are placed into the hands of large firms or national governments. For example, centralized recycling plants can be under the control of large manufacturers with the lowest level of consumer involvement. On the other hand, decentralized governance focuses on expanding local economic and political autonomy by transferring responsibilities and power away from large incorporations and national administrative and political institutions (Bauwens et al., 2020). In Fig. 5.6, the vertical axis represents the extent of the centrality of governance in the transition to CE.

5.5.3.1 Planned circularity
Transition to a CE is guided through strong coercive measures by the government. For example, the government can set regulations on manufacture and consumption to urge companies as well as consumers to get involved in recycling activities (Bauwens et al., 2020).

5.5.3.2 Circular modernism
In this scenario, the transition is based on technological advancement on the one hand and political and economic centralized governance under the control of a handful of large corporations and the government on the other hand (Bauwens et al., 2020). This approach is close to the eco-modernism perspective in which it is argued that nature can be protected from human activities by boosting technology (Mol & Spaargaren, 2000). This scenario is focused on the production or supply side rather than altering the consumption patterns of consumers. It is also the dominant conception and current narrative of CE (Lowe & Genovese, 2022).

5.5.3.3 Bottom-up sufficiency
Small-scale and decentralized manufacturing in a small local community makes a basis for this scenario. The emphasis is on the reduction of resource consumption while promoting democratic participation, emotional consumer

Table 5.7 Two-dimensional stratified game table for SSCM (Stratified decision-making models).

| Reverse integration | Circularity scenarios | SDG8: Decent Work and Economic Growth | | |
		Good	Moderate	Poor
High supplier integration and high customer integration (H)	1. PCI 2. CIM 3. BUS 4. PPC	SE_1	SE_2	SE_3
Medium supplier integration and medium customer integration (M)	1. PCI 2. CIM 3. BUS 4. PPC	SE_4	SE_5	SE_6
Low supplier integration and low customer integration (L)	1. PCI 2. CIM 3. BUS 4. PPC	SE_7	SE_8	SE_9

engagement, and voluntary behavioral changes in consumption patterns (Bauwens et al., 2020; Lowe & Genovese, 2022).

5.5.3.4 Peer-to-peer circularity

This scenario promotes the idea that the implementation of modern technologies, such as Industry 4.0, will contribute to a highly decentralized economy while promoting possible sustainability benefits and consequently leading to a CE. This scenario can be related to the sharing economy, gig economy, collaborative consumption, or access economy (Asian et al., 2019; Bauwens et al., 2020). On-demand access to various types of resources is a core vision among these concepts (Table 5.9).

5.5.4 Stratified decision-making models analysis

The web application proposed by Vafadarnikjoo et al. (2023) can be used for the SDMM analysis[1]. Table 5.8 and Fig. 5.7 illustrate the transition between states (Tables 5.7 and 5.10).

Status and outcome transition probabilities are generated by simulation and a decision maker's judgment, as shown in Table 5.11. In Table 5.12, all the utility values are presented, which are derived from a decision maker's judgments.

---

[1] The web app can be accessed at: https://amvaf.shinyapps.io/SDMM/.

Table 5.8 CST (concept of stratification) for state transitions for $N = 3$ and $M = 3$ (Vafadarnikjoo et al., 2023).

SE_t	Status	Outcome	SE_{t+1}
1	1	1	1
2	1	2	1, 2
3	1	3	1, 2, 3
4	2	1	1, 4
5	2	2	1, 2, 4, 5
6	2	3	1, 2, 3, 4, 5, 6
7	3	1	1, 4, 7
8	3	2	1, 2, 4, 5, 7, 8
9	3	3	1, 2, 3, 4, 5, 6, 7, 8, 9

Table 5.9 The verbal scale for obtaining utility values (Vafadarnikjoo et al., 2023).

Linguistic term for effectiveness	Score	Expected utility
Not at all	0	0.00
Low	1	0.26
Fairly low	2	0.38
Medium	3	0.50
Fairly high	4	0.68
High	5	0.90
Absolutely high	6	1.00

The analysis steps in the web application tool are illustrated in Figs. 5.8–5.11.

Based on the input data in this case example, the obtained results (see Fig. 5.11) from the SDMM analysis indicate that in the long-term period considering two dimensions of reverse integration and levels for decent work and economic growth (SDG8), planned circularity is the best plausible future strategy followed by circular modernism, peer-to-peer circularity, and bottom-up sufficiency.

5.6 Managerial implications and future research opportunities

The case example in this chapter was not based on generalizable real-world data and was computed using simulation and decision-maker's judgments. The web application tool proposed by Vafadarnikjoo et al. (2023)

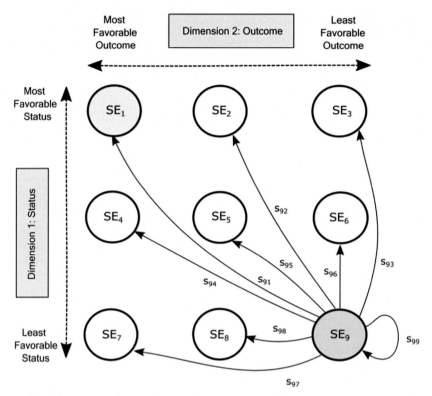

Figure 5.7 State transitions beginning at state 9 for $N = 3$ and $M = 3$. Source: *Adapted from Vafadarnikjoo et al. (2023).*

Table 5.10 The rating scale used for acquiring probability estimations (Vafadarnikjoo et al., 2023).

Linguistic term	Score	Expected probability
Almost zero (AZ)	0	0.00
Very small (VS)	1	0.15
Small (S)	2	0.25
Moderate (M)	3	0.45
Large (L)	4	0.65
Very large (VL)	5	0.85
Almost certain (AC)	6	1.00

was utilized to show how the model can be applied in a numerical example case in sustainable supply chain management to identify the most plausible circular future scenario. For future managerial implications, the proposed model can be applied and tested in real-world cases by obtaining

Table 5.11 Status and outcome transition probabilities.

P	$p_{11} = 1$	$p_{12} = 0$	$p_{13} = 0$	Q	$q_{11} = 1$	$q_{12} = 0$	$q_{13} = 0$
	$p_{21} = 0.38$	$p_{22} = 0.62$	$p_{23} = 0$		$q_{21} = 0.48$	$q_{22} = 0.52$	$q_{23} = 0$
	$p_{31} = 0.30$	$p_{32} = 0.45$	$p_{33} = 0.25$		$q_{31} = 0.36$	$q_{32} = 0.40$	$q_{33} = 0.24$

Table 5.12 Utility values.

		SDG8: Decent work and economic growth		
RI	Circularity scenarios	Good	Moderate	Poor
H	1. *PCI*	0.68	0.68	0.68
	2. *CIM*	0.90	0.68	0.50
	3. *BUS*	0.68	0.68	0.38
	4. *PPC*	0.90	0.68	0.50
M	1. *PCI*	0.90	0.90	0.68
	2. *CIM*	0.68	0.68	0.38
	3. *BUS*	0.50	0.38	0.38
	4. *PPC*	0.50	0.50	0.38
L	1. *PCI*	0.90	0.68	0.50
	2. *CIM*	0.50	0.38	0.26
	3. *BUS*	0.68	0.50	0.38
	4. *PPC*	0.50	0.38	0.26

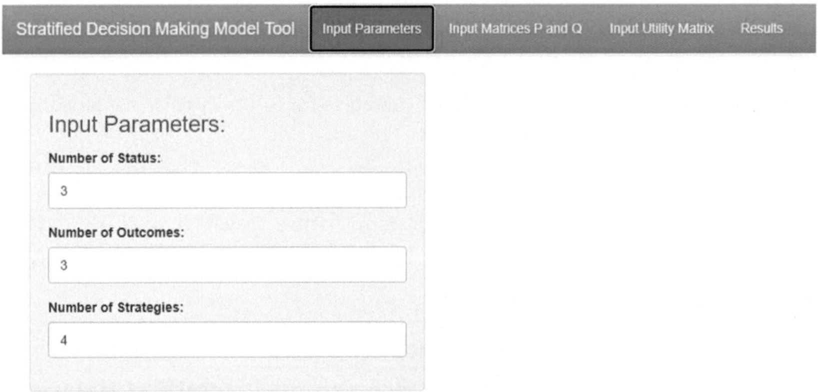

Figure 5.8 Parameters input in the SDMM web application tool. *SDMM*, stratified decision-making models.

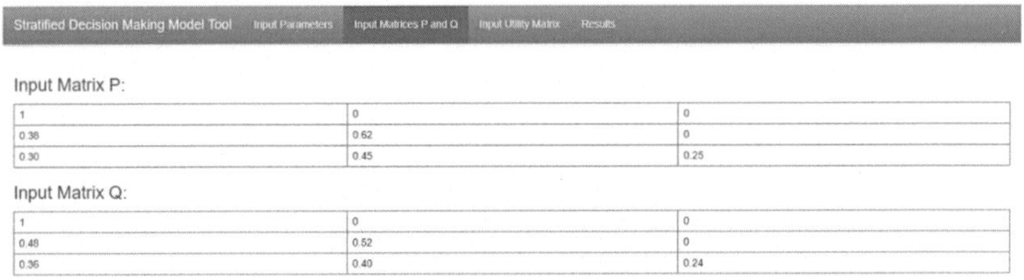

Figure 5.9 Matrices P and Q input in the SDMM web app. *SDMM*, stratified decision-making models.

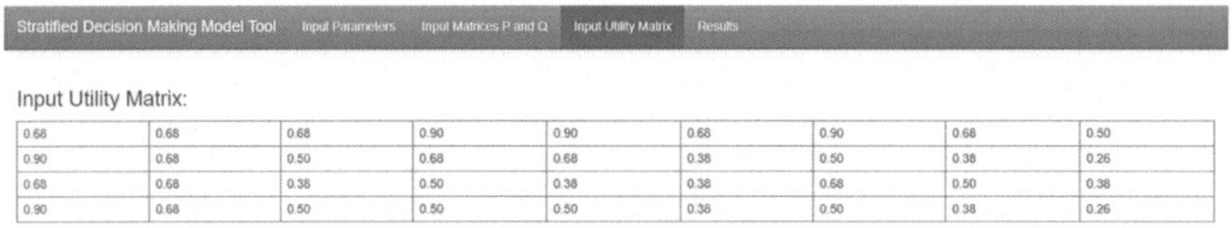

Figure 5.10 Utility matrix input in the SDMM web app. *SDMM*, stratified decision-making models.

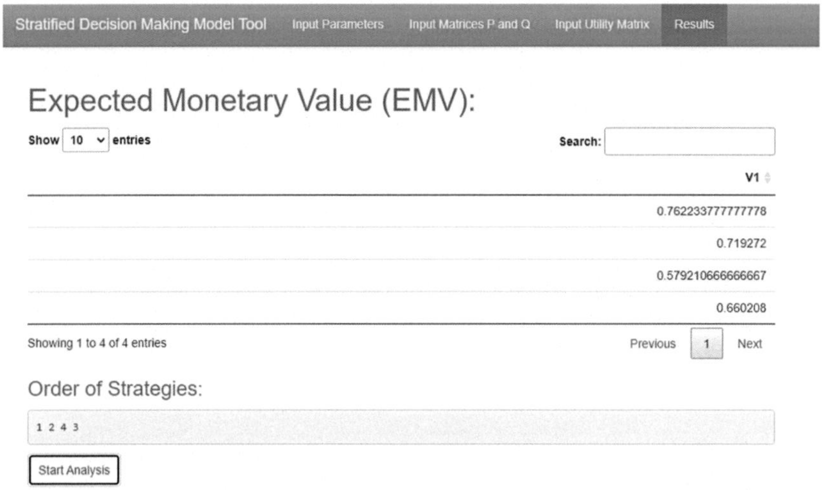

Figure 5.11 EMV and the final ranking of strategies for future circular scenarios. *EMV*, expected monetary/utility value.

empirical data, involving subject experts in future research, and integrating with other artificial intelligence (AI) tools. A few potential future research directions are discussed as follows:

5.6.1 Explainable artificial intelligence and stratified decision-making models

Future extensions of the model, for instance, can integrate explainable artificial intelligence (XAI) within SDMM to measure the overall equipment effectiveness (OEE) and provide a comparative analysis with the traditional deep learning and ML methods. Xu et al. (2019) discussed that AI systems should be able to assist humans in making decisions in crucial missions tasks by being more transparent and trustworthy AI such as in XAI. This opens up an opportunity to integrate traditional decision-making models with XAI to achieve a higher level of result reliability in the era of big data by being more transparent in the algorithm (Janssen et al., 2022).

5.6.2 Horizon scanning techniques and stratified decision-making models

Horizon scanning is a useful practice for businesses that strive not only to survive in a competitive business environment but also to flourish in the long run of their functionality. In doing so, understanding change and monitoring the business environment are critical tasks (Brown, 2007). For example, organizations can do environmental scanning, risk assessment, or trend analysis and horizon scanning is not limited to a specific business context. SDMM can facilitate the modeling process to incorporate the transition between states applicable for horizon scanning in long-term planning. Computational intelligence techniques for scenario analysis can be useful in this regard by implementing computational intelligence algorithms.

5.6.3 Three-dimensional or multi-dimensional models in stratified decision-making models

The SDMM is flexible by incorporating a third dimension. In this chapter, the SDMM has been proposed in two-dimensional settings in Vafadarnikjoo et al. (2023) and the circularity futures in sustainable supply chains. Vafadarnikjoo et al. (2023) studied two dimensions of the SDMM. In the current chapter, the reverse integration status and levels for the SDGs (i.e., decent work and economic growth-SDG8) have been considered as two dimensions of the SDMM. However, SDMM has the potential to have more than two dimensions to deal with the higher complexity of real-world decision-making problems. This would add a higher computational and data processing challenge for the decision modeling process which can be overcome by using novel AI/ML and computational techniques and can be an interesting future research.

5.6.4 Application of stratified decision-making models in other decision-making contexts

The SDMM can be applied in other decision-making contexts, such as disaster management in case of flooding, earthquakes, or droughts, modern slavery risk assessment within the socio-economic environment, and pandemic containment strategy prioritization, just to name a few.

5.7 Conclusion

In this chapter, SDMM and its applicability in a sustainable supply chain context within the CE were explained. The SDMM was described as a suitable tool to explore the potential future scenarios for CE implementation within sustainable supply chain management under two dimensions of reverse integration and decent work and economic growth (SDG8). The SDMM model can be applied in similar decision-making scenarios such as environmental or disaster management. The case results indicate that in the long-term period considering two dimensions of reverse integration and levels for decent work and economic growth (SDG8), planned circularity would be the best plausible future strategy followed by circular modernism, peer-to-peer circularity, and bottom-up sufficiency. The results have a few limitations that cannot be generalized as the analysis was carried out based on the obtained data from one decision maker and simulated numerical data for transition probabilities were randomly generated via simulation to show how the model can be applied.

References

Arsova, S., Genovese, A., & Ketikidis, P. H. (2022). Implementing circular economy in a regional context: A systematic literature review and a research agenda. *Journal of Cleaner Production, 368*(September)133117. Available from https://doi.org/10.1016/j.jclepro.2022.133117.

Asadabadi, M. R. (2018). The stratified multi-criteria decision-making method. *Knowledge-Based Systems, 162*(December), 115−123. Available from https://doi.org/10.1016/j.knosys.2018.07.002.

Asadabadi, M. R., Ahmadi, H. B., Gupta, H., & Liou, J. J. H. (2022). Supplier selection to support environmental sustainability: The stratified BWM TOPSIS method. *Annals of Operations Research*. Available from https://doi.org/10.1007/s10479-022-04878-y, August.

Asadabadi, M.R., Saberi,M., & Chang,E. (2017). Logistic Informatics Modelling Using Concept of Stratification (CST). In *2017 IEEE International Conference on Fuzzy Systems (FUZZ-IEEE)*, 1−7. Naples, Italy: IEEE. Available from https://doi.org/10.1109/FUZZ-IEEE.2017.8015510.

Asadabadi, M. R., & Zwikael, O. (2021). Integrating risk into estimations of project activities' time and cost: A stratified approach. *European Journal of Operational Research, 291* (2), 482−490. Available from https://doi.org/10.1016/j.ejor.2019.11.018.

Asian, S., Hafezalkotob, A., & John, J. J. (2019). Sharing economy in organic food supply chains: A pathway to sustainable development. *International Journal of Production Economics, 218*(December), 322−338. Available from https://doi.org/10.1016/j.ijpe.2019.06.010.

Bauwens, T. (2021). Are the circular economy and economic growth compatible? A case for post-growth circularity. *Resources, Conservation and Recycling, 175*(December) 105852. Available from https://doi.org/10.1016/j.resconrec.2021.105852.

Bauwens, T., Hekkert, M., & Kirchherr, J. (2020). Circular futures: What will they look like? *Ecological Economics*, *175*(September)106703. Available from https://doi.org/10.1016/j.ecolecon.2020.106703.

Bimpizas-Pinis, M., Calzolari, T., & Genovese, A. (2022). Exploring the transition towards circular supply chains through the arcs of integration. *International Journal of Production Economics*, *250*(August)108666. Available from https://doi.org/10.1016/j.ijpe.2022.108666.

Brown, D. (2007). Horizon scanning and the business environment—The implications for risk management. *BT Technology Journal*, *25*(1), 208–214. Available from https://doi.org/10.1007/s10550-007-0022-8.

Calzolari, T., Genovese, A., & Brint, A. (2022). Circular economy indicators for supply chains: A systematic literature review. *Environmental and Sustainability Indicators*, *13*(February)100160. Available from https://doi.org/10.1016/j.indic.2021.100160.

Carter, C. R., Kaufmann, L., & Ketchen, D. J. (2020). Expect the unexpected: Toward a theory of the unintended consequences of sustainable supply chain management. *International Journal of Operations & Production Management ahead-of-print (ahead-of-print)..* Available from https://doi.org/10.1108/IJOPM-05-2020-0326.

Castro, C. G., Trevisan, A. H., Pigosso, D. C. A., & Mascarenhas, J. (2022). The rebound effect of circular economy: Definitions, mechanisms and a research agenda. *Journal of Cleaner Production*, *345*(April)131136. Available from https://doi.org/10.1016/j.jclepro.2022.131136.

Colman, A. M. (1982). *Game Theory and Experimental Games: The Study of Strategic Interaction* (1st ed.). Pergamon Press.

Ecer, F., & Torkayesh, A. (2022). A stratified fuzzy decision-making approach for sustainable circular supplier selection. *IEEE Transactions on Engineering Management*, 1–15. Available from https://doi.org/10.1109/TEM.2022.3151491.

Flynn, B. B., Koufteros, X., & Lu, G. (2016). On theory in supply chain uncertainty and its implications for supply chain integration. *Journal of Supply Chain Management*, *52*(3), 3–27. Available from https://doi.org/10.1111/jscm.12106.

Geissdoerfer, M., Morioka, S. N., de Carvalho, M. M., & Evans, S. (2018). Business models and supply chains for the circular economy. *Journal of Cleaner Production*, *190*(July), 712–721. Available from https://doi.org/10.1016/j.jclepro.2018.04.159.

Ghisellini, P., Cialani, C., & Ulgiati, S. (2016). A review on circular economy: The expected transition to a balanced interplay of environmental and economic systems. *Journal of Cleaner Production*, *114*(February), 11–32. Available from https://doi.org/10.1016/j.jclepro.2015.09.007.

Hák, T., Janoušková, S., & Moldan, B. (2016). Sustainable development goals: A need for relevant indicators. *Ecological Indicators*, *60*(January), 565–573. Available from https://doi.org/10.1016/j.ecolind.2015.08.003.

Hickel, J., & Kallis, G. (2020). Is green growth possible? *New Political Economy*, *25*(4), 469–486. Available from https://doi.org/10.1080/13563467.2019.1598964.

ILO. (2023). 'Relevant SDG Targets Related to Global Supply Chains'. *International Labour Organization* (blog). 2023. Available at: https://www.ilo.org/global/topics/dw4sd/themes/supply-chains/WCMS_558569/lang-en/index.htm.

Janssen, M., Hartog, M., Matheus, R., Yi Ding, A., & Kuk, G. (2022). Will algorithms blind people? The effect of explainable ai and decision-makers' experience on AI-supported decision-making in government. *Social Science Computer Review*, *40*(2), 478–493. Available from https://doi.org/10.1177/0894439320980118.

Krausmann, F., Gingrich, S., Eisenmenger, N., Erb, K. H., Haberl, H., & Fischer-Kowalski, M. (2009). Growth in global materials use, GDP and population during the 20th century. *Ecological Economics*, *68*(10), 2696–2705. Available from https://doi.org/10.1016/j.ecolecon.2009.05.007.

Küfeoğlu, S. (2022). SDG-8: Decent work and economic growth. In S. Küfeoğlu (Ed.), *Sustainable development goals series* (pp. 331–348). Cham: Springer International Publishing. Available from https://doi.org/10.1007/978-3-031-07127-0_10.

Lowe, B. H., & Genovese, A. (2022). What theories of value (Could) underpin our circular futures? *Ecological Economics*, *195*(May)107382. Available from https://doi.org/10.1016/j.ecolecon.2022.107382.

Luce, R., Duncan., & Raiffa, H. (1989). *Games and decisions: Introduction and critical survey*. New York: Dover Publications.

MahmoumGonbadi, A., Genovese, A., & Sgalambro, A. (2021). Closed-loop supply chain design for the transition towards a circular economy: A systematic literature review of methods, applications and current gaps. *Journal of Cleaner Production*, *323*(November) 129101. Available from https://doi.org/10.1016/j.jclepro.2021.129101.

Merton, R. K. (1936). The unanticipated consequences of purposive social action. *American Sociological Review*, *1*(6), 894. Available from https://doi.org/10.2307/2084615.

Mol, A. P. J., & Spaargaren, G. (2000). Ecological modernisation theory in debate: A review. *Environmental Politics*, *9*(1), 17–49. Available from https://doi.org/10.1080/09644010008414511.

Rodriguez-Anton, J. M., Rubio-Andrada, L., Celemín-Pedroche, M. S., & Alonso-Almeida, M. D. M. (2019). Analysis of the relations between circular economy and sustainable development goals. *International Journal of Sustainable Development & World Ecology*, *26*(8), 708–720. Available from https://doi.org/10.1080/13504509.2019.1666754.

Schroeder, P., Anggraeni, K., & Weber, U. (2019). The relevance of circular economy practices to the sustainable development goals. *Journal of Industrial Ecology*, *23*(1), 77–95. Available from https://doi.org/10.1111/jiec.12732.

Selvaraj, G., & JeongHwan, J. (2022). Decision-making technique to achieve stratified target performance: Analyze science and technology innovation policy investment of South Korea. *International Journal of Intelligent Systems*, *37*(8), 4670–4714. Available from https://doi.org/10.1002/int.22736.

Siderius, T., & Poldner, K. (2021). Reconsidering the circular economy rebound effect: Propositions from a case study of the dutch circular textile valley. *Journal of Cleaner Production*, *293*(April)125996. Available from https://doi.org/10.1016/j.jclepro.2021.125996.

Torkayesh, A. E., Malmir, B., & Rajabi Asadabadi, M. (2021). Sustainable waste disposal technology selection: The stratified best-worst multi-criteria decision-making method. *Waste Management*, *122*(March), 100–112. Available from https://doi.org/10.1016/j.wasman.2020.12.040.

Torkayesh, A. E., & Simic, V. (2022). Stratified hybrid decision model with constrained attributes: Recycling facility location for urban healthcare plastic waste. *Sustainable Cities and Society*, 77(February)103543. Available from https://doi.org/10.1016/j.scs.2021.103543.

Ulansky, V., & Raza, A. (2021). Generalization of minimax and maximin criteria in a game against nature for the case of a partial a priori uncertainty. *Heliyon*, *7*(7) e07498. Available from https://doi.org/10.1016/j.heliyon.2021.e07498.

Vafadarnikjoo, A. (2020). Decision analysis in the uk energy supply chain risk management: Tools development and application'. PhD Thesis, The University of East Anglia. https://ueaeprints.uea.ac.uk/id/eprint/77909.

Vafadarnikjoo, A., Chalvatzis, K., Botelho, To, & Bamford, D. (2023). A stratified decision-making model for long-term planning: Application in flood risk management in Scotland. *Omega*, *116*102803. Available from https://doi.org/10.1016/j.omega.2022.102803.

Webb, A. (2019). 'How to Do Strategic Planning like a Futurist'. *Harvard Business Review*. Available at: https://hbr.org/2019/07/how-to-do-strategic-planning-like-a-futurist.

Xu, F., Uszkoreit, H., Du, Y., Fan, W., Zhao, D., & Zhu, J. (2019). Explainable AI: A brief survey on history, research areas, approaches and challenges. In J. Tang, M. Y. Kan, D. Zhao, S. Li, & H. Zan (Eds.), *Natural Language Processing and Chinese Computing* (11839, pp. 563—574). Cham: Springer International Publishing, Lecture Notes in Computer Science. Available from https://doi.org/10.1007/978-3-030-32236-6_51.

Zadeh, L. A. (2016). Stratification, target set reachability and incremental enlargement principle. *Information Sciences*, *354*(August), 131—139. Available from https://doi.org/10.1016/j.ins.2016.02.047.

CHAPTER SIX

Machine learning techniques for sustainable industrial process control

Imtiaz Ahmed and Ahmed Shoyeb Raihan

Department of Industrial and Management Systems Engineering, West Virginia University, Morgantown, WV, United States

6.1 Introduction

The manufacturing industry is currently experiencing an unprecedented surge in data availability, which encompasses various types of data in different formats, quality, and semantics, such as data collected from various sensors in production or assembly lines, including parameters controlling the operations of machine tools, and environmental data (Davis et al., 2015). The specific terminology for this trend (influx of data) differs based on the region: in Germany, it is known as Industry 4.0; in the USA, it is called smart manufacturing; and in South Korea, it is termed Smart Factory (Wuest et al., 2016). The rise in data availability and quantity is commonly known as "Big Data," and it presents significant opportunities for sustainable improvements in process and product quality, especially with quality-related data availability (Elangovan et al., 2015; Lee et al., 2013). As a result, Industry 4.0 is gaining ground in the manufacturing industry, signifying a substantial transformation in the management and operation of factories and supply chains. This transition can revolutionize conventional manufacturing methods, resulting in improved efficiency, productivity, and cost-effectiveness. Industry 4.0, often referred to as the fourth phase of the industrial revolution, encompasses the fusion of multiple digital technologies such as Artificial Intelligence (AI), cloud computing, and the Internet of Things (IoT) within production processes. This integration enables machines and systems to communicate with each other, making it possible to achieve greater levels of automation and flexibility. AI enables machines to learn from data, make predictions, and perform tasks that would typically require

Computational Intelligence Techniques for Sustainable Supply Chain Management.
DOI: https://doi.org/10.1016/B978-0-443-18464-2.00014-5

human intelligence. In the manufacturing domain, AI can be utilized to enhance production line optimization, anticipate maintenance requirements, minimize waste, improve quality control, and identify defects in real-time. It can also improve quality control and detect defects in real-time. By leveraging AI and Industry 4.0, manufacturing companies can optimize their operations, leverage data more effectively, and create more adaptable and efficient production systems, leading to a significant transformation in the industry.

Traditional industrial process control (IPC) methodologies, including control charts, design of experiments (DOE), root cause analysis, failure modes and effects analysis (FMEA), and other statistical methods such as ANOVA, have long been the bedrock of quality assurance and operational efficiency in manufacturing and production settings. However, in today's dynamic and data-rich industrial environments, these process control methods face several limitations and are not as effective as they used to be a couple of decades ago. Traditional methods often struggle to adapt to non-linear and dynamic process behaviors, and they can be limited in handling intricate interactions between variables. Moreover, these methods might not effectively accommodate real-time adjustments or unpredictable disturbances. It is within these gaps that machine learning (ML) demonstrates its potential to revolutionize IPC. In recent times, as industries strive to enhance their efficiency, reduce resource consumption, and minimize environmental impact, the integration of AI and its branch, ML into sustainable IPC has emerged as a pivotal area of research and innovation successfully overcoming the drawbacks of traditional IPC methods. For example, one of the pressing challenges in industrial processes is optimizing energy consumption without compromising product quality. ML algorithms can analyze historical energy usage data, production output, and environmental conditions to develop predictive models that optimize energy consumption. These models are capable of dynamically adjusting process parameters in response to real-time data, which can largely reduce energy waste while meeting production targets. Another real-life example is the significant waste reduction in chemical industries. Chemical manufacturing processes often involve intricate reactions where minor deviations can lead to significant waste generation. ML models can predict reaction outcomes based on input variables such as reactant concentrations, temperature, and catalyst usage. Through proper adjustment and optimization of these parameters using online sensor data and historical process information, these models can minimize unwanted byproducts, thereby enhancing resource efficiency and sustainability.

This chapter delves into the dynamic intersection of ML and IPC and illustrates how sustainability in industries can be achieved in the era of Industry 4.0. To this end, in Section 6.1, we discuss elaborately the concepts of sustainability, and Industry 4.0, and how they are interconnected. In Section 6.2, we describe the importance of IPC, traditional approaches to process control in industry, and their limitations. Section 6.3 provides a detailed description of ML and its classification whereas an elaborate discussion of the applications of different ML techniques employed to address and solve a multitude of problems in industries is provided in Section 6.4. The crucial problem of anomaly detection in industry is described in Section 6.5 with an in–depth classification of various anomaly detection techniques. In Section 6.6, a case study is presented where anomalies are detected and root causes behind these anomalies are identified in a hydropower generation plant. Finally, we conclude the findings, limitations, and direction for future research in Section 6.7.

6.1.1 Sustainability in supply chains

Sustainable development, an important strategic goal in today's world, refers to advancement that fulfills the current requirements while safeguarding the capability of upcoming generations in the future to satisfy their necessities (Chalmeta & Santos-deLeón, 2020; Ortiz et al., 2009). To obtain the equilibrium among economic growth, societal progress, and environmental conservation, sustainable development mandates the amalgamation and contemplation of economic, social, cultural, political, and ecological factors in decision-making processes (Maes et al., 2019; Roy et al., 2020). In recent times, with rapid industrialization, digitalization, and globalization coming into play, manufacturing firms are considering sustainable development in their supply chains more often than ever (Azevedo et al., 2012; Raihan et al., 2022). An important factor driving industries to implement a sustainable supply chain into their operations is the growing pressure from stakeholders to lessen the detrimental environmental effects that manufacturing companies are having on the environment (Srivastava, 2007). In addition, with the current global market being more competitive than ever, a company can promote itself and thus get ahead of its competitors in terms of brand value by embracing sustainability in its supply network (Chalmeta & Santos-deLeón, 2020). However, there are significant challenges that need to be dealt with before a company can incorporate sustainability into its supply chain. According to a study, the current understanding of Sustainable Supply Chain Management (SSCM) is insufficient to construct supply chains that are sustainable in their operations and there exist issues with assumptions, regulations, organizations, measures, and techniques that need to be addressed by future study (Pagell & Shevchenko, 2014). In this regard, it is important to know the concepts of Industry 4.0 which can greatly impact the inclusion of sustainability in the supply chain networks of the modern industries.

6.1.2 Industry 4.0

The ongoing digitization of manufacturing, coupled with the utilization of advanced technologies that generate a constant flow of data, has led to a significant transformation in the way products are manufactured, and operations are managed. This shift is so profound that it has been labeled Industry 4.0, marking the fourth industrial revolution. In the first industrial revolution, water and steam power were used to run machines which were replaced by electricity for mass production and assembly lines in the second industrial revolution. In this fourth industrial revolution, smart and autonomous systems powered by data and ML will improve what was started in the previous industrial revolution with the use of computers and automation.

Industry 4.0 is transforming the production, enhancement, and distribution of goods. Manufacturing industries are integrating advancements like the IoT, cloud solutions, and data analytics, alongside AI and ML into their production sites and overall operational processes (Islam et al., 2023). The advanced technologies in Industry 4.0 have provided us with unprecedented levels of automation, predictive maintenance, and self-optimization of process improvements in the so-called smart factories. These technologies which are driving this fourth industrial revolution are illustrated in Fig. 6.1 (Akhtar, 2022).

In this chapter, we will not go into the details of these technologies. However, in later sections, we will discuss ML which is considered as a subset of AI.

6.1.3 Attaining sustainable supply chains through Industry 4.0

Sustainability is a multidimensional notion that encompasses the economic, environmental, and social aspects of a company. By achieving a balance between costs and revenues in sourcing, manufacturing, and dispatch of goods and services, the primary objective of every business

Figure 6.1 Some major Industry 4.0 technologies.

organization is to generate profit for long-term economic sustainability. Global pressure, climate change, and pollution have placed environmental sustainability at the forefront of modern business (Boons & Lüdeke-Freund, 2013). The focus is on minimizing the consumption of natural resources, waste, and emissions (affecting air, water, and soil) while maximizing the adoption of renewable energy in both production and distribution (Akhtar, 2022). Workplace conditions, employee morale, equity, and community social integration are all aspects of social sustainability. By shortening lead times, offering customized products, enhancing product quality, and improving the working environment, and staff morale, Industry 4.0 technologies assist manufacturing organizations in achieving long-term goals (Kamble et al., 2018).

Attaining a high degree of process integration is made simpler by Industry 4.0, which boosts organizational performance in the three areas of sustainability. Braccini and Margherita (2019) conducted a case study on a ceramics manufacturing company to examine the impact of implementing Industry 4.0. Their study revealed enhancements in product quality and productivity, efficient energy monitoring resulting in decreased energy consumption, and a safer workplace, leading to a more contented workforce. From an economic standpoint, integrating Industry 4.0 into the supply chain boosts its efficiency and robustness, improves synchronization between production and logistics, guarantees punctual deliveries and order precision, facilitates optimal supplier choices, anticipates downtime, streamlines repair and upkeep tasks, and shortens lead time. The pursuit of economic sustainability is expected to result in reduced costs and increased profits. In terms of environmental sustainability, Industry 4.0-driven technologies will allow less consumption of energy resources, and reduce wastages and GHG emissions leaving a low carbon footprint in the process. Finally, Industry 4.0 contributes to promoting social sustainability in a company's supply chain by providing a

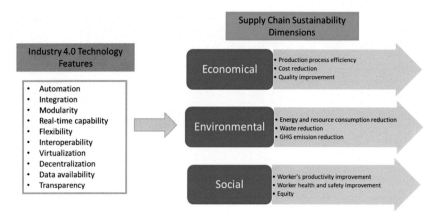

Figure 6.2 Influence of Industry 4.0 innovations on the sustainability aspects of the supply chain.

safer and less stressful work environment that offers greater flexibility and eliminates hazardous tasks. It will foster a better learning environment which will positively impact the growth of the employees. The crucial impact of Industry 4.0 on supply chain sustainability across its three dimensions is summarized in Fig. 6.2. All the stages of a supply chain starting from the supplier and passing through the manufacturer, distributor, retailer, and finally reaching out to the customers, are influenced by Industry 4.0. In this chapter, we will concentrate on the **manufacturing** stage of the supply chain where one of the crucial activities is **industrial process control**.

6.2 Industrial process control

IPC is a set of principles and tools employed in manufacturing settings to supervise, regulate, and optimize the various processes involved in producing goods from raw materials. An effective IPC system can ensure the production of effective, dependable, and high-quality products while reducing waste, downtime, and energy consumption. To this end, sensors, actuators, and control systems are commonly utilized in IPC to monitor and manage different aspects of the manufacturing process. These systems collect real-time data, analyze it using algorithms and models, and determine how to control the process. For instance, temperature sensors may be used to track the reaction's temperature, and a control system may modify the reactant's flow rate to keep the temperature within a predetermined range. Additional sensors can gather data on various process variables, including pressure, humidity, flow rate, and others, utilizing individual sensors. This data is analyzed to detect patterns, recognize problems, and make process adjustments as necessary to achieve optimal performance. IPC encompasses safety systems and alarms that alert operators to potential issues and ensure that processes operate within acceptable parameters. By providing real-time process information, IPC empowers operators to make informed decisions, improve overall efficiency and consistency, and minimize waste and downtime.

6.2.1 Importance of industrial process control

IPC is critical to the success of manufacturing companies, as it provides several benefits that can improve the efficiency, consistency, and quality of their operations. The significance of IPC in manufacturing companies can be comprehended by examining the following advantages:

- Increased Efficiency: By continuously monitoring and controlling key process variables, IPC can help reduce downtime, reduce energy consumption, and minimize waste, leading to increased efficiency and reduced costs.
- Improved Consistency: IPC helps to ensure that processes run consistently and within acceptable limits, resulting in improved quality and reduced variability. This, in turn, leads to higher customer satisfaction and a stronger reputation for the company.
- Enhanced Product Quality: By controlling the key parameters that affect the quality of a product, IPC helps to ensure that products meet specifications and are of consistent quality.
- Improved Safety: IPC includes safety systems and alarms that alert operators to potential problems, ensuring that processes run within safe limits and reducing the risk of accidents and equipment damage.
- Better Data Analysis: IPC involves collecting and analyzing data in real-time, providing valuable information about the process that can be used to identify areas for improvement and make data-driven decisions.

6.2.2 Traditional process control methods

Statistical process control (SPC) and statistical quality control (SQC) are methodologies utilized in IPC to enhance the quality and efficiency of processes. SPC involves the monitoring and control of a process by analyzing data and detecting patterns to identify and correct any potential issues. It requires statistical analysis of data collected from a process to determine if it is producing results within acceptable limits and in a state of statistical control. On the other hand, SQC involves utilizing statistical methods to ensure that the quality of a product or service meets specified standards and requirements. This includes the use of tools such as control charts and sampling plans to monitor the quality of the product or service over time. Both SPC and SQC are frequently used in conjunction as traditional process control tools to provide a comprehensive quality control system in an industrial setting, resulting in increased efficiency, reduced waste, and enhanced customer satisfaction. Fig. 6.3 shows the seven quality control tools that are popularly used in industrial settings. These basic tools are the same for both SPC and SQC. Control charts are the most frequently used SPC/SQC tools. Because of their widespread adoption and efficacy as a quality control tool, we will discuss it briefly.

6.2.3 Control chart

A control chart, also known as a Shewhart chart, is a statistical tool used in IPC to monitor and control process performance. The control chart is a graphical representation of the process data over time, which is used to

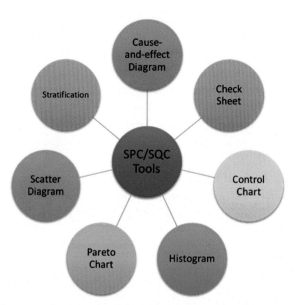

Figure 6.3 The seven basic tools of SPC/SQC. *SPC*, Statistical process control; *SQC*, statistical quality control.

detect any changes in the process and evaluate its stability. A control chart comprises an average line representing the process data's mean, and upper and lower control limits that are derived from the process data. Process data is plotted on the control chart, and any data points that fall beyond the control limits are regarded as out-of-control points and are examined to identify the possible causes of the deviation. Fig. 6.4 shows a simple control chart. Despite being a major tool for IPC, control charts have some limitations which are discussed in the following:

- Control charts assume that the process data follows a normal distribution, which may not always be the case in real-world processes.
- These charts can generate false alarms when small random fluctuations in the data are interpreted as real signals.
- Control charts use control limits that are pre-determined based on historical data or other factors. This can be problematic if the underlying process changes over time, as the control limits may no longer be appropriate.
- The charts are typically used to detect changes in the process over time, but they can have a significant time lag in detecting changes in real-time processes.
- They are designed to detect random patterns in the data, but they may have difficulty detecting non-random patterns that indicate a problem with the process.
- Control charts indicate when a process is out-of-control, but they do not explain the reason behind the deviation.

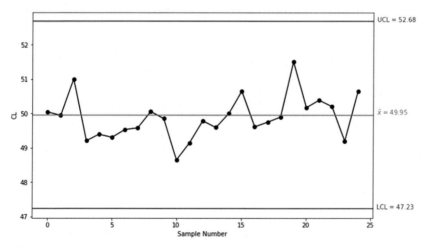

Figure 6.4 A simple control chart.

ML techniques can enhance control charts by providing more precise predictions, better identification of anomalies, and dynamic adjustments to evolving process conditions. Decision trees, random forests, and gradient boosting are examples of ML algorithms that can model intricate relationships in the process data, without assuming a normal distribution. These algorithms can rapidly adapt to changes in the process conditions, enabling quick adjustments. Furthermore, ML algorithms can continuously monitor the process, minimizing the delay in detecting changes. Additionally, ML algorithms can effectively identify non-random patterns in data, making it easier to detect anomalies or unusual data points.

6.3 Machine learning in industrial process control

Conventional process control methods, such as SPC and SQC, have several limitations including the assumption of normality, lack of flexibility, insensitivity to outliers, limited ability to deal with multivariate processes, reliance on pre-determined control limits, difficulty in identifying non-random patterns and time lag in detecting process changes. ML can overcome the limitations of SPC and SQC and provide more adaptable and resilient approaches to process control, which are commonly referred to as advanced process control methods. In this section, we provide a brief introduction to ML and discuss its applications in the manufacturing industries as a process control tool.

6.3.1 Machine learning

ML involves the creation of algorithms that mimic human intelligence by allowing computer programs to learn and adapt based on their exposure to data and experiences (El Naqa & Murphy, 2015). ML has been defined in various ways in the academic literature. Arthur Samuel defined it as allowing computers to learn without explicit programming (Samuel, 2000). Tom Mitchell viewed it as a computer program's ability to improve its performance on a task through experience (Mitchell, 1997). Ethem Alpaydin described it as the optimization of a performance criterion through the use of example data or experience (Alpaydin, 2020). All these definitions emphasize the concept of instructing computers to perform tasks that exceed basic data processing by learning from repeated exposure to examples in their environment (El Naqa & Murphy, 2015).

6.3.2 Classification of machine learning

After an extensive review of the literature, it is evident that the authors of several books and scientific articles have proposed diverse classification techniques for ML algorithms. Typically, ML algorithms fall into three main categories: supervised learning, unsupervised learning, and reinforcement learning (RL) (Jordan & Mitchell, 2015).

6.3.2.1 Supervised learning

Supervised learning is an ML approach in which the algorithm learns from data with existing labels, and the desired outcome or label is predetermined (Alpaydin, 2020). The algorithm aims to understand the connection between the input attributes and the output label, enabling it to forecast results on unfamiliar data. Supervised learning can again be classified into two main classes: regression and classification (Sen et al., 2020).

- Regression: Regression is a technique used for predicting continuous data, such as temperature or stock prices. Regression seeks to establish a connection between input parameters and the desired variable by training the algorithm using a collection of input characteristics and a continuous target value. For example, a regression algorithm could be trained on the plot area, and number of bedrooms, bathrooms, and garages in a group of houses to estimate the approximate cost of a newly constructed house that has a different plot area and number of bedrooms, bathrooms, and garages.
- Classification: Classification is a method for grouping data into one of several predetermined categories, such as determining whether a credit card transaction is normal or not. In classification, an algorithm is trained on a set of input features and a categorical target value to discover a relationship between the inputs and the target category. For instance, a classification system could use data on the color, shape, and texture of fruit to predict whether a new fruit is an apple, banana, or orange.

In both regression and classification, the training procedure entails presenting the algorithms with several instances of inputs and their associated outputs and modifying the parameters of these algorithms to reduce the difference between their predictions and the actual outputs. After the training process of these algorithms is completed, they can be employed to forecast data that is new and not seen before. Supervised learning is an effective method for handling a wide range of issues, from predicting stock prices to identifying handwritten numbers.

6.3.3 Unsupervised learning

Unsupervised learning is an ML technique that does not involve labeled data. Instead, the algorithm must autonomously discover patterns or structures within the data (El Naqa & Murphy, 2015). The objective of this type of ML method is to extract meaningful information from the data and to learn the underlying structure or distribution of the data. Unsupervised learning can also be classified into two main sub categories: clustering and dimensionality reduction (Jordan & Mitchell, 2015).

- Clustering: Clustering is an approach in unsupervised learning that categorizes alike data points into groups. The primary objective of clustering is to segment a collection of data points into separate groups, ensuring that data within a group share similarities and are distinct from those in other groups. For example, if provided with data on the height and weight of a group of people, a clustering algorithm could learn to categorize them into groups such as children, adults, and seniors based on their height and weight.
- Dimensionality Reduction: Dimensionality reduction, on the other hand, is a method used to minimize the number of data features while retaining the most important information. The objective of dimensionality reduction is to transform a set of data points with numerous characteristics into a lower-dimensional space while preserving the most critical information. For example, if given data on the color, shape, texture, and size of a group of fruits, a dimensionality reduction algorithm could learn to transform the data into a two-dimensional space where the most important features, such as color and texture, are retained.

The algorithm must learn the inherent structure of the data without the help of labeled data in both clustering and dimensionality reduction. This makes unsupervised learning a powerful tool for exploratory data analysis and data visualization, as well as for discovering hidden patterns in the data.

6.3.3.1 Reinforcement learning

RL is a special form of ML in which an entity, better known as an agent, learns the decision-making process through its interactions with the surrounding environment. In RL, the goal is to train an agent to take actions that will maximize some notion of cumulative reward (Jordan & Mitchell, 2015). The process of RL can be depicted as a Markov Decision Process (MDP), encompassing the subsequent elements (van Otterlo & Wiering, 2012).

- States: The states represent the environment of the agent. For instance, in a manufacturing assembly line, the state could be the current position or status of items being assembled.
- Actions: Actions symbolize the choices available to the agent. In the context of a manufacturing assembly line, an action could be adjusting the speed of a conveyor belt or redirecting a robotic arm.
- Rewards: Rewards are numerical values assigned to the agent based on each state-action combination. The agent's objective is to optimize the total reward it accrues. In a manufacturing setting, the reward might be $+1$ for completing an assembly correctly and -1 for any defect.
- Policy: This represents a guide from states to actions, delineating the agent's operations. RL aims to determine the best policy that elevates the overall cumulative reward.

RL algorithms are usually divided into two categories: model-based and model-free (Doody et al., 2022). In model-based RL, the agent constructs a representation of the environment, encompassing the transition likelihoods and reward mechanisms. The agent then utilizes this model to select its actions. For instance, in a game of chess, a model-based RL algorithm could use the rules of the game to build a model of the environment and plan its moves. On the other hand, in model-free RL, the agent does not necessarily build a model of the environment but instead learns a policy directly from experience. In a chess game, for example, a model-free RL algorithm could learn a policy by playing many games and updating its policy based on the rewards it receives.

RL has been implemented in a variety of fields, including robotics, gaming, and autonomous vehicles. Self-driving cars, which use RL to learn how to drive safely and efficiently are an example (Kiran et al., 2022). RL can be a difficult form of ML since the agent must learn from its environment through trial and error, and the ideal policy might be tricky to determine. Nonetheless, it has proved successful in a variety of applications and remains a hot topic of research within the subject of ML.

These are the primary categories of ML techniques, but many others have been developed, including semi-supervised learning, transfer learning, and multi-task learning, which are hybrid approaches that integrate components of various categories. ML seeks to empower computers to enhance their task performance autonomously through accumulated experience and training, regardless of the specific approach employed. The following figure shows a classification of ML techniques (Fig. 6.5).

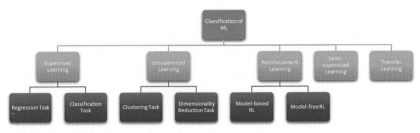

Figure 6.5 Classification of machine learning.

6.3.4 Deep learning

Deep learning (DL) is a field of AI that focuses on developing artificial neural networks capable of learning and making decisions based on input data, similar to ML. Nevertheless, DL algorithms are crafted to autonomously identify patterns and characteristics in extensive datasets, rendering them apt for intricate endeavors such as image categorization, voice recognition, and processing natural language. Unlike other ML algorithms, DL models have multiple hidden layers in their neural networks, which enable them to learn increasingly complex representations of the input data (Shrestha & Mahmood, 2019). The table below highlights some key differences between DL and ML (Janiesch et al., 2021) (Table 6.1).

6.3.5 Sequential learning

Sequential learning, also called active learning or incremental learning refers to a ML method in which a model is trained sequentially on data, with new data being continually provided over time. In this approach, the model updates its parameters and makes predictions based on each new data point as it is presented, in contrast to batch learning where the model is trained on a fixed dataset using a single training session. Sequential learning is commonly used in cases where the data is vast, intricate, or rapidly evolving, such as in financial forecasting, fraud detection, and natural language processing. Sequential learning has applications in the fields of material discovery, drug discovery, autonomous manufacturing, neuroscience, and robotics (Ahmed et al., 2021; Raihan and Ahmed, 2023b).

6.3.6 Offline and online learning

Based on the way the algorithms process the data, there are two types of learning processes: Offline learning and online learning. Fig. 6.6 illustrates the process of offline and online learning (Zheng et al., 2017).

Table 6.1 Differences between deep learning and machine learning.

Criterion	Deep learning	Machine learning
Complexity	Deep learning algorithms are more complex than traditional machine learning algorithms, as they are based on artificial neural networks that can have multiple hidden layers. This added complexity allows deep learning algorithms to learn more abstract and sophisticated relationships between the inputs and outputs	Traditional machine learning algorithms are simpler and less capable of modeling complex relationships
Automation	Deep learning algorithms are structured to assimilate data without direct programming instructions. This allows them to automatically identify patterns and features in the data and make predictions	Traditional machine learning algorithms often require manual feature engineering, where the input data is preprocessed and transformed into a format that can be used by the algorithm
Data Requirements	Deep learning algorithms mostly require massive amounts of data to train effectively. This is because deep learning algorithms are designed to automatically learn patterns in the data	Traditional machine learning algorithms can often be trained on smaller datasets. Oftentimes, these algorithms necessitate handcrafted feature design, a task that can be intricate and lengthy
Applications	Deep learning is well-suited to tasks such as classifying images, recognizing speeches and voices, and processing natural language, where the input data is highly structured and there are many complex relationships between the inputs and outputs	Traditional machine learning algorithms, on the other hand, are better suited to tasks such as regression and classification, where the relationships between the inputs and outputs are relatively simple

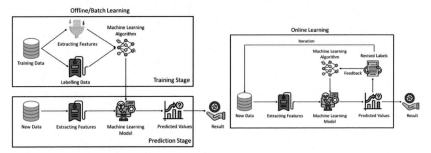

Figure 6.6 Process of offline (left) and online (right) machine learning methods.

- Offline learning: In offline learning or batch learning, data is processed in batches. The algorithm is first trained on a dataset and then applied to make predictions on a new dataset. Offline ML algorithms require all the data to be available at the time of training. It is best suited for applications where the data is relatively static and can be processed in bulk. Examples of offline ML include batch processing of log data, image or speech recognition, and large-scale data mining applications.
- Online learning: Online learning is an approach where data is processed sequentially as it arrives. The algorithm is updated continuously, and predictions are made in real-time. Online ML algorithms require the data to be available in a stream or small batches. Online ML algorithms are typically used when the data is continuously evolving or when there is a need for immediate response to changing conditions. It is also useful in situations where it is impractical or impossible to store all the data at once, such as in real-time systems. Examples of online ML include fraud detection, recommendation systems, and adaptive control systems.

The decision of choosing between offline and online ML depends on the specific requirements of the application. Online ML is preferable for applications that require instantaneous responses. Offline ML is better appropriate for situations where the data are relatively static. In many instances, a hybrid strategy that blends offline and online ML may deliver a superior solution overall.

6.4 Role of machine learning in manufacturing sustainability

The application of AI algorithms is transforming both service and manufacturing industries, enhancing productivity while decreasing costs related to maintenance, repairs, and errors in manufacturing (Chien et al., 2020; Li et al., 2017). The manufacturing sector has seen a significant increase in the use of several traditional and novel ML and DL-based approaches in recent years, largely due to the enormous amounts of data generated by activities on the shop floor, in the supply chain, and during manufacturing processes (Lee et al., 2018; Rawat et al., 2021). However, many industries have not been able to fully leverage this data. By utilizing AI-based techniques such as ML and DL, significant and valuable information can be extracted and interpreted, leading to increased productivity and efficiency in the manufacturing industry (Jamwal et al., 2022). A literature survey on Industry 4.0 and its applications has demonstrated that 82% of manufacturing-related organizations have improved their performance and consequently, realized an increase in revenue, through the adoption of Industry 4.0 technologies such as solutions based on ML and leveraging IoT (Akter et al., 2021). ML-based tools are essential for promoting sustainable development and shaping the future. Advanced ML-based data analytical tools are now aiding industries in making more efficient use of resources, reducing waste, and lowering carbon emissions (Carvalho et al., 2019). The incorporation of ML in supply chain operations is contributing to sustainability efforts. ML's integration into supply chain practices is resulting in reduced carbon footprints and improved performance of supply chains (Belhadi et al., 2022). The application of ML-powered tools in supply chain design reduces movements and mitigates the supply chain's environmental impact (Abdirad & Krishnan, 2021).

In contemporary manufacturing operations, the adoption of data-centric technologies has become imperative. These advanced technologies, pillars of Industry 4.0, drive innovation and champion sustainable practices in manufacturing entities (Raj et al., 2020). To enhance decision-making and achieve optimal outcomes, manufacturers are progressively integrating AI-driven solutions like DL and ML into eco-friendly manufacturing methodologies (Hansen & Bøgh, 2021). This is particularly important because of the resource, energy, and carbon emission constraints that manufacturing organizations face (Enyoghasi & Badurdeen, 2021). ML methods have found extensive applications across multiple facets of manufacturing, encompassing design phases, lifecycle management of products, production endeavors, operational tasks, and practices related to sustainability (Jung et al., 2021). In the next subsections, we will discuss the applications of ML-based approaches in the manufacturing industry, focusing on Industry 4.0 and sustainability.

6.4.1 Machine learning in predictive maintenance

Predictive maintenance is an essential part of the manufacturing business since it enables firms to anticipate and prevent equipment breakdowns. In the industrial business, ML and DL have become significant enablers for predictive maintenance, allowing organizations to accurately predict equipment breakdowns and undertake maintenance in a timely and effective manner. Some of the applications of ML in predictive maintenance are discussed in the following:

- Predicting Equipment Failures: Utilizing ML for predictive maintenance offers a significant advantage in assessing vast amounts of information from various sources, including maintenance logs, sensors, and prior performance data. By processing this data, ML algorithms can identify patterns and associations that may indicate potential equipment failures. For example, temperature readings from sensors could be analyzed by an ML algorithm to identify any abnormal temperature changes that may indicate impending mechanical problems.
- Condition-based Monitoring: This is another significant application of ML in predictive maintenance. This entails employing sensors to collect data on the health of equipment and feeding this data into an ML algorithm to detect performance changes or equipment degradation over time. For instance, an algorithm may be taught to recognize variations in vibration patterns that indicate an imminent breakdown. This enables firms to prioritize maintenance tasks and reduce the impact of equipment failures on production.
- Optimizing Maintenance Schedules: In addition, ML can be utilized to optimize maintenance schedules. By examining previous maintenance data, ML algorithms can determine the best frequency and timing for maintenance tasks, thereby minimizing downtime and enhancing equipment reliability.

6.4.2 Machine learning in quality control

ML algorithms are playing a critical role in improving the quality control processes in the manufacturing industry. By leveraging large amounts of data and advanced algorithms, ML and DL are enabling organizations to automate quality control processes and make data-driven decisions.

- Defect Detection: ML algorithms can be trained on product images to detect defects rapidly and accurately such as scratches, cracks, and other irregularities. This reduces the chance of human error and enables manufacturers to identify faults early in the production process.

- Root Cause Analysis: By examining data from various sources, including sensors and production logs, multiple ML algorithms can identify patterns and correlations between factors that may contribute to quality control issues, facilitating root cause analysis. This helps manufacturers identify the primary cause of problems and implement corrective actions more effectively.
- Quality Prediction: Advanced ML algorithms can predict the quality of future products by analyzing data from previous manufacturing cycles. This assists manufacturers in anticipating potential quality control issues and taking preventative measures.
- SPC: ML-based methods are used to conduct SPC, which involves monitoring and managing a process to ensure it stays within predefined boundaries. By automating SPC, ML can improve the accuracy and efficiency of the quality control process.

6.4.3 Machine learning in supply chain optimization

Several ML-based techniques play a key role in optimizing supply chain operations in the manufacturing business, as shown below.

- Predictive Analytics: Based on historical data, ML algorithms can forecast future demand and supply. These forecasts can be utilized to optimize the supply chain by reducing lead times and boosting the percentage of on-time deliveries. For instance, an ML model can be used to forecast the demand for a product based on historical sales patterns, weather patterns, and economic factors. The manufacturing company can alter its production schedule and the flow of items along the supply chain based on these forecasts.
- Route Optimization: To improve delivery routes and minimize the carbon footprint of the supply chain, ML and DL algorithms can both be utilized. For example, a DL model can be trained to determine the most efficient delivery route based on factors such as traffic conditions, delivery schedules, and fuel consumption, resulting in reduced delivery times, costs, and carbon emissions.
- Inventory Optimization: ML algorithms can optimize inventory levels, decreasing waste and enhancing supply chain efficiency. An ML model can predict product demand based on historical sales data and adjust inventory levels accordingly. This can reduce inventory levels and increase order fulfillment speed.

6.4.4 Machine learning in energy efficiency

ML and DL algorithms can be implemented to lower energy consumption in manufacturing facilities, hence reducing costs, and enhancing sustainability. Some notable applications of ML techniques in efficient energy management are explained in the following.

- Predictive Energy Management: Energy usage can be predicted using ML methods based on production schedules, weather conditions, and energy prices. This data can be applied to reduce energy consumption and associated costs and enhance sustainability.
- Equipment Monitoring: ML methods can be used to monitor equipment performance, identifying anomalies and inefficiencies that may contribute to increased energy usage. This data can be used to identify and address energy efficiency problems.
- Energy-saving Recommendations: Using ML algorithms, energy usage data may be analyzed and recommendations for reducing energy consumption can be generated. For instance, computers can recognize times when energy consumption exceeds expectations and offer solutions to cut consumption during these times.
- Load Forecasting: ML algorithms can be used to forecast energy demand, allowing manufacturers to optimize energy generation and storage to meet demand efficiently and cost-effectively.
- Energy Portfolio Optimization: An organization's energy portfolio, comprising the selection of energy sources, the scheduling of energy usage, and the management of energy storage, can be optimized using many ML algorithms.

6.4.5 Machine learning in carbon footprint and waste reduction

ML, methods can play a major role in reducing carbon footprint in various ways.

- Energy Management: To decrease energy waste and utility costs while reducing carbon emissions, ML algorithms can analyze energy consumption patterns, recognize high-consumption areas, and offer real-time suggestions for optimizing energy consumption in energy management.
- Renewable Energy: By utilizing ML-based methods to analyze meteorological and energy data, energy providers can anticipate renewable energy generation from sources such as wind and solar. This allows

them to adjust their operations to utilize more renewable energy, decrease their reliance on fossil fuels, and reduce carbon emissions.

- Transportation: ML algorithms can assist in optimizing transportation routes, reducing fuel usage, and limiting emissions. By examining traffic patterns, weather data, and other variables, ML algorithms can reduce the number of miles driven and fuel consumed, thereby lowering carbon emissions.
- Industrial Processes: ML algorithms can help optimize industrial processes to reduce energy consumption and carbon emissions. By evaluating data from sensors and other sources, ML algorithms may discover improvement opportunities, forecast outcomes, and make modifications in real time.
- Resource Allocation: ML algorithms can also optimize resource allocation problems in manufacturing, logistics, and other industries. By forecasting demand and modifying resource allocation accordingly, these algorithms can help reduce waste caused by overproduction, underutilization of resources, and inefficient processes.

The manufacturing industry has been driven toward adopting Industry 4.0 by recent advancements in AI, leading to a more sustainable production process from start to finish. ML applications have had an impact on every sector of the manufacturing industry. In the following section, we will examine two case studies that apply ML and DL algorithms in anomaly detection tasks. First, we will provide a brief overview of anomaly detection before delving into the case studies.

6.5 Anomaly detection

Many countries have embraced the concept of Industry 4.0 as a strategic blueprint to address the needs of advanced manufacturing that is more efficient, cost-effective, robust, and capable of producing customized products and services (Hsieh et al., 2019). This initiative seeks to transform the industrial manufacturing landscape by incorporating digital advancements and leveraging cutting-edge technologies such as cyber-physical systems (CPS), IoT, cloud computing, and AI. Smart manufacturing is one such example that strives to build an integrated, collaborative manufacturing system that can adapt to widely volatile demands and circumstances in the production floor, supply chain, and highly customized customer demands in real-time (Lu et al., 2016). Anomaly detection is an essential challenge in smart manufacturing because it facilitates predictive maintenance, minimizing maintenance frequency and conserving resources (Lopez et al., 2017; Wang et al., 2015). This becomes increasingly crucial as manufacturing processes become more complex and scaled up, increasing the likelihood of machine and equipment malfunctions, and defective products. As a result, to minimize the frequency and severity of such unwanted situations in the industry, it is crucial to detect any abnormal behavior of mechanical devices on the production line as early as possible, allowing potential problems to be predicted and production plans to be adjusted accordingly to avoid any failures (Hsieh et al., 2019). Early detection also helps prevent any further issues that may arise as a result of such failures.

6.5.1 What are anomalies?

Anomalies are data points that deviate substantially from the typical data distribution. Anomaly detection involves identifying uncommon occurrences, objects, or observations that are suspicious due to their deviation from normal behaviors or patterns.

In the literature, anomaly detection is often considered an unsupervised learning problem, as there is typically no pre-labeled training dataset with normal observations or outliers (Ahmed, Dagnino et al., 2019). Anomaly detection techniques aim to identify anomalies in large volumes of data that may initially appear uniform. This can offer valuable insights for a range of practical applications, such as detecting credit card fraud, ensuring cybersecurity, analyzing medical images, monitoring surveillance, and ensuring safety in industrial processes (Ahmed et al., 2019; Raihan & Ahmed, 2023a). The ability to accurately and quickly identify anomalies can help prevent losses, including those related to human, financial, and informational factors.

6.5.2 Categories of anomaly detection

The literature describes three main classes of anomaly detection techniques based on the labels or output targets of the data in a training set (Ahmed et al., 2019).

6.5.2.1 Supervised anomaly detection

This category of anomaly detection approach involves using labeled data, where anomalies have been previously identified, to train an ML model that can predict if a new observation is an anomaly. The algorithm is provided with both normal and abnormal observations to learn from the labeled data and make predictions on new, unseen data. For example, if a credit card company has a dataset of known fraudulent transactions, a supervised ML model can be trained to identify similar transactions as anomalies.

6.5.2.2 Semisupervised anomaly detection

This approach involves training an ML algorithm using a dataset that mainly consists of regular observations, with only a small number of identified anomalies. The algorithm utilizes this knowledge to detect abnormalities in new data. This technique is advantageous when obtaining tagged data is difficult or expensive. For instance, in medical imaging, it may be easier to collect normal images with only a few tagged images of anomalies, making a semi-supervised method a useful tool for detecting novel anomalies.

6.5.2.3 Unsupervised anomaly detection

This technique examines the data without any previous knowledge of what an abnormality may look like. By utilizing statistical or ML techniques, the system recognizes observations that differ significantly from the rest of the data and identifies them as anomalies. When obtaining labeled data is challenging or when abnormalities are uncommon and difficult to predict, this approach is frequently used (Ahmed et al., 2022; Ahmed, Hu, et al., 2021). For example, in IPC, it may be challenging to predict all potential equipment failures; hence, an unsupervised method could be useful in detecting and reporting unexpected anomalies.

6.5.3 Anomaly detection techniques

When identifying anomalous data that are rare, it is common to come across high levels of noise that might resemble anomalous behavior. This is due to the imprecise boundary between normal and abnormal behavior, which can change over time. Additionally, because many data patterns are time-based, with seasonality included, it can be even more complex to detect anomalies. For example, it requires sophisticated methods to distinguish between actual changes in seasonality versus noise or anomalous data. Therefore, there are multiple techniques for detecting anomalies, and the most suitable one for a particular user or dataset may vary

Table 6.2 A brief description of different anomaly detection methods.

Anomaly detection techniques	Description	Some algorithms used
Statistical methods	Involve using statistical models to identify anomalies	Gaussian mixture models, principal component analysis (PCA), histogram-based outlier score (HBOS), ARIMA
Clustering methods	Involve grouping similar data points into clusters and identifying outliers that do not belong to any of the clusters	K-means, DBSCAN, LDCOF
Density-based methods	Involve identifying areas of high density in the data and labeling points outside these areas as anomalies	Local outlier factor (LOF), minimum covariance determinant (MCD), connectivity-based outlier factor (COF)
Neural network-based methods	Involve using deep learning algorithms to learn the normal patterns in the data and identify deviations from these patterns as anomalies	Autoencoders, recurrent neural networks (RNNs), and convolutional neural networks (CNNs)
Support vector machine-based methods	Involve using a supervised learning algorithm to train a model on a dataset that is labeled and then using the same model to identify anomalies in new data	SVM, one-class SVM
Ensemble methods	Involve combining multiple anomaly detection algorithms to improve the accuracy of the results	Feature bagging, isolation forest

depending on the circumstances. The following table mentions a list of anomaly detection approaches with the algorithms used (Table 6.2).

6.5.4 Offline and online anomaly detection

Anomaly detection approaches can be online or offline. In online anomaly detection, the algorithm processes the data as it arrives and updates the model in real time (Ahmed, Galoppo, et al., 2019). In contrast, an offline

Table 6.3 Differences between offline and online anomaly detection.

Criterion	Offline anomaly detection	Online anomaly detection
Data processing	The entire dataset is processed at once	Data is processed sequentially as it arrives
Model training	A complete dataset is required beforehand for model training	Model can be trained with a small dataset and then updated as new data arrives
Computational resources	Significant computational resources are required to process large datasets	Fewer resources and less computational cost as data is processed incrementally
Accuracy	Offline anomaly detection generally produces more accurate results because it processes the entire dataset	Online anomaly detection must sacrifice some accuracy to provide real-time processing
Time sensitivity	Offline anomaly detection is better for applications that can tolerate a longer processing time	Online anomaly detection is better suited for applications that require an immediate response

anomaly detection algorithm processes the entire dataset at once and generates a static model that is used to identify anomalies. The key differences between these two approaches are mentioned in Table 6.3.

6.6 Anomaly detection and root cause identification in a hydropower dataset (a case study)

In this section, we present a case study (Ahmed et al., 2019) focused on anomaly detection and subsequent root cause identification in a hydropower plant. The hydropower facility is an intricate structure made up of many physical elements, and detecting anomalous behavior is necessary to monitor the health of these components. Developing performance guidelines and identifying critical variables responsible for abnormal behavior can assist maintenance personnel in identifying potential process shifts on time. To achieve this goal, a system must be in place to differentiate regular data points from anomalies. In this case study, we utilized three separate methods to identify unusual observations and assessed their performance with a historical dataset from the hydropower facility. Domain specialists confirmed the detected anomalies, and we pinpointed the key variables linked to these irregularities through a decision tree and a feature evaluation procedure. Furthermore, we created a one-class classifier utilizing the dataset without outliers to establish standard operating conditions. This has allowed us to identify anomalous observations in future operations.

6.7 Problem description

The hydropower generation plant comprises several functional areas such as generators, turbines, and bearings, each containing components that generate real-time data from multiple sensors. Outliers can originate from multiple causes, resulting in issues like bearing oil and metal components getting excessively hot, vibrations originating from bearings, and diminished active or reactive power production. Detecting anomalies at the earliest is crucial, but it is challenging due to the high dimensionality of the data.

This case study examines three anomaly identification methods, each grounded in a unique approach. Local outlier factor (LOF) is rooted in density making it a density-based technique, Feature Bagging for Outlier Detection (FBOD) is termed as an ensemble-based method, and Subspace Outlier Degree (SOD) is a technique based on subspace methodology. The dataset (Ahmed et al., 2019), which has been employed in this work, is collected from a plant generating hydropower, and the results from the outlier detection methods are compared to identify any commonalities. The study further investigates the factors that have the most influence on the detected anomalies and their respective value ranges. Moreover, we trained a one-class support vector machine (SVM) classifier on the dataset after excluding anomalies, establishing a standard boundary to pinpoint prospective anomalies in the future.

6.7.1 Description of the dataset

We sourced the data directly from a power generation plant's Distributed Control System. This data, which spanned several months, came in a time-stamped format, and was segmented into various operational zones. The data were collected every 10 minutes daily but had some gaps. Initially, there were 9508 observations and 222 variables without missing values, mainly consisting of temperature, vibration, pressure, and active power. After conducting some pre-processing tasks and statistical analyses, the number of observations (expressed by the rows in the data table) decreased from 9508 to 9219.

6.7.2 Solution methodology

After preliminary data processing, it is essential to pinpoint anomalies which are harder to identify than those detected by basic clustering or statistical methods. With high-dimensional data, it is almost impossible to visually detect anomalies through graphical analysis. Moreover, the clusters formed by data with varying densities indicate the possible presence of local outliers.

To address these real-world challenges, we require outlier detection methods capable of handling unsupervised, high-dimensional data with clustering tendencies. Our objective is to identify these points and understand the reasons and conditions that caused them, aiding engineering decision-making. Additionally, we aim to develop a classifier to flag outlier observations in the future.

To achieve these goals, three outlier detection methods were selected and employed in the hydropower dataset. Multiple methods from various anomaly detection categories were chosen because no single method can effectively handle the challenging unsupervised outlier detection problem. A brief explanation of how these three approaches work is provided below.

6.7.2.1 Local outlier factor

Breunig et al. introduced the LOF algorithm for outlier detection (Breunig et al., 2000). The algorithm measures the local density of a data point by assessing its proximity to nearby points. Points with lesser density compared to their surrounding ones (signified by a higher LOF score) are marked as anomalies. The dataset used in this research is composed of clusters of different densities, which makes it apt for the LOF technique. Even though there have been recent modifications to the LOF method, we have utilized its original form for this data. A notable limitation of LOF, however, is its suboptimal performance on high-dimensional datasets, especially when the anomalies differ from the norm in just a single or a few dimensions.

6.7.2.2 Subspace outlying degree

To address the challenge of handling high-dimensional data, it may be necessary to use dimension reduction techniques that consider a subset of the original features. When data points are transformed into a subspace, their uniform distribution in the original full-dimensional space may change, potentially amplifying any anomalous behavior and separating it from the normal points. However, using the subspace approach can be risky since an inappropriate choice of subspace can eliminate the margin or gap between a potential anomaly and a normal data point. Therefore, in this study, the SOD technique was utilized, which is one of several subspace methods recently introduced (Kriegel et al., 2009). The SOD method does not rely on certain assumptions commonly used in other subspace approaches, such as monotonicity, which can be restrictive.

6.7.2.3 Feature bagging in anomaly detection

While subspace methods can manage data with high dimensions, there lies an inherent danger of choosing an inappropriate subspace, which can complicate or even hinder anomaly detection. In this case study, a technique known as FBOD was adopted. FBOD randomly selects subsets of the feature space to calculate outlier scores for observations, and the cumulative scores from multiple iterations are considered as the final outlier score (Lazarevic & Kumar, 2005). For this analysis, only the LOF method was employed, and it was executed across several cycles. Since density and distance-based methods may suffer from the curse of dimensionality, employing LOF within FBOD combines density and subspace-based methods, which takes full advantage of both.

6.7.3 Results and discussion

The three approaches show a reasonable level of consistency. In the process of pinpointing potential outliers, 22 of the top 30 timestamps identified by these techniques overlap (marked in green in Table 6.4). This overlap expands to 33 for the top 50 timestamps and rises to 55 for the top 100 timestamps. To save space, events beyond the top 30 are not displayed here.

Upon examining Table 6.4, a crucial finding is that specific time intervals on particular days (e.g., September 14, January 11 and 12) are more prone to outliers. Although there are minor differences in the timestamps provided by each method beyond the top 30 timestamps, they are quite similar (within a range of 10−50 minutes). The third observation indicates that the SOD method identifies some timestamps from July 4th as outliers, while the other two methods do not. The SOD technique can detect an anomaly in a subspace with as few as one variable, a capability not found

Table 6.4 Most anomalous 30 timestamps identified by the anomaly detection techniques (Ahmed et al., 2019).

Subspace outlier degree (SOD)	Feature bagging for outlier detection (FBOD)	Local outlier factor (LOF)
Jul-04-2015 04:20	Apr-16-2015 16:00	Apr-16-2015 16:00
Jul-04-2015 04:30	Apr-16-2015 23:10	Apr-16-2015 23:10
Jul-04-2015 04:40	Sep-13-2015 19:00	Sep-13-2015 19:00
Jul-04-2015 05:40	Sep-14-2015 01:50	Sep-14-2015 01:50
Jul-04-2015 05:50	Sep-14-2015 02:00	Sep-14-2015 02:00
Jul-04-2015 06:30	Sep-14-2015 08:00	Sep-14-2015 08:00
Jul-04-2015 08:20	Sep-14-2015 08:10	Sep-14-2015 08:10
Jul-04-2015 08:30	Sep-14-2015 13:00	Sep-14-2015 13:00
Sep-13-2015 19:00	Sep-14-2015 13:10	Sep-14-2015 13:10
Sep-14-2015 01:50	Sep-16-2015 10:50	Sep-16-2015 10:50
Sep-14-2015 02:00	Oct-03-2015 14:40	Sep-17-2015 11:30
Sep-14-2015 08:00	Oct-04-2015 03:10	Oct-03-2015 14:40
Sep-14-2015 08:10	Oct-13-2015 08:15	Oct-04-2015 03:10
Sep-14-2015 13:00	Oct-13-2015 08:25	Oct-13-2015 08:15
Sep-14-2015 13:10	Nov-02-2015 09:56	Oct-14-2015 23:15
Sep-16-2015 10:50	Jan-02-2016 21:10	Oct-14-2015 23:35
Oct-03-2015 14:40	Jan-02-2016 21:10	Nov-02-2015 09:56
Oct-13-2015 08:15	Jan-02-2016 21:30	Jan-02-2016 21:10
Nov-02-2015 09:56	Jan-02-2016 21:40	Jan-02-2016 21:20
Jan-02-2016 13:30	Jan-09-2016 18:00	Jan-02-2016 21:30
Jan-02-2016 21:10	Jan-09-2016 18:30	Jan-02-2016 21:40
Jan-02-2016 21:20	Jan-09-2016 18:40	Jan-09-2016 18:30
Jan-02-2016 21:40	Jan-09-2016 18:50	Jan-09-2016 18:50
Jan-09-2016 18:50	Jan-11-2016 12:00	Jan-11-2016 12:00
Jan-11-2016 12:00	Jan-11-2016 13:30	Jan-11-2016 13:00
Jan-11-2016 13:00	Jan-11-2016 14:40	Jan-11-2016 13:30
Jan-11-2016 13:30	Jan-12-2016 11:20	Jan-11-2016 13:50
Jan-11-2016 14:40	Jan-12-2016 11:30	Jan-11-2016 14:40
Jan-12-2016 11:20	Mar-07-2016 09:40	Jan-12-2016 11:20
Jan-12-2016 11:30	Mar-11-2016 12:30	Jan-12-2016 11:30

in the other two methods. Thus, it is likely that the timestamps from July 4 showed significant variations in a low-dimensional subspace, explaining why they were not identified by the other two techniques. To pinpoint the factors most influencing the outliers, emphasis was put on the regular events among the top 100 anomalies identified by all three techniques. These were deemed the definitive outliers. After consulting with a domain specialist, a threshold of 100 was established. It was observed that 55 out of the top 100 anomalies were consistent across all methods.

These were given a label of 1, while the rest of the data entries received a label of 0, signaling regular conditions. This action effectively converted the original dataset from unclassified to a binary classification. Using the R package "rpart," a classification and regression tree (CART) was constructed, depicted in Fig. 6.7. This tree suggests that certain conditions, specifically pertaining to air pressure and the temperature variance between the oil and air in bearing F4, can precisely flag 24 out of the 55 shared anomalies. Specifically, when the air pressure exceeds 945 mbar and the temperature difference between the oil and air of bearing F4 falls below 10.582°C, the generator tends to malfunction, as highlighted by 9 recurrent readings which were anomalous. Such insights are invaluable for professionals aiming to swiftly identify malfunctions and manage alerts during the system's operation.

Another important finding was uncovered when a decision tree model was applied to the 100 anomalies detected by the LOF model. Threshold values for two important variables in the turbine system, the oil temperature, and the harmonic readings for bearing F4, were determined, as depicted in Fig. 6.8. It was a significant discovery as it was confirmed

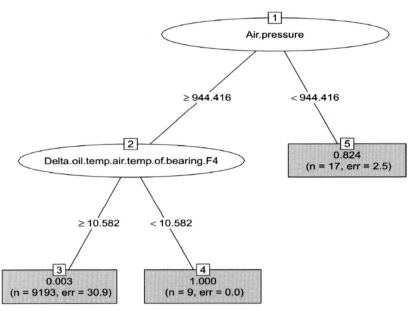

Figure 6.7 Construction of a decision tree from 55 common anomalies identified by the anomaly detection techniques (Ahmed et al., 2019).

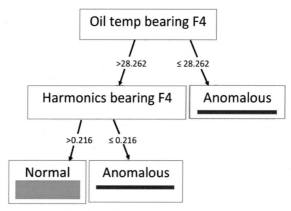

Figure 6.8 Construction of a decision tree from 100 anomalies identified by the LOF technique (Ahmed et al., 2019). *LOF*, Local outlier factor.

during the maintenance procedure of the turbine under investigation that corrective repairs were necessary for bearing F4 to prevent future damage or costly disruptions to the turbine's operation.

Utilizing a real-world dataset obtained from a hydropower generation facility, we discussed the real-world application of anomaly detection and root cause analysis in this section. The data obtained from the hydro power-plant is at first subjected to preprocessing before employing three distinct anomaly detection techniques, all of which identify similar and comparable anomalies. Verification of these anomalies' accuracy is established through validation by domain experts and maintenance personnel. Despite their absence in the physical system's current state, the anomalies become evident during turbine preventive maintenance procedures. The root causes and threshold values for pivotal attributes contributing to these anomalies are ascertained through the construction of a decision tree. Moreover, the identification of crucial variables driving the anomalous instances is also pinpointed. This new-found knowledge offers practitioners the means to effectively supervise and diagnose the power plant's activities.

6.8 Conclusion

ML, a pivotal subset of AI, has a profound impact on the operations of the manufacturing sector. ML is increasingly being integrated into every stage of the supply chain, from sourcing to production to distribution. As sustainability concerns increase and data becomes more available, the incorporation of ML's computational intelligence is an obvious choice. Many factories in developed countries have already adopted advanced AI-based approaches that focus on various ML techniques and are experiencing benefits such as improved product quality, reduced costs, and faster and more efficient production processes. It is only a matter of time before AI reaches industries worldwide.

This book chapter highlights the significance of Industry 4.0 technologies and ML techniques in helping manufacturing companies maintain a sustainable supply chain and remain competitive in an ever-changing market. Traditional process control methods suffer from major drawbacks such as lacking adaptivity to a continuously growing dataset and inability to determine the underlying sophisticated relationship between the process variables and outcomes. These conventional methods perform poorly under dynamic environments and are no longer efficient in this new era of AI. The chapter highlights how data-driven ML-based approaches have impacted IPC, surpassing the limitations of traditional process control tools. Advanced concepts in ML, including DL and sequential learning, are also discussed in detail. Anomaly detection, one of the most critical tasks in IPC, is explained comprehensively. A case study using data from a hydropower generation plant is presented to demonstrate the practical application of anomaly detection techniques and root cause analysis in detecting abnormal patterns and identifying system faults.

In the pursuit of sustainable industrial processes, the integration of ML and process control presents a promising trajectory. Through the proper identification of existing gaps and harnessing the potential of data-driven intelligence, industries can pave the way for more efficient, eco-friendly, and resilient operations. This chapter attempts to shed light on this transformative journey, empowering the practitioners to navigate the evolving landscape of sustainable IPC.

References

Abdirad, M., & Krishnan, K. (2021). Industry 4.0 in logistics and supply chain management: A systematic literature review. *Engineering Management Journal, 33*(3), 187–201. Available from https://doi.org/10.1080/10429247.2020.1783935.

Ahmed, I., Bukkapatnam, S., Botcha, B., & Ding, Y. (2021). Towards futuristic autonomous experimentation—A surprise-reacting sequential experiment policy. arXiv preprint arXiv:2112.00600. Available from https://doi.org/10.48550/arXiv.2112.00600.

Ahmed, I., Dagnino, A., & Ding, Y. (2019). Unsupervised anomaly detection based on minimum spanning tree approximated distance measures and its application to hydropower turbines. *IEEE Transactions on Automation Science and Engineering, 16*(2), 654−667. Available from https://doi.org/10.1109/TASE.2018.2848198.

Ahmed, I., Galoppo, T., & Ding, Y. (2019). O-LoMST: An online anomaly detection approach and its application in a hydropower generation plant. In: Proceedings of the IEEE 15th international conference on automation science and engineering (CASE), 762−767. https://doi.org/10.1109/COASE.2019.8843317.

Ahmed, I., Galoppo, T., Hu, X., & Ding, Y. (2022). Graph regularized autoencoder and its application in unsupervised anomaly detection. *IEEE Transactions on Pattern Analysis and Machine Intelligence, 44*(8), 4110−4124. Available from https://doi.org/10.1109/TPAMI.2021.3066111.

Ahmed, I., Hu, X., Ben., Acharya, M. P., & Ding, Y. (2021). Neighborhood structure assisted non-negative matrix factorization and its application in unsupervised point-wise anomaly detection. *The Journal of Machine Learning Research, 22*(1), 1624−1655.

Akhtar, M. (2022). In T. Bányai, Á. Bányai, & I. Kaczmar (Eds.), *Industry 4.0 technologies impact on supply chain sustainability.* IntechOpen, p. Ch. 5. Available from https://doi.org/10.5772/intechopen.102978.

Akter, S., McCarthy, G., Sajib, S., Michael, K., Dwivedi, Y. K., D'Ambra, J., & Shen, K. N. (2021). Algorithmic bias in data-driven innovation in the age of AI. *International Journal of Information Management, 60*, 102387. Available from https://doi.org/10.1016/j.ijinfomgt.2021.102387.

Alpaydin, E. (2020). *Introduction to machine learning.* MIT Press.

Azevedo, S. G., Carvalho, H., Duarte, S., & Cruz-Machado, V. (2012). Influence of green and lean upstream supply chain management practices on business sustainability. *IEEE Transactions on Engineering Management, 59*(4), 753−765.

Belhadi, A., Kamble, S., Fosso Wamba, S., & Queiroz, M. M. (2022). Building supply-chain resilience: an artificial intelligence-based technique and decision-making framework. *International Journal of Production Research, 60*(14), 4487−4507. Available from https://doi.org/10.1080/00207543.2021.1950935.

Boons, F., & Lüdeke-Freund, F. (2013). Business models for sustainable innovation: State-of-the-art and steps towards a research agenda. *Journal of Cleaner Production, 45*, 9−19. Available from https://doi.org/10.1016/j.jclepro.2012.07.007.

Braccini, A. M., & Margherita, E. G. (2019). Exploring organizational sustainability of industry 4.0 under the triple bottom line: The case of a manufacturing company. *In Sustainability, 11*(1), 36. Available from https://doi.org/10.3390/su11010036.

Breunig, M. M., Kriegel, H. P., Ng, R. T., & Sander, J. (2000). LOF: Identifying density-based local outliers. In: *Proceedings of the 2000 ACM SIGMOD International Conference on Management of Data*, 93−104.

Carvalho, T. P., Soares, F. A. A. M. N., Vita, R., Francisco, R., da, P., Basto, J. P., & Alcalá, S. G. S. (2019). A systematic literature review of machine learning methods applied to predictive maintenance. *Computers & Industrial Engineering, 137*, 106024. Available from https://doi.org/10.1016/j.cie.2019.106024.

Chalmeta, R., & Santos-deLeón, N. J. (2020). Sustainable supply chain in the era of industry 4.0 and big data: A systematic analysis of literature and research. *Sustainability, 12*(10), 4108. Available from https://doi.org/10.3390/su12104108.

Chien, C.-F., Dauzère-Pérès, S., Huh, W. T., Jang, Y. J., & Morrison, J. R. (2020). Artificial intelligence in manufacturing and logistics systems: algorithms, applications, and case studies. *International Journal of Production Research, 58*(9), 2730−2731. Available from https://doi.org/10.1080/00207543.2020.1752488.

Davis, J., Edgar, T., Graybill, R., Korambath, P., Schott, B., Swink, D., Wang, J., & Wetzel, J. (2015). Smart manufacturing. *Annual Review of Chemical and Biomolecular Engineering, 6*, 141−160. Available from https://doi.org/10.1146/annurev-chem-bioeng-061114-123255.

Doody, M., Van Swieten, M. M. H., & Manohar, S. G. (2022). Model-based learning retrospectively updates model-free values. *Scientific Reports, 12*(1), 2358. Available from https://doi.org/10.1038/s41598-022-05567-3.

El Naqa, I., & Murphy, M. J. (2015). *What is machine learning?* Springer International Publishing. Available from https://doi.org/10.1007/978-3-319-18305-3_1.

Elangovan, M., Sakthivel, N. R., Saravanamurugan, S., Nair, B. B., & Sugumaran, V. (2015). Machine learning approach to the prediction of surface roughness using statistical features of vibration signal acquired in turning. *Procedia Computer Science, 50*, 282−288.

Enyoghasi, C., & Badurdeen, F. (2021). Industry 4.0 for sustainable manufacturing: Opportunities at the product, process, and system levels. *Resources, Conservation and Recycling, 166*, 105362. Available from https://doi.org/10.1016/j.resconrec.2020.105362.

Hansen, E. B., & Bøgh, S. (2021). Artificial intelligence and internet of things in small and medium-sized enterprises: A survey. *Journal of Manufacturing Systems, 58*, 362−372. Available from https://doi.org/10.1016/j.jmsy.2020.08.009.

Hsieh, R. J., Chou, J., & Ho, C. H. (2019). Unsupervised online anomaly detection on multivariate sensing time series data for smart manufacturing. In: *2019 IEEE 12th Conference on Service-Oriented Computing and Applications (SOCA)*, 90−97. https://doi.org/10.1109/SOCA.2019.00021.

Islam, F., Raihan, A. S., & Ahmed, I. (2023). Applications of federated learning in manufacturing: identifying the challenges and exploring the future directions with industry 4.0 and 5.0 visions. *Proceedings of the International Conference on Industrial Engineering and Operations Management.*

Jamwal, A., Agrawal, R., & Sharma, M. (2022). Deep learning for manufacturing sustainability: Models, applications in Industry 4.0 and implications. *International Journal of Information Management Data Insights, 2*(2), 100107. Available from https://doi.org/10.1016/j.jjimei.2022.100107.

Janiesch, C., Zschech, P., & Heinrich, K. (2021). Machine learning and deep learning. *Electronic Markets, 31*(3), 685−695. Available from https://doi.org/10.1007/s12525-021-00475-2.

Jordan, M. I., & Mitchell, T. M. (2015). Machine learning: Trends, perspectives, and prospects. *Science (New York, N.Y.), 349*(6245), 255−260. Available from https://doi.org/10.1126/science.aaa8415.

Jung, H., Jeon, J., Choi, D., & Park, J. Y. (2021). Application of machine learning techniques in injection molding quality prediction: Implications on sustainable manufacturing industry. *Sustainability, 13*(8), 4120.

Kamble, S. S., Gunasekaran, A., & Gawankar, S. A. (2018). Sustainable Industry 4.0 framework: A systematic literature review identifying the current trends and future perspectives. *Process Safety and Environmental Protection, 117*, 408−425. Available from https://doi.org/10.1016/j.psep.2018.05.009.

Kiran, B. R., Sobh, I., Talpaert, V., Mannion, P., Al Sallab, A. A., Yogamani, S., & Pérez, P. (2022). Deep reinforcement learning for autonomous driving: A survey. *IEEE Transactions on Intelligent Transportation Systems, 23*(6), 4909−4926. Available from https://doi.org/10.1109/TITS.2021.3054625.

Kriegel, H.-P., Kröger, P., Schubert, E., & Zimek, A. (2009). Outlier Detection in Axis-Parallel Subspaces of High Dimensional Data BT - Advances in Knowledge Discovery and Data Mining (pp. 831−838). Springer Berlin Heidelberg.

Lazarevic, A., & Kumar, V. (2005). Feature bagging for outlier detection. *Proceedings of the Eleventh ACM SIGKDD International Conference on Knowledge Discovery in Data Mining*, 157−166.

Lee, J., Davari, H., Singh, J., & Pandhare, V. (2018). Industrial artificial intelligence for industry 4.0-based manufacturing systems. *Manufacturing Letters, 18*, 20−23. Available from https://doi.org/10.1016/j.mfglet.2018.09.002.

Lee, J., Lapira, E., Bagheri, B., & Kao, H. (2013). Recent advances and trends in predictive manufacturing systems in big data environment. *Manufacturing Letters, 1*(1), 38−41. Available from https://doi.org/10.1016/j.mfglet.2013.09.005.

Li, B., Hou, B., Yu, W., Lu, X., & Yang, C. (2017). Applications of artificial intelligence in intelligent manufacturing: a review. *Frontiers of Information Technology & Electronic Engineering, 18*(1), 86−96. Available from https://doi.org/10.1631/FITEE.1601885.

Lopez, F., Saez, M., Shao, Y., Balta, E. C., Moyne, J., Mao, Z. M., Barton, K., & Tilbury, D. (2017). Categorization of anomalies in smart manufacturing systems to support the selection of detection mechanisms. *IEEE Robotics and Automation Letters, 2* (4), 1885−1892.

Lu, Y., Morris, K. C., & Frechette, S. (2016). Current standards landscape for smart manufacturing systems. *National Institute of Standards and Technology, 8107*(3).

Maes, M. J., Jones, K. E., Toledano, M. B., & Milligan, B. (2019). Mapping synergies and trade-offs between urban ecosystems and the sustainable development goals. *Environmental Science & Policy, 93*, 181−188.

Mitchell, T. M. (1997). *Machine learning* (1). New York: McGraw-hill.

Ortiz, O., Castells, F., & Sonnemann, G. (2009). Sustainability in the construction industry: A review of recent developments based on LCA. *Construction and Building Materials, 23*(1), 28−39.

Pagell, M., & Shevchenko, A. (2014). Why research in sustainable supply chain management should have no future. *Journal of Supply Chain Management, 50*(1), 44−55.

Raihan, A. S., & Ahmed, I. (2023a). A Bi-LSTM autoencoder framework for anomaly detection − A case study of a wind power dataset. *arXiv preprint arXiv, 2303*, 09703. Available from https://doi.org/10.48550/arXiv.2303.09703.

Raihan, A.S., & Ahmed, I. (2023b). Guiding the sequential experiments in autonomous experimentation platforms through EI-based bayesian optimization and bayesian model averaging. arXiv:2302.13360. Available from https://doi.org/10.48550/arXiv.2302.13360.

Raihan, A. S., Ali, S. M., Roy, S., Das, M., Kabir, G., & Paul, S. K. (2022). Integrated model for soft drink industry supply chain risk assessment: Implications for sustainability in emerging economies. *International Journal of Fuzzy Systems, 24*(2), 1148−1169. Available from https://doi.org/10.1007/s40815-020-01039-w.

Raj, A., Dwivedi, G., Sharma, A., Lopes de Sousa Jabbour, A. B., & Rajak, S. (2020). Barriers to the adoption of industry 4.0 technologies in the manufacturing sector: An inter-country comparative perspective. *International Journal of Production Economics, 224*, 107546. Available from https://doi.org/10.1016/j.ijpe.2019.107546.

Rawat, S., Rawat, A., Kumar, D., & Sabitha, A. S. (2021). Application of machine learning and data visualization techniques for decision support in the insurance sector. *International Journal of Information Management Data Insights, 1*(2), 100012. Available from https://doi.org/10.1016/j.jjimei.2021.100012.

Roy, S., Das, M., Mithun, S., Shoyeb, A., Kumar, S., & Kabir, G. (2020). Evaluating strategies for environmental sustainability in a supply chain of an emerging economy. *Journal of Cleaner Production, 262*, 121389. Available from https://doi.org/10.1016/j.jclepro.2020.121389.

Samuel, A. L. (2000). Some studies in machine learning using the game of checkers. *IBM Journal of Research and Development, 44*(1.2), 206−226.

Sen, P. C., Hajra, M., & Ghosh, M. (2020). Supervised classification algorithms in machine learning: A survey and review. *Emerging Technology in Modelling and Graphics: Proceedings of IEM Graph, 2018*, 99−111.

Shrestha, A., & Mahmood, A. (2019). Review of deep learning algorithms and architectures. *IEEE Access, 7*, 53040−53065. Available from https://doi.org/10.1109/ACCESS.2019.2912200.

Srivastava, S. K. (2007). Green supply-chain management: A state-of-the-art literature review. *International Journal of Management Reviews*, *9*(1), 53–80.

van Otterlo, M., & Wiering, M. (2012). *Reinforcement learning and markov decision processes. Reinforcement learning: State-of-the-art* (pp. 3–42). Springer Berlin Heidelberg.

Wang, K.-S., Li, Z., Braaten, J., & Yu, Q. (2015). Interpretation and compensation of backlash error data in machine centers for intelligent predictive maintenance using ANNs. *Advances in Manufacturing*, *3*(2), 97–104. Available from https://doi.org/10.1007/s40436-015-0107-4.

Wuest, T., Weimer, D., Irgens, C., & Thoben, K. D. (2016). Machine learning in manufacturing: Advantages, challenges, and applications. *Production and Manufacturing Research*, *4*(1), 23–45. Available from https://doi.org/10.1080/21693277.2016.1192517.

Zheng, S., Lu, J., Ghasemzadeh, N., Hayek, S., Quyyumi, A., & Wang, F. (2017). Effective information extraction framework for heterogeneous clinical reports using online machine learning and controlled vocabularies. *JMIR Medical Informatics*, *5*(2), e7235. Available from https://doi.org/10.2196/medinform.7235.

CHAPTER SEVEN

Intelligent sustainable infrastructure for procurement and distribution

Pawan Whig[1], Sandeep Kautish[2,3], Rahul Reddy Nadikattu[4] and Yusuf Jibrin Alkali[5]

[1]Vivekananda Institute of Professional Studies, New Delhi, India
[2]LBEF Campus (APU Malaysia) Kathmandu, Nepal
[3]Model Institute of Engineering and Technology, Jammu, Jammu and Kashmir, India
[4]Department of IT, University of Cumbersome, Williamsburg, KY, United States
[5]Federal Inland Revenue Service, Abuja, Nigeria

7.1 Introduction

Metropolitan infrastructure systems are vital to the smooth operation of urban regions, but they also have enormous financial costs, human consequences, and contentious shape and function. Thus, it is vital and frequently disputed how these systems are defined, used, and assessed within industrial societies (Whig et al., 2022d). The things that are built—and the way they are built—clearly show the goals and values of a society, and they will be crucial in deciding how well global nations adhere to regional and global sustainability demands and obligations (Alkali et al., 2022).

The language of value is frequently used in infrastructure commissioning and decision-making, both to discuss what is valuable (i.e., worthy) and which values (i.e., principles) are most pertinent in certain settings. A rising corpus of writing has been produced on the importance of infrastructure in sustainability as the connection between sustainability and infrastructure has drawn more attention in academic literature and public debate (Whig et al., 2022g). The word "value" has been employed in this literature to communicate various ideas in a variety of ways (Whig, Kouser, Velu, et al., 2022a).

The possibility of building sustainable infrastructure is arguably most important at the selection phase of the large focus cycle. Authorities, the

Computational Intelligence Techniques for Sustainable Supply Chain Management.
DOI: https://doi.org/10.1016/B978-0-443-18464-2.00015-7

project's instigators, issue public bids at this time to find bidding coalitions that provide the most value for their funds (Whig et al., 2022f). Value for Money (VfM) has a completely new meaning when trying to build sustainable infrastructure as shown in Fig. 7.1.

VfM is one of the four main principles guiding public procurement, along with openness, competition, and fairness. Public procurers must make sure that public spending is properly focused toward choices that maximize VfM for residents and taxpayers because they are the custodians and bursars of public cash. The problem with the conventional method is that VfM is frequently taken to mean the lowest bid (Whig et al., 2022c). The drawback of choosing the lowest-priced offer is that it frequently sacrifices quality, durability, and sustainability, leading to assets that may end up costing more to finance, develop, manage, maintain, and dispose of.

As developers look for every way to save costs, such assets frequently also have more negative effects on the environment and society. This may lead to concessions in safeguarding compliance and even shortcuts in technical design and quality. protocols for environmental, health, and safety considerations, among many other things (Whig et al., 2022e). Making purchasing selections based only on price greatly hinders the potential to

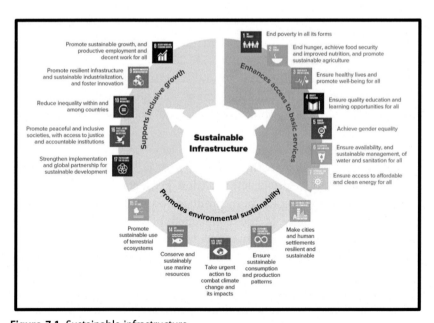

Figure 7.1 Sustainable infrastructure.

use resources that integrate sustainability. environmental design, resource and material efficiency, building techniques, environmental, health, and safety regulations, and many other methods (Alnoaimi & Rahman, 2019; Busari & Ndlovu, 2012). Therefore, it is preferable to base public procurement choices on the total cost of ownership (TCO): solutions that maximize VfM throughout an asset's life cycle rather than only at the time of purchase. TCO refers to procedures that account for all direct and indirect costs related to the acquisition of an asset throughout its life. As a result, it makes it possible for investors and procurers to calculate the asset's overall cost, which includes finance, planning, design, construction, operation, maintenance, and management expenses as well as, if applicable, decommissioning costs (Jupalle et al., 2022).

This justification is the foundation of sustainable public procurement, which bases decisions on how well assets perform in terms of the environment and society as well as how much it costs to plan, design, construct, manage, and maintain them as opposed to just how much it would cost to buy them outright (Tomar et al., 2021).

7.1.1 Sustainable infrastructure

The New Climate Economy predicts that by 2030, sustainable infrastructure will require an investment of $90 trillion worldwide as shown in Fig. 7.2. These investments are necessary to support green economic growth in emerging markets and developing nations, as well as to update outdated machinery in industrialized nations and put it in line with the battle against climate change (Whig et al., 2022b).

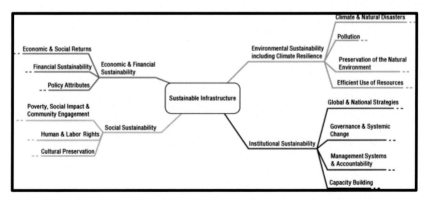

Figure 7.2 Various factors for sustainable infrastructure.

With only 2000 residents, the West Bank settlement of Jerico rose to become the most densely inhabited place on Earth some 9000 years ago (Anand et al., 2022). The number was stunning back then, but it now looks trivial in light of today's enormous megacities like Tokyo, Delhi, and Shanghai (China). The United Nations' (UN) most recent population study indicates that by 2030, there will be more cities and megacities than ever before, with 60% of the world's people living in them (Cass & Mukherjee, 2011; Ibrahim et al., 2019).

Many city residents benefit from opportunity, affluence, and well-being as cities expand, but this expansion also seriously disturbs the balance of the social, economic, and environmental systems (Kattel et al., 2019). For instance, the UN has noted that metropolitan areas, the majority of which are poorly built, devoid of public transportation, and energy-intensive, account for 70% of all greenhouse gas emissions (GHG) (Khera et al., 2021).

An alternative kind of city, one that is more compact, sustainable, and resistant to the consequences of climate change, is yet possible on our planet. These new cities, as envisioned by the UN in its Sustainable Development Goals (SDG9) are based on areas including innovation, research, sustainable infrastructure, and a more inclusive and environmentally responsible industry as shown in Fig. 7.3. These sectors support economic growth and citizen well-being (Nadikattu et al., 2020).

Figure 7.3 Various sustainable goals.

Roads, bridges, telephone pylons, hydroelectric power plants, and other structures that are built based on generally sustainable principles and are intended to provide the public with necessary services are included in the definition of sustainable infrastructure (Kivilä et al., 2017; Lee, 2011; Meng, 2015). This indicates that all aspects of the infrastructure, including those related to the economy, finances, society, and institutions, are ecologically friendly (Whig & Ahmad, 2014).

Urban areas are expanding rapidly, particularly in developing nations, and sustainable infrastructure is proving to be a more effective, fruitful, and ecologically friendly alternative (Su & Lee, 2010; Van Eck & Waltman, 2014; Xue et al., 2018; Zhou & Liu, 2015). The World Bank also claims that these facilities are more lucrative since they improve services, increase resilience to harsh weather, and minimize the impact of natural risks on people and the economy (Shen et al., 2010).

7.1.2 Considerations of sustainable structure

Cities will become more livable and inclusive if the outdated urban infrastructure is replaced with new, contemporary, and sustainable components (Whig & Ahmad, 2012). Over the following ten years, this would necessitate an investment of several trillion dollars worldwide. However, if we act responsibly, it will also put us on the path to economic expansion. The following is a list of the principal benefits of sustainable infrastructure.

7.1.2.1 Lowering our carbon and environmental footprint

According to The New Climate Economy, improved urban design combined with more environmentally friendly infrastructure will save the earth 3.7 gigatons of CO_2 annually over the next 15 years as shown in Fig. 7.4.

1. Quit purchasing water in plastic bottles. Purchase a reusable water bottle, fill it, and carry it with you at all times. Both money and the environment will benefit.
2. Include some of your frequent short-trip locations for biking or walking. You can often do a mile of walking in less than 20 minutes. This is a fantastic way to get some exercise into your hectic day.
3. When not in use, disconnect the electronics and turn off the lights. Every small deed counts!
4. Maintain adequate tire pressure on your car's tires and get frequent tune-ups. Low tire pressure forces your automobile to work harder to

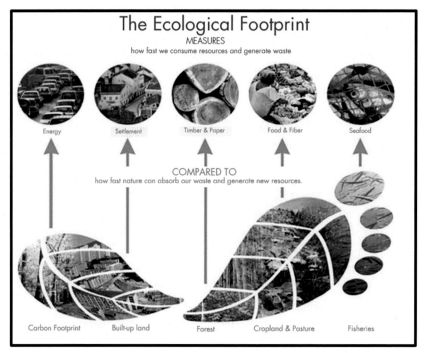

Figure 7.4 Ecological carbon footprint.

go from point A to point B, using more petrol and producing more pollution.

7.1.2.2 Promoting renewable energy

The creation of a decentralized, digital electric infrastructure and the reduction of economic carbon emissions may allow the billion people who lack access to power today (Whig & Ahmad, 2011).

More work than merely implementing decentralized systems is necessary to bring about this dramatic shift. To fully reap the rewards of energy access, it is necessary to invest in an "ecology" that places the variety of people's livelihoods at the center of energy access activities and provides specialized energy solutions, the funding, capacity, and skills, market access, and policy support as shown in Fig. 7.5.

7.1.2.3 Generating green jobs

By 2030, the number of green employment in the renewable energy sector might increase from the current 2.3 to 20 million. Twenty-four

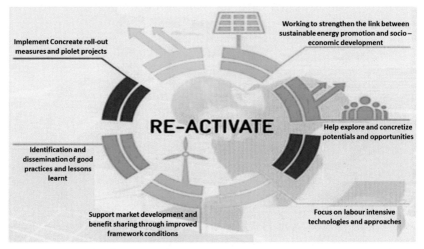

Figure 7.5 Specialized energy solutions

million new employment would be produced globally by 2030, according to International Labor Organization (ILO) research, if sustainable practices are embraced and put into effect.

Millions of people will be able to escape poverty and enjoy a better quality of life as a result, which is in keeping with one of the goals of the Paris Agreement, which highlights the commitment of countries to a just transition and the development of decent employment and green jobs (Velu & Whig, 2021; Whig & Ahmad, 2013). According to the analysis, while there will be some job losses, particularly in the petroleum business, these will be more than compensated by jobs being generated in the renewable energy sector and the shift to a circular economy as shown in Fig. 7.6.

The International Labor Organization (ILO) works to advance and mainstream environmentally sound social and economic development in the developing world, where the poor are most impacted by the effects of climate change. This is done by promoting green jobs and expanding opportunities for income generation. The Green Jobs Programme in rural Zambia enabled impoverished subsistence farmers to live new lives and boost their output. Women learned how to construct homes using green technologies, constructing 18 in 5 months. Many individuals in Zambia never have power since they live off the grid and the country is experiencing an energy crisis. Through the initiative, women were given

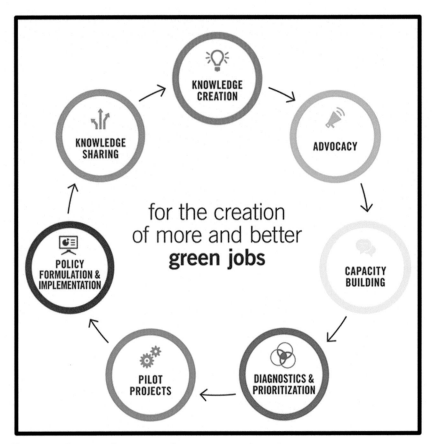

Figure 7.6 Process of green employment.

the skills to assemble and install solar panels, which they then did for the newly constructed homes. These are vital life skills that improve the environment and benefit the entire community.

7.1.2.4 Promoting sustainable economic growth

An important component of the new economy focused on climate and sustainability action is the construction of sustainable infrastructure.

Local governments have several prospects for growth as a result of the green economy. Even though the recession has affected all businesses, spending on green technology, including federal and state financing, has performed better than other industries. Furthermore, between 1998 and 2007, the rise of green occupations outpaced that of all other job types as shown in the flowchart given in Fig. 7.7.

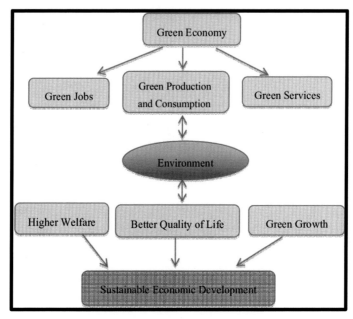

Figure 7.7 Flowchart for sustainable economic development.

The fusion of economic growth, worker development, and environmental care is represented by a green economy. Green economic practices are distinctive in that they encourage workforce and development departments of cities and counties to consider the environment when making decisions, while environmental departments are encouraged to take the economic growth and job creation of their policies into consideration. Both economic growth and environmental protection are benefits of a green economy.

7.1.2.5 Reducing disparities
In contrast to the sustainable alternative, the current infrastructure is not equipped to provide even the most fundamental requirements of developing nations, such as access to running water, sanitary facilities, transportation networks, etc.

7.1.3 Sustainable urban development's importance
Future city expansion should consider the New Urban Agenda's suggestions (NUA). In this strategic plan created by the UN, cities are advised to change via planning, development, governance, and administration

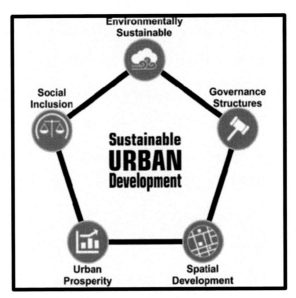

Figure 7.8 Importance of sustainable development.

based on innovations in design, law, and economic and urban policies. The plan also specifies sustainable urban development criteria as shown in Fig. 7.8.

To meet the Paris Agreement's 2050 deadline, sustainable urban development also promotes the economy's progressive transition to a renewable energy model and the decarbonization of the economy. According to the Global Commission on the Economy and Climate of the New Climate Economy, by then sustainable cities would have saved the globe $17 trillion (NCE).

7.2 Participants in the infrastructure workflow

7.2.1 Sponsors

The project company's owners, equity holders, or shareholders, are known as the sponsors. While all sponsors have a stake in the project's success, their reasons for being involved and the extent to which they participate in the various stages of the project life cycle might vary greatly. The ownership structure often stays the same during a project. This is

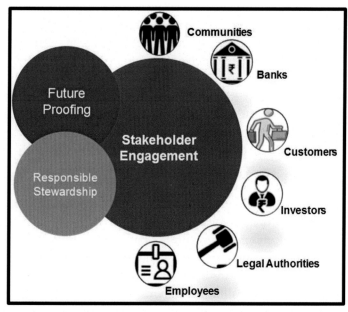

Figure 7.9 Various stakeholders for sustainable development.

especially valid for strategic sponsors, who are essential to the project's success (e.g., developers, government). Frequently, the departure of a strategic sponsor from the project might cause a default. Changes in the project's sole financial sponsors, such as investors and multilateral development banks, are more frequent and have a significantly lesser effect. Various stakeholders for sustainability are shown in Fig. 7.9.

7.2.2 Governments

Governments (national or municipal) commission infrastructure to carry out their economic goals, fulfill their obligations as sovereigns, and reap other significant social and environmental advantages. Because it is the project's major sponsor (also known as the promoter), the government also plays a unique role among sponsors. This responsibility includes starting the tender, defining what requirements and criteria the bidders (i.e., developers) must fulfill during the bidding process, and selecting the winning bidder.

7.2.3 Banks for development

By contributing risk capital (by becoming sponsors) and other credit enhancement strategies, development banks (both national and international) take part in infrastructure agreements to influence development in their target areas and nations. Development banks (DBs) might actively participate in the project's design and execution by providing technical help, particularly if the host nation lacks the requisite resources to create bankable transactions. However, during the operating phase, DBs often play a supporting role. They often have an exit strategy in place at the time of investing, so they may sell their interest to the other sponsors or through an IPO and use their funds for other initiatives.

7.2.4 Equity traders

Private equity funds, sovereign wealth funds, insurance corporations, banks, and pension funds are examples of equity investors. These institutional investors are among those with the knowledge and risk tolerance to invest in infrastructure projects. Their engagement in the asset's design, development, and operation is minimal even if they are sponsors. Finding bankable, financially appealing agreements is their only goal to increase their risk-adjusted return and pay for their long-term responsibilities. In addition, put another way, they invest money if the project's internal rate of return is favorable for the project's inherent risks and is consistent with their investment mission.

7.2.5 Developers

Developers include building firms and other service providers who are in charge of building, running, and maintaining the asset. In essence, they are the bidders or a significant participant throughout the tendering procedure, a bidding consortium. Depending on how the project is set up, their precise role may vary greatly. By the conventional procurement approach, these businesses may only be accountable for the construction and prompt asset delivery.

However, under public-private partnership (PPP) agreements, their duties may also extend to the asset's engineering design, finance, operation, and maintenance. Developers can provide equity funding and take on the role of sponsor, demonstrating their long-term dedication to the undertaking. This might be a crucial factor for other investors to take into account when determining the project's viability, particularly if the underlying technological and construction risks are extremely significant and would raise the total cost of funding.

Engineering-procurement-construction (EPC) contractual agreements are often used by contractors. In this situation, the developer assumes liability for the asset's design, the acquisition of necessary materials, and turn-key construction (i.e., taking on the construction risk by being financially liable for any construction delays and cost overruns).

7.2.6 Specialized automobiles

When a project financing structure is employed, a special purpose vehicle (SPV) specifically intended for the project must be established. The project company, or SPV, is the legal entity that obtains non-recourse loans from lenders and investors. It is in charge of project management and also owns the asset. The sponsors mentioned above are SPV's shareholders. When borrowing without recourse, the SPV's assets and projected future cash flows act as the only form of security. The sponsors are not legally obligated to reimburse any possible losses suffered by the SPV beyond their equity investment. Various stakeholder engagement is shown in the flowchart given in Fig. 7.10.

7.2.7 Debt traders

Banks continue to be the primary financiers of infrastructure, but other institutional investors and MDBs are becoming more significant players as well. As they rank higher in the capital structure (i.e., are more senior) than equity, debt investments (including loans and project bonds) can offer steady long-term returns and are thus safer than equity if the project defaults. Even less than sponsors, private lenders are involved in the planning and execution of the project. They have a more indirect impact on project design since the project must be designed such that debt investors will find it bankable and have enough equity to cover any possible losses.

7.3 Case study

7.3.1 Stakeholder responsibilities and duties in the implementation of sustainable procurement

Any infrastructure project must successfully manage social and environmental hazards. Non-compliance with regional and global standards might have a major financial impact at the time of not only during construction

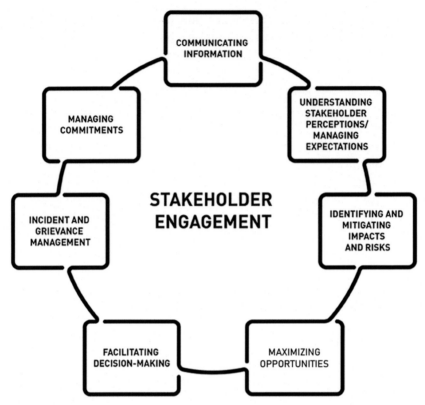

Figure 7.10 Various stakeholder engagement flowcharts.

but throughout the asset's life. The respect for pertinent laws and the reduction of the asset's sustainability effect is, in fact, in the best interests of all parties involved. The World Bank forecasts that the advantages may offset the potential rise in project capital costs. The World Bank included greater risk management for the environment, society, health, and safety as an environmental advantage, along with better community and government connections, financial availability, increased reputation, and brand awareness value, as well as the prospective market for increased sustainability. Relevant social advantages, on the other hand, include a reduction in accidents or health damages, increased population safety standards, enhanced developmental chances, protection of use rights from common property resources, and improved livelihood prospects (Figs. 7.11 and 7.12).

Procurement Framework's **Core Procurement Principles** provided the opportunity to modernize procurement of MDI equipment

Identified a need for customized buyer's guidance on how to procure MDI equipment

Applying the tools, techniques and approaches in the Procurement Framework to support and optimize procurement solutions and implementation

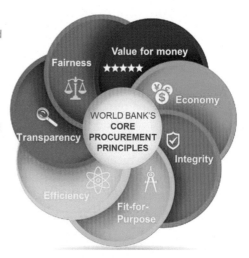

Figure 7.11 Sustainable procurement principles.

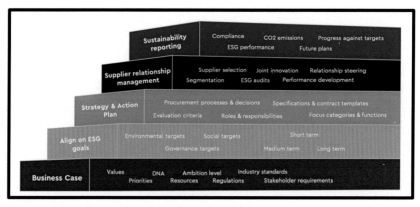

Figure 7.12 Sustainable procurement policy.

7.3.2 The government's role

To promote effective, efficient, and sustainable infrastructure development, governments should constantly strive to plan infrastructure projects that produce the maximum VfM over the life cycle. VfM comprises all expenses, revenues, and associated risks and uncertainties that occur during the various stages of the project: design, construction, operation and maintenance, and asset disposal and decommissioning. When planning and implementing infrastructure projects, governments, and their procurement agencies have a mission to maximize VfM for taxpayers.

7.3.3 The function of development banks

DBS (both national and international) was established to give financial and technical aid to promote development, with a heavy emphasis on infrastructure. They have also taken the lead since 2015 in carrying out the sustainable development goals multilateral development banks (MDBs) have adopted environmental and social safeguard policies. The cornerstones of these protection programs are land acquisition and environmental impact assessments. The World Bank amended its safeguard policies, as well as its procurement regulations and guidelines, in 2016. The latter's changes are centered on sustainable public procurement, VfM, and fitness-for-purpose (World Bank, 2016a). Many other MDBs are also implementing similar programs.

7.3.4 The role of debt and equity investors

Because project prices are sensitive to environmental risk, compliance with safeguards is vital to both stock and loan investors. However, as the asset's owners, equity investors' interests go beyond the minimal compliance with safeguards and may undoubtedly involve exploring ways to reduce the asset's total environmental imprint. The World Bank's environmental and social protection policies are acknowledged as the approved basis for significant international projects. A handful of investors will no longer support initiatives that do not match these criteria.

The Equator Principles are another important risk management methodology (EP). Financial institutions utilize it to determine, analyze, and manage environmental and social risk in projects, as well as to provide a standard for due diligence to assist responsible investment decision-making. EP is particularly essential for infrastructure because it is intended for use in project finance and associated activities. The EP has been accepted by 84 financial institutions globally, including the largest participants in project finance.

7.3.5 The developer's role

As previously noted, developers can play various roles in the project, determining how much influence and interest they have in the asset satisfying local and international environmental standards. If developers are simply responsible for the construction and do not become sponsors, their obligations are restricted to following the contract agreement's environmental criteria.

In this instance, the project company (i.e., the SPV) is in charge of overseeing the asset's long-term performance.

7.3.6 The special purpose vehicle's function

The project company, or SPV, is in charge of monitoring the project's compliance with environmental and social safeguards, as stated in the asset's design requirements throughout the procurement process. However, the SPV's monitoring obligations extend beyond the building phase to the operational phase, when it must detect and handle any unanticipated concerns until the asset's life cycle is completed. Because environmental and social risks can have a significant influence on the project's overall financial viability, the SPV's goal is to assure compliance and establish internal rules that reduce the asset's footprint.

7.4 Conclusion

Sourcing is critical in the deployment of sustainable infrastructure. It captures the moment when governments go to the market to declare their desire to acquire and deploy. It encompasses all of the methods by which the government communicates with markets to tell them what they want to buy and how they intend to maximize the best VfM for the public purse. Given the large quantities spent on infrastructure, the public procurement process represents a significant opportunity to guide whole value chains toward sustainable growth. Authorities, however, need to become aware of this potential and duty. Sustainable building cannot be achieved unless they design the acquisition process to demand green assets, provide the legal and regulatory frameworks to enable the supply of sustainable infrastructure financially possible, and offer incentives and rewards to creative front-runners.

Governments need to actively pursue VfM across the asset life cycles as they become more aware of this potential and duty rather than merely managing projects and selecting the lowest bidder. The fact that many governments have failed to recognize the commercial case for sustainable public procurement creates further difficulties. The civil service as a whole, the policy and procurement cadre, and the laws themselves all require significant improvement. The writers have also noted that even when legislators support the idea, turning that support into actual action is still quite difficult.

The durability of the assets that financiers invest in will eventually define their risk and reward profiles, thus they have an important role to play. They have a lot of power over governments since they are the ones that give the capital. The challenge for bankers is to utilize their collective power to demand predictability and stability in the application of laws and practices relating to the transition to a green economy and, more broadly, the achievement of Sustainable Development Goals. This will perhaps be the most effective catalyst for sustainable infrastructure that generates VfM for both investors and residents.

7.5 Future scope

Project management in sustainable infrastructure projects is now one of the burgeoning academic topics as a result of the necessity to produce sustainable projects. Previous research indicates that to improve the sustainability of infrastructure projects, new project management methodologies must be developed. Although developed countries like the US, the UK, and Australia are dominating this field of study, the review revealed that developing nations like China are becoming increasingly interested in this subject. Community impact, sustainability measurements, processes, and factors, as well as the sustainability triangle made up of cost, assessment, and factor, have been recognized as the main study areas in this area.

References

Alnoaimi, A., & Rahman, A. (2019). Sustainability assessment of sewerage infrastructure projects: A conceptual framework. *International Journal of Environmental Science and Development, 10*(1).

Alkali, Y., Routray, I., & Whig, P. (2022). Strategy for reliable, efficient, and secure IoT using artificial intelligence. *IUP Journal of Computer Sciences, 16*(2).

Anand, M., Velu, A., & Whig, P. (2022). Prediction of loan behaviour with machine learning models for secure banking. *Journal of Computer Science and Engineering (JCSE), 3*(1), 1−13.

Busari, O., & Ndlovu, J. (2012). Leveraging water infrastructure for transformative social-economic development in South Africa. *WIT Transactions on Ecology and the Environment, 162*, 435−446.

Cass, D., & Mukherjee, A. (2011). Calculation of greenhouse gas emissions for highway construction operations by using a hybrid life-cycle assessment approach: A case study for pavement operations. *Journal of Construction Engineering and Management*, *137*(11), 1015–1025.

Ibrahim, M. N., Thorpe, D., & Mahmood, M. N. (2019). Risk factors affecting the ability for earned value management to accurately assess the performance of infrastructure projects in Australia. *Construction Innovation*.

Jupalle, H., Kouser, S., Bhatia, A. B., Alam, N., Nadikattu, R. R., & Whig, P. (2022). Automation of human behaviors and their prediction using machine learning. *Microsystem Technologies*, *28*, 1879–1887.

Kattel, G. R., et al. (2019). China's south-to-north water diversion project empowers sustainable water resources system in the North. *Sustainability*, *11*(13), 3735.

Khera, Y., Whig, P., & Velu, A. (2021). efficient effective and secured electronic billing system using AI. *Vivekananda Journal of Research*, *10*, 53–60.

Kivilä, J., Martino, M., & Vuorinen, L. (2017). Sustainable project management through project control in infrastructure projects. *International Journal of Project Management*, *35*(6), 1167–1183.

Lee, J. C., et al. (2011). Evaluation of variables affecting sustainable highway design with BE2ST-in-Highways system. *Transportation Research Record*, *2233*(1), 178–186.

Meng, J., et al. (2015). Relationships between top managers' leadership and infrastructure sustainability: A Chinese urbanization perspective. *Engineering, Construction and Architectural Management*, *22*(6), 692–714.

Nadikattu, R.R., Mohammad, S.M., & Whig, P. (2020). Novel economical social distancing smart device for covid-19. *International Journal of Electrical Engineering and Technology (IJEET)*.

Shen, L.-Y., et al. (2010). Project feasibility study: The key to the successful implementation of sustainable and socially responsible construction management practice. *Journal of Cleaner Production*, *18*(3), 254–259.

Su, H.-N., & Lee, P.-C. (2010). Mapping knowledge structure by keyword co-occurrence: A first look at journal papers in Technology Foresight. *Scientometrics*, *85*(1), 65–79.

Tomar, U., Chakroborty, N., Sharma, H., & Whig, P. (2021). AI-based smart agriculture system. *Transactions on Latest Trends in Artificial Intelligence*, *2*(2).

Van Eck, N. J., & Waltman, L. (2014). *Visualizing bibliometric networks. Measuring scholarly impact* (pp. 285–320). Springer.

Velu, A., & Whig, P. (2021). Protect personal privacy and wasting time using NLP: A comparative approach using AI. *Vivekananda Journal of Research*, *10*, 42–52.

Whig, P., & Ahmad, S. N. (2011). On the performance of ISFET-based device for water quality monitoring. *International Journal of Communications, Network and System Sciences*, *4*(11), 709.

Whig, P., & Ahmad, S. N. (2012). A CMOS-integrated CC-ISFET device for water quality monitoring. *International Journal of Computer Science Issues*, *9*(4), 1694–1814.

Whig, P., & Ahmad, S. N. (2013). A novel pseudo-PMOS integrated ISFET device for water quality monitoring. *Active and Passive Electronic Components*, *2013*.

Whig, P., & Ahmad, S.N. (2014). Simulation of a linear dynamic macro model of the photocatalytic sensor in SPICE. In: *COMPEL The International Journal for Computation and Mathematics in Electrical and Electronic Engineering*.

Whig, P., Kouser, S., Velu, A., & Nadikattu, R. R. (2022a). *Fog-IoT-assisted-based smart agriculture application. Demystifying federated learning for blockchain and industrial internet of things* (pp. 74–93). IGI Global.

Whig, P., Nadikattu, R. R., & Velu, A. (2022b). *COVID-19 pandemic analysis using the application of AI, . Healthcare monitoring and data analysis using IoT: Technologies and applications* (1). IET,.

Whig, P., Velu, A., & Bhatia, A. B. (2022c). *Protect nature and reduce the carbon footprint with an application of blockchain for IoT. Demystifying federated learning for blockchain and industrial internet of things* (pp. 123−142). IGI Global.

Whig, P., Velu, A., & Naddikatu, R. R. (2022d). *The economic impact of AI-enabled blockchain in 6G-based industry. AI and blockchain technology in 6G wireless network* (pp. 205−224). Singapore: Springer.

Whig, P., Velu, A., & Nadikattu, R. R. (2022e). *Blockchain platform to resolve security issues in IoT and smart networks. AI-enabled agile internet of things for sustainable FinTech ecosystems* (pp. 46−65). IGI Global.

Whig, P., Velu, A., & Ready, R. (2022f). *Demystifying federated learning in artificial intelligence with human-computer interaction. Demystifying federated learning for blockchain and industrial internet of things* (pp. 94−122). IGI Global.

Whig, P., Velu, A., & Sharma, P. (2022g). *Demystifying federated learning for blockchain: A case study. Demystifying federated learning for blockchain and industrial internet of things* (pp. 143−165). IGI Global.

Xue, B., Liu, B., & Sun, T. (2018). What matters in achieving infrastructure sustainability through project management practices: A preliminary study of critical factors. *Sustainability*, *10*(12), 4421.

Zhou, J., & Liu, Y. (2015). The method and index of sustainability assessment of infrastructure projects based on system dynamics in China. *Journal of Industrial Engineering and Management*, *8*(3), 1002−1019.

CHAPTER EIGHT

Machine learning techniques for route optimizations and logistics management description

A. Reyana[1] and Sandeep Kautish[2,3]

[1]Department of Computer Science and Engineering, Karunya Institute of Technology and Sciences, Coimbatore, Tamilnadu, India
[2]LBEF Campus (APU Malaysia) Kathmandu, Nepal
[3]Model Institute of Engineering and Technology, Jammu, Jammu and Kashmir, India

8.1 Introduction

With the development of new technologies, the volume, delivery, and customer expectation in supply chain management has increased—for example, logistics in real-time threat for Amazon Prime. On ordering a product, it moves from the warehouse shelf to the truck and reaches the customer's door. Gaining customer satisfaction is a challenge. Getting into rural delivery is another challenge. However, minimizing logistics complications in improving customer-demanding service is necessary.

In an age when internet shopping is the norm, last-mile delivery is more vital than ever. Fortunately, artificial intelligence (AI) is redefining logistics management with last-mile logistics and cost reduction. Considering the last-mile delivery, the rising cost is one of the significant issues for businesses. Customer needs to have an interest in paying additional expenses. Another area for improvement is meeting the delivery deadline due to shipping delays, traffic congestion, etc. In extreme cases, the customer cancels the delivery. Today customer tracks their shipment, and regular update is a real-time challenge. Another is route planning and optimization. For the above challenges, AI use in business can fulfill the expectations of the customer to a large extent.

Similarly, another rapidly expanding industry is grocery delivery. Grocery delivery orders in the United States are predicted to increase by 20% annually until the 2020s. Because of the growing demand for food

Computational Intelligence Techniques for Sustainable Supply Chain Management.
DOI: https://doi.org/10.1016/B978-0-443-18464-2.00010-8

delivery, last-mile logistics service providers are under tremendous pressure to enhance efficiency and cut prices. Delivery demand varies greatly, and it is necessary to manage timely delivery (Ding et al., 2021). ML algorithms can improve the complex process by predicting customer preferences and planning routes accordingly.

- This chapter discusses how ML technology can support the optimization of routes and schedules.
- It discusses the recent developments in logistics and their applications.
- Briefs out the usage of AI in the logistics industry for route planning and optimization.
- The role of ML in intelligent transportation, warehouse management, and a case study on last-mile delivery with a real-time scenario and its results is herewith discussed.

Last-mile logistics is a complicated procedure that leads to the high cost of last-mile delivery. Furthermore, because of fuel and labor costs, the last mile is sometimes the maximum cost component in the service delivery process (Deng, 2018). When planning shipments, it is necessary to consider the logistical expenses of missing deliveries, returns, and repatriation. These considerations make optimizing low-cost last-mile deliveries challenging. The rise of e-commerce has made last-mile deliveries even more difficult. As online sales expand, so does the demand for rapid and accurate delivery (Zhao et al., 2019). Customers anticipate their delivery immediately after their purchase.

8.2 Development in logistics

Woschank et al. (2020) explored the ongoing evolution of micro- and macro-economic entities in the context of automation and digitalization, with a particular focus on the significant role of AI in the realm of smart logistics. Their research, which encompassed studies conducted between 2014 and 2019, revealed a widespread adoption of AI within the logistics sector. The rapid advancements in this field have propelled AI to the forefront, ushering in a new era of possibilities in logistics and transportation. Consequently, the integration of AI in logistics has led to cost reductions, reduced emissions, and faster delivery times, all while maintaining the quality of supply chain services through the intelligent allocation of resources and efficient transport flow coordination. Liu et al.

(2020) introduced a forward-thinking E-commerce business model called "Cloud Laundry," which harnessed the power of the Internet of Things (IoT) to revolutionize large-scale laundry services. This innovative model leveraged big data analytics, intelligent logistics management, and ML techniques to optimize various aspects of the laundry service. It calculated the most efficient transportation routes and provided real-time updates and re-routing options for logistic terminals, making it an intelligent and dynamically responsive solution for laundry needs. Cloud Laundry enabled smartphone-based control by capitalizing on the rapid growth of the big data industry, user interest modeling, and prioritizing information security and privacy concerns. It utilized sophisticated big data models for safeguarding customer security requirements. This approach departed from traditional laundry services, resulting in higher capital turnover, increased liquidity, and enhanced profitability for Cloud Laundry companies. Ding et al. (2021) conducted a comprehensive review of contemporary research and practical applications of smart logistics within the IoT framework, encompassing areas such as intelligent freight transportation, warehouse management, and delivery systems. Their examination of publications spanning from 2008 to 2019 shed light on persistent challenges in the evolution of smart logistics. These ongoing issues encompass technical hurdles related to radio frequency identification and wireless sensor networks, limitations in the scalability and technical capabilities of the IoT, concerns regarding IoT standardization, data acquisition and processing challenges, and the critical matter of security and privacy within the IoT landscape. Consequently, there is a pressing need for robust research efforts aimed at tackling these IoT-related challenges head-on. Tang and Meng (2021) introduced a groundbreaking fusion structure that combines data analytics and optimization within the realm of smart industrial engineering. This model, meticulously crafted through theoretical and technological research, establishes a four-level framework for smart industry. Within this framework, data analytics plays a pivotal role in perceiving and analyzing industrial production and logistics processes. Furthermore, it showcases its intelligent capabilities in the domains of planning, scheduling, operation optimization, and optimal control. The model addresses critical challenges that frequently afflict manufacturing, resource management, energy consumption, and logistics systems by seamlessly integrating data analytics and system optimization technologies across these four levels. These challenges encompass issues such as high energy consumption, elevated costs, low energy efficiency, underutilized resources, and environmental pollution. The fusion of data analytics and optimization empowers enterprises to enhance their predictive and control capabilities in uncharted territories, unearthing hidden insights to bolster decision-making efficiency. Consequently, the significance of industrial intelligence in the context of industrial advancement cannot be overstated. Sulova (2021) conducted a comprehensive analysis of the technological facets underpinning the digital transformation in logistics. Furthermore, Sulova put forth a conceptual framework tailored for efficient big data management within the logistics sector. This framework harnessed the power to process structured, semi-structured, and unstructured data, aiming to optimize the logistics industry. The result was an enhanced capacity to predict and respond to events, perform real-time route optimization, preempt unforeseen delivery delays, introduce innovative solutions, and facilitate a seamless adaptation to the digitalization trends reshaping the industry. Lee et al. (2020) pioneered a traffic control system driven by ML predictions. They also devised a dynamic routing approach that autonomously selected routes for vehicle systems, effectively reducing congestion rates. This system exhibited the ability to anticipate traffic jams in critical areas and underwent a rigorous evaluation involving the assessment of four widely used algorithms. The outcomes of this research demonstrated that incorporating ML into traffic control can significantly improve the efficiency and performance of existing autonomous vehicle systems.

8.2.1 Intelligent traffic systems

The development of intelligent transportation systems (ITS) merges sensor technology and vehicle connectivity to monitor traffic conditions in real time. Vehicle distribution, transportation coordination, and interactive information sharing are all aspects of intelligent transportation in logistics. Communication technologies such as IEEE 802.11n Wi-Fi and cellular-based LTE are currently developed for transmission.

8.2.2 Internet of things-enabled logistics

Operations transformation could result in determinants such as viability, networks, and agency. IT-enabled skills improve product and process efficiency, which aids knowledge exploitation. Any organization, especially those involved in supply chains, currently requires faster transactions and processing. RFID and other IoT solutions are commonly employed. Firms' incremental innovation skills at all levels remain a challenge in the context of the transformation of Industry 4.0 (Lu et al., 2020). With autonomous coordination, the IoT delivers a paradigm shift in logistics. However, decision-making in an IoT setting necessitates the development of new tools. Logistics information may be extracted automatically using barcode technology and electronic tag systems. Both wired and wireless transmissions are used to access logistics processes. Real-time communication, processing, and response are now possible because of rapid advancements in information technology (James et al., 2019).

Furthermore, modern logistics technology links logistics activities with mechanization, automation, and information technology. Unpredictable events can create delays at times. Current conditions and travel plans can be changed using real-time vehicle location tracking. As a result, unnecessary delays are avoided. Logistics 4.0 meets individualized consumer expectations without rising costs and assists by boosting relief operations performance.

8.2.3 Travel time prediction

It entails gathering data on road users and traffic conditions. Because of the varying traffic flow circumstances, predicting travel time is difficult. A wide range of methodologies have been used to study traffic. However, there are still concerns about the predictability of travel times. The components of travel time variation include regular, irregular, and random fluctuations. Predictive analytics classifies short-term traffic conditions. Regular variations provide information on remarkably similar traits that can be used to support predictions. K-nearest neighbor, multi-step forecasting, and other techniques are used. However, when dealing with large datasets, these are computationally expensive.

8.3 Applications of artificial intelligence in transport

AI can potentially revolutionize the transport industry by improving efficiency, safety, and sustainability (Abduljabbar et al., 2019). Here are some applications of AI in transport:

8.3.1 Autonomous vehicles

AI enables self-driving cars, trucks, and buses to navigate roads and make decisions without human intervention. AI-powered sensors, computer vision, and ML algorithms allow vehicles to perceive their surroundings, interpret traffic patterns, and respond accordingly. Autonomous vehicles have the potential to reduce accidents, optimize traffic flow, and enhance mobility for people with limited mobility.

8.3.2 Traffic management

AI can optimize traffic flow by analyzing real-time data from sensors, cameras, and connected vehicles. ML algorithms can predict traffic patterns, congestion, and accidents, enabling authorities to make informed decisions about signal timing, lane management, and route planning (Abduljabbar et al., 2019). This helps reduce travel times, fuel consumption, and greenhouse gas emissions.

8.3.3 Intelligent transportation systems

AI can enhance the overall efficiency of transportation systems. AI algorithms can analyze data from various sources, such as weather conditions, public transportation schedules, and historical traffic data, to provide real-time information to travelers. This includes predicting arrival times, suggesting alternate routes, and notifying passengers about delays or disruptions.

8.3.4 Predictive maintenance

AI can monitor the condition of vehicles and predict maintenance needs. By analyzing sensor data and historical performance, AI algorithms can detect potential faults or failures before they occur. This allows for proactive maintenance, reducing downtime, and optimizing the lifecycle of vehicles, thereby improving operational efficiency.

8.3.5 Smart infrastructure

AI can be used to optimize the operation of infrastructure elements such as traffic lights, toll booths, and parking systems. By analyzing real-time data and traffic patterns, AI algorithms can adjust traffic signal timings, optimize toll collection processes, and provide real-time parking availability information to drivers. This reduces congestion, improves traffic flow, and enhances user experience.

8.3.6 Supply chain management

AI can optimize logistics and supply chain operations. ML algorithms can analyze data from various sources, including customer demand, inventory levels, and transportation networks, to optimize delivery routes, minimize transportation costs, and improve inventory management (Abduljabbar et al., 2019). This helps reduce delays, improve customer satisfaction, and increase overall operational efficiency.

8.3.7 Risk assessment and safety

AI can enhance safety in transportation by predicting and preventing accidents. By analyzing historical data and real-time inputs, AI algorithms can identify high-risk areas, predict potential hazards, and suggest proactive safety measures. AI can monitor driver behavior and provide real-time feedback to promote safer driving practices.

These are some applications where AI is being applied in the transport industry. As technology advances, AI is expected to play an increasingly important role in transforming the travel and transportation of goods. Therefore, transport systems appear to be a strong fit for AI. The continual increase in rural and urban transportation exacerbates the difficulty. AI approaches are more adaptive to achieve a more reliable transportation system.

8.4 Benefits of artificial intelligence usage in the logistics industry

The logistics industry is the backbone of global trade and is vital in ensuring the efficient movement of goods from manufacturers to consumers. In recent years, the application of AI has emerged as a game-changer in the logistics sector. By harnessing the power of AI, logistics companies can revolutionize their operations, enhance efficiency, and deliver superior customer experiences (Janarden, 2017). The significant benefits of AI usage in the logistics industry include:

8.4.1 Enhanced supply chain visibility

AI enables logistics companies to achieve real-time visibility and control over their supply chains. Through advanced data analytics and ML algorithms, AI can analyze vast amounts of data from multiple sources, such as

GPS trackers, RFID tags, and sensors, to provide accurate and up-to-date information about shipment location, status, and condition. This enhanced visibility allows for proactive decision-making, timely interventions, and improved coordination across the supply chain.

8.4.2 Optimized route planning and resource allocation

AI algorithms lead to cost savings and increased efficiency. By considering factors such as traffic conditions, weather forecasts, historical data, and delivery constraints, AI can determine the most optimal routes for transportation, reducing fuel consumption, travel time, and vehicle wear and tear (James et al., 2019). Additionally, AI can optimize the allocation of resources, such as trucks, drivers, and warehouses, ensuring efficient utilization and reducing idle time and unnecessary expenses.

8.4.3 Predictive maintenance and reduced downtime

AI-powered predictive maintenance can help logistics companies identify and address maintenance issues before they cause equipment failures or disruptions. AI algorithms can detect patterns and anomalies by analyzing sensor data, historical performance, and maintenance records, allowing for timely maintenance interventions (Janarden, 2017). This proactive approach reduces downtime, increases equipment reliability, and minimizes the risk of unexpected breakdowns, thereby improving operational efficiency and customer satisfaction.

8.4.4 Efficient inventory management

AI can optimize inventory management processes by providing accurate demand forecasting, stock level optimization, and dynamic replenishment strategies. By analyzing historical sales data, market trends, and external factors, AI algorithms can predict future demand patterns more accurately. This enables logistics companies to optimize inventory levels, reduce stockouts, and avoid overstocking, resulting in cost savings, improved order fulfillment rates, and enhanced customer satisfaction.

8.4.5 Intelligent warehouse operations

AI-powered automation and robotics can significantly enhance warehouse operations. AI algorithms can optimize storage layouts, predict demand patterns, and automate inventory management processes. Robots with AI capabilities can efficiently pick, pack, and sort goods, reducing human errors, improving speed and accuracy, and enabling 24/7 operations. Furthermore, AI-powered warehouse management systems can track inventory, manage workflows, and optimize labor allocation, resulting in increased productivity and reduced operational costs.

8.4.6 Risk management and security

AI can be crucial in risk management and security within the logistics industry. By analyzing data from various sources, including historical incident records, weather patterns, and real-time inputs, AI algorithms can identify potential risks and develop proactive strategies to mitigate them. AI-powered security systems can monitor facilities, vehicles, and personnel, detecting anomalies and unauthorized activities. This enhances safety and security, protects valuable cargo, and reduces the risk of theft, damage, and accidents. Thus the benefits of AI usage in the logistics industry are undeniable.

8.5 Role of artificial intelligence in route planning and route optimization

In today's fast-paced world, efficient route planning and optimization are crucial for various industries, including logistics, transportation, and delivery services. The ability to find the most optimal routes can significantly impact operational costs, fuel consumption, delivery times, and customer satisfaction. Route optimization, aided by advanced technologies such as AI and Geographic Information Systems (GIS), has become an essential strategy for businesses to maximize efficiency and savings. Route optimization enables businesses to minimize fuel consumption and operational costs (Jiang & Ma, 2020). By considering factors such as distance, traffic patterns, road conditions, and vehicle characteristics, AI algorithms can determine the most efficient routes for transportation. Companies can reduce travel distances and fuel consumption by avoiding congested routes, minimizing detours, and optimizing delivery sequences. This leads to significant cost savings and lower carbon emissions, promoting economic and environmental sustainability.

Also, Efficient route planning allows for faster and more accurate deliveries, improving customer satisfaction. Enhanced customer satisfaction can lead to increased loyalty, positive word-of-mouth, and a competitive edge in the market. Ensuring that resources are utilized efficiently, minimizing idle time, and maximizing productivity. Companies can streamline operations, reduce unnecessary costs, and achieve higher operational efficiency. For businesses with a fleet of vehicles, route optimization plays a vital role in fleet management. AI algorithms can analyze real-time data, including vehicle locations, traffic conditions, and delivery requirements, to make informed decisions on route adjustments and dispatching. By dynamically reassigning vehicles and adapting routes based on changing circumstances, businesses can optimize fleet utilization, minimize empty miles, and improve overall fleet efficiency. This leads to reduced operational costs and increased responsiveness to customer demands. Thus contributing to improved safety and regulatory compliance. AI algorithms can consider factors such as road conditions, speed limits, and hazardous areas to plan routes that ensure driver safety. By avoiding high-risk areas and adhering to regulations. Businesses can reduce the likelihood of accidents, fines, and penalties. Compliance with safety standards also enhances the reputation and trustworthiness of the company. Whether a company is expanding its operations, adding new delivery locations, or facing fluctuating demand, AI algorithms can quickly adapt and optimize routes accordingly. This flexibility allows businesses to respond effectively to market dynamics, maximize operational efficiency, and maintain a high level of service quality. This route optimization is a powerful tool that can significantly impact the efficiency and profitability of businesses in various industries. In an increasingly competitive landscape, businesses prioritizing route optimization gain a competitive edge, achieve sustainability goals, and deliver exceptional customer experiences. Embracing route optimization is a strategic decision that propels businesses toward increased efficiency, savings, and success.

8.6 Machine learning and internet of things in smart transportation

ML and the IoT are two transformative technologies that are vital in enabling ITS. ML leverages algorithms and statistical models to enable machines to learn from data and make predictions or decisions. At the same time, IoT connects various devices and sensors to gather and exchange data over a network (Chen et al., 2021). In the context of intelligent transportation, ML and IoT together create an ecosystem that enhances safety, efficiency, and sustainability. ML algorithms can analyze real-time traffic data from IoT sensors, such as traffic cameras, GPS devices, and vehicle sensors. ML models can identify traffic patterns, predict congestion, and optimize traffic flow by processing this data. This information enables traffic management authorities to make data-driven decisions, such as adjusting traffic signal timings or re-routing vehicles, to alleviate congestion and improve traffic efficiency (Kuo et al., 2015). The integration of ML and IoT enables the development of Intelligent Transportation Systems (ITS). ITS uses ML algorithms to analyze real-time data from IoT devices, including vehicle, road, and weather sensors. This data predicts traffic conditions, identifies hazardous road conditions, and provides real-time information to drivers and transportation management systems. By continuously monitoring and analyzing sensor data, ML models can detect anomalies, identify potential failures, and schedule maintenance proactively. This predictive maintenance approach reduces vehicle downtime, improves fleet efficiency, and saves costs associated with unscheduled repairs. They are also used in the operation of autonomous vehicles. IoT sensors, including LiDAR, cameras, and radar, collect massive amounts of data about the vehicle's surroundings. The continuous interaction between IoT sensors and ML models allows autonomous vehicles to navigate safely and efficiently in various driving conditions. IoT sensors embedded in parking spaces can gather data on occupancy and availability (Valadarsky et al., 2017). ML algorithms can process this data to predict parking availability, guide drivers to vacant spaces, and optimize parking resource utilization.

Further, IoT sensors can collect data on air quality, noise levels, and other environmental parameters. ML algorithms can analyze this data to monitor pollution levels, identify high-emission areas, and optimize transportation routes to minimize environmental impact. ML models can also provide personalized transportation options, routes, or service recommendations.

8.7 Machine learning in logistics—warehouse management

ML is significantly transforming various aspects of logistics, including warehouse management. ML algorithms and techniques enable warehouses to optimize operations, improve efficiency, and enhance decision-making processes in the below-mentioned contexts.

Demand Forecasting: ML algorithms can analyze historical sales data, customer behavior, market trends, and other relevant factors to predict future demand accurately. By understanding demand patterns and fluctuations, warehouses can optimize inventory levels, allocate resources efficiently, and minimize stockouts or excess inventory. ML-based demand forecasting helps warehouses achieve better inventory management and improve customer satisfaction.

Inventory Optimization: ML algorithms can analyze real-time data on inventory levels, order history, lead times, and other variables to optimize inventory management. ML models can recommend optimal inventory replenishment strategies by considering various factors, such as demand variability, supplier performance, and storage capacity. This helps warehouses maintain optimal stock levels, reduce carrying costs, and improve inventory performance.

Route Optimization: ML algorithms can optimize picking and packing routes within the warehouse. ML models can generate optimized routes for warehouse personnel or autonomous robots by analyzing order profiles, product locations, and traffic patterns. This reduces travel time, minimizes congestion, and improves efficiency in order fulfillment processes (Liu, 2021). Route optimization algorithms enable warehouses to fulfill orders quickly and accurately.

Quality Control: ML algorithms can analyze sensor data, images, and other inputs to detect product anomalies or defects. By training ML models on labeled data, warehouses can automate quality control processes, identifying non-conforming items in real time. ML-powered quality control systems improve efficiency, reduce errors, and enhance product quality, ensuring that only high-quality products are shipped to customers.

Resource Allocation: ML algorithms can optimize the allocation of warehouse resources, such as labor and equipment. ML models can allocate resources effectively by analyzing historical data, order volumes, and other variables, ensuring that the right personnel and equipment are available when needed. This improves resource utilization, reduces idle time, and enhances overall operational efficiency in the warehouse.

Predictive Maintenance: ML algorithms can analyze sensor data from warehouse equipment, such as conveyor belts or forklifts, to predict maintenance needs and prevent breakdowns. ML models can proactively schedule maintenance by monitoring equipment health, detecting anomalies, and identifying potential failures. This reduces equipment downtime, improves operational continuity, and saves costs associated with unscheduled repairs.

Performance Analytics: ML algorithms can analyze various performance metrics and operational data to provide valuable insights for continuous improvement. By monitoring key performance indicators (KPIs) and analyzing data, ML models can identify bottlenecks, inefficiencies, or areas for improvement in warehouse operations. This enables warehouses to make data-driven decisions, optimize processes, and enhance performance.

8.8 Supply chain planning using machine learning

Supply chain planning is a strategic and tactical process involving coordinating and optimizing all activities within a supply chain network. It aims to align supply and demand, optimize inventory levels, minimize costs, and meet customer requirements efficiently (Stampa et al., 2017). The goal of supply chain planning is to ensure the smooth flow of products, information, and resources from the suppliers to the end customers.

8.8.1 Key elements
8.8.1.1 Demand planning
This involves forecasting customer demand for products or services based on historical data, market trends, and other relevant factors. Demand planning helps businesses anticipate future demand patterns, understand customer preferences, and plan production and inventory levels accordingly.

8.8.1.2 Inventory planning
Inventory planning focuses on determining the optimal inventory levels to maintain at different supply chain stages. It involves analyzing demand forecasts, lead times, supplier capabilities, and cost considerations to determine the correct inventory quantity to hold (Xie et al., 2018). Effective inventory planning ensures sufficient stock is available to meet customer demand while minimizing carrying costs and the risk of stockouts or obsolescence.

8.8.1.3 Production planning
Production planning involves determining the optimal production capacity, scheduling production activities, and allocating resources to meet the

anticipated demand. It considers factors such as production capacities, lead times, labor availability, and raw material availability. By aligning production capabilities with demand forecasts, production planning helps optimize production efficiency, reduce costs, and ensure timely product delivery.

8.8.1.4 Procurement planning

Procurement planning focuses on determining the quantity and timing of raw material and component purchases to support production and meet customer demand. It involves assessing supplier capabilities, negotiating contracts, managing supplier relationships, and optimizing the procurement process (Zantalis et al., 2019). Effective procurement planning helps ensure a reliable supply of materials, reduces procurement costs, and minimizes supply chain disruptions.

8.8.1.5 Distribution planning

Distribution planning involves determining the most efficient way to transport products from manufacturing facilities or distribution centers to end customers. It includes network design, transportation mode selection, route optimization, and warehouse management. Distribution planning aims to minimize transportation costs, reduce delivery lead times, and improve customer service.

8.8.1.6 Risk management

Supply chain planning also involves identifying and mitigating risks that could disrupt the flow of goods or impact supply chain performance. This includes assessing risks related to demand variability, supplier reliability, transportation disruptions, natural disasters, and other factors. Effective risk management strategies help minimize the impact of disruptions and enhance supply chain resilience.

8.8.2 Benefits

- Improved customer service levels and satisfaction.
- Enhanced supply chain visibility and transparency.
- Optimal inventory levels and reduced carrying costs.
- Increased operational efficiency and cost savings.
- Minimized stockouts and improved product availability.
- Efficient allocation of resources and improved productivity.

- Faster response times to changing market conditions and customer demands.
- Enhanced collaboration and coordination across the supply chain network.

8.9 Objectives and main specificities of reinforcement learning-based routing protocols

Reinforcement Learning (RL) is a ML approach that involves an agent learning to make decisions in an environment to maximize a reward signal. RL-based routing protocols apply RL techniques to the network routing domain, where the objective is to find optimal routes for data packets to traverse a network (Liu et al., 2017). RL-based routing protocols aim to find routes that optimize specific performance criteria, such as minimizing delay, maximizing throughput, or reducing network congestion (Waschneck et al., 2018). RL agents learn to make routing decisions based on observed states and rewards to achieve these objectives. RL-based routing protocols can dynamically adapt to changing network conditions and traffic patterns. The RL agent continuously interacts with the network environment, observing states (e.g., network topology, traffic load) and actions (e.g., selecting next-hop routers) based on its learned policy. This adaptability enables routing protocols to respond to network dynamics effectively. RL-based routing protocols balance exploration and exploitation (Stampa et al., 2017). During the learning phase, the RL agent explores different routing options to gather information about the network and learn the best actions. As the agent gains knowledge and experience, it exploits its learned policy to make optimal routing decisions. The design of the reward function is crucial in RL-based routing protocols (Liu et al., 2019). The reward signal guides the RL agent to learn actions that lead to desirable outcomes. The reward can be based on various performance metrics, such as delay, throughput, or fairness. Designing an appropriate reward function is essential to incentivize the agent to achieve the desired routing objectives. RL-based routing protocols typically involve a training phase where the RL agent learns to make routing decisions through interactions with the network environment. This training phase can be offline or online, depending on the availability of historical data or the ability to learn in real time (Mammeri, 2019).

The RL agent updates its policy based on feedback received from the environment. RL-based routing protocols face challenges related to scalability and complexity. As the network size and complexity increase, the number of possible states and actions also grows exponentially, making the learning process computationally intensive. Techniques such as function approximation, value iteration, or deep learning can address these challenges and make RL-based routing protocols more scalable (Waschneck et al., 2018). RL-based routing protocols can be designed to work alongside traditional routing protocols, leveraging the strengths of both approaches (Mammeri, 2019). By integrating with existing protocols, RL-based solutions can inherit the stability and robustness of traditional routing algorithms while introducing RL's adaptability and optimization capabilities. For example, optimizing for throughput may lead to increased network congestion (Stampa et al., 2017). RL agents can learn to make decisions that strike an appropriate balance between these trade-offs, considering the specific requirements and constraints of the network environment (Liu et al., 2017). By effectively integrating RL techniques with traditional routing protocols, RL-based routing protocols can potentially enhance network efficiency and performance.

8.10 Path planning strategies

Path planning strategies are approaches used to determine the optimal or near-optimal path for a mobile agent, such as a robot, vehicle, or unmanned aerial vehicle (UAV), to navigate an environment from a starting point to a goal location. These strategies consider obstacle avoidance, efficiency, safety, and feasibility factors. Here are some commonly used path planning strategies.

8.10.1 Dijkstra's algorithm

Dijkstra's algorithm is a classic path planning algorithm that finds the shortest path between two points in a graph. It explores the graph by iteratively selecting the node with the lowest cost until the goal node is reached (Mao et al., 2018). Dijkstra's algorithm guarantees finding the optimal path but may be computationally expensive for large-scale environments.

8.10.2 A* Algorithm

The A* algorithm is an extension of Dijkstra's algorithm, incorporating heuristics to guide the search process. It estimates the cost of reaching the goal by considering the actual cost from the start node and a heuristic function that estimates the cost to the goal. A* algorithm is widely used due to its efficiency and ability to find near-optimal paths.

8.10.3 Probabilistic roadmap

Probabilistic roadmap (PRM) is a sampling-based approach where random samples are taken from the configuration space of the agent and connected to form a roadmap. It then searches for a path by connecting the start and goal configurations to the roadmap. PRM can handle complex and high-dimensional spaces but may require significant preprocessing time for roadmap construction.

8.10.4 Rapidly exploring random trees

Rapidly exploring Random Trees (RRT) is another sampling-based algorithm that builds a tree structure by iteratively expanding toward unexplored areas of the configuration space. RRT is particularly suitable for dynamic and uncertain environments as it adapts well to changing conditions. It tends to produce feasible paths quickly but may only sometimes find the optimal solution.

8.10.5 Potential fields

The Potential fields approach uses forces to guide the agent through the environment. It assigns attractive forces towards the goal location and repulsive forces from obstacles, creating a virtual potential field. The agent navigates by following the resultant force vector toward the goal while avoiding obstacles. Potential fields are computationally efficient but can suffer from local minima and oscillations.

8.10.6 Model predictive control

Model predictive control (MPC) is a control-based approach that predicts the future behavior of the agent over a finite time horizon and optimizes a cost function. It considers dynamic constraints, such as velocity and acceleration limits, to generate a trajectory. MPC can handle complex dynamics and constraints but requires real-time computations.

8.10.7 Swarm intelligence

Swarm intelligence algorithms, inspired by collective behaviors in natural systems like ant colonies or bird flocks, use a population of agents to explore and search for paths collaboratively. Algorithms such as Ant Colony Optimization (ACO) and Particle Swarm Optimization (PSO) leverage the collective knowledge of the swarm to find paths efficiently.

These are just a few examples of path planning strategies, and many variations and hybrid approaches combine different techniques. The path planning strategy choice depends on the environment's specific characteristics, the agent's dynamics, computational requirements, and the desired trade-offs between optimality, efficiency, and adaptability.

8.11 Path planning and obstacle avoidance algorithms

Path planning refers to finding a path that satisfies specific criteria, such as minimizing travel time, avoiding obstacles or collisions, maximizing efficiency, or considering other constraints (Radmanesh et al., 2018). The choice of path planning algorithm depends on factors such as the complexity of the environment, the agent's dynamics, computational resources, real-time requirements, and the specific objectives and constraints of the application. Path planning algorithms often incorporate obstacle detection and perception techniques, sensor fusion, mapping, localization, and decision-making to ensure safe and effective navigation (Lu et al., 2020). Path planning is crucial in various applications, including autonomous driving, robotic navigation, logistics and transportation, virtual reality, gaming, and more. It enables autonomous agents to move efficiently and accomplish tasks in their respective environments.

Obstacle avoidance algorithms enable autonomous agents, such as robots or vehicles, to navigate an environment while avoiding obstacles or collisions. These algorithms utilize sensor data and perception techniques to detect obstacles and generate safe paths. The following are some commonly used obstacle avoidance algorithms (Radmanesh et al., 2018).

8.11.1 Potential fields

The potential fields algorithm creates a virtual potential field around obstacles, where obstacles exert repulsive forces on the agent, while the goal location exerts an attractive force. The agent navigates by following the resultant force vector, moving away from obstacles and towards the goal. Potential fields are simple and computationally efficient but can suffer from local minima and oscillations.

8.11.2 Vector field histogram

Vector Field Histogram (VFH) algorithm divides the agent's sensor range into angular sectors and creates a histogram representing the presence of obstacles in each sector. It then selects a collision-free heading direction based on the sectors with the fewest obstacles. VFH allows for real-time obstacle avoidance and is robust to sensor noise, but it may struggle in cluttered environments.

8.11.3 Rapidly exploring random trees

RRT algorithms, such as RRT* or RRT-connect, use a tree structure to explore the configuration space. They generate a tree by iteratively expanding toward unexplored areas while avoiding collisions with obstacles. RRT-based algorithms are suitable for dynamic environments and can handle complex spaces but may only sometimes find the optimal path.

8.11.4 Dynamic-Window Approach

Dynamic-Window Approach algorithm calculates the achievable velocities of the agent based on its dynamic constraints (e.g., maximum speed, acceleration). It generates a set of reachable trajectories and evaluates their safety and proximity to obstacles. The agent selects the best trajectory that maximizes both progress toward the goal and obstacle avoidance.

8.11.5 Elastic band

The elastic band algorithm models the path as an elastic band that can stretch and contract. It iteratively adjusts the band by pushing it away from obstacles and pulling it toward the goal. The agent follows the deformed band, representing a safe and feasible path. Elastic band algorithms suit nonholonomic agents and provide smooth and continuous paths.

8.11.6 Reciprocal Velocity Obstacles

Reciprocal Velocity Obstacles algorithm focuses on collision avoidance in multi-agent scenarios. It computes the preferred velocities of each agent and predicts potential collisions based on their velocities and positions.

The algorithm then calculates reciprocal velocity obstacles and adjusts the velocities of the agents to avoid collisions while maintaining their preferences.

8.11.7 Sensor-based local navigation

Sensor-based algorithms like LiDAR or cameras utilize real-time sensor data to perceive the environment and plan collision-free paths. Techniques like occupancy grids, sensor fusion, or point cloud processing are used to represent and interpret the sensor data. Local navigation algorithms generate safe trajectories based on the sensed obstacles.

Each algorithm has its strengths and limitations, and the choice depends on factors such as the agent's dynamics, environment complexity, available sensors, computational resources, and specific requirements of the application. Often, a combination of algorithms or hybrid approaches may be used to achieve robust and efficient obstacle avoidance.

8.12 Operations with route optimization

Route optimization involves finding the most efficient and cost-effective routes for vehicles or delivery drivers to follow when making multiple stops or deliveries (Liu, 2021). Several operations and processes are associated with route optimization, including:

Route Planning involves determining the best sequence of stops or destinations for a vehicle or driver. It considers distance, traffic conditions, time windows, customer priorities, and specific constraints (Liu et al., 2020). Route planning aims to minimize travel distance, reduce fuel consumption, and optimize time utilization.

8.12.1 Vehicle assignment

Vehicle assignment involves assigning specific vehicles to routes based on their capacity, availability, and compatibility with the types of deliveries or stops (Snoeck et al., 2020). It ensures that suitable vehicles are allocated to the corresponding routes to optimize efficiency and resource utilization.

8.12.2 Time window management

Time windows refer to specific intervals within which deliveries or stops must be made. Managing time windows involves scheduling deliveries to accommodate customer preferences, service level agreements, and any operational constraints. An efficient time window management minimizes waiting times, reduces customer dissatisfaction, and improves overall route performance.

8.12.3 Load balancing

Load balancing distributes the workload evenly among vehicles or drivers to avoid overloading or underutilization. It ensures that the capacity of vehicles is optimized, leading to balanced routes, reduced idle time, and improved overall operational efficiency.

8.12.4 Dynamic route optimization

Dynamic route optimization involves adjusting real-time or near real-time routes based on changing conditions. It considers traffic congestion, road closures, weather conditions, and unexpected events (Liu, 2021). Dynamic route optimization allows agile adjustments, minimizing delays and improving customer service.

8.12.5 Constraint handling

Constraints are specific limitations or requirements that must be considered during route optimization. These include vehicle capacity limits, delivery time windows, road restrictions, and other operational constraints. Effective constraint handling ensures routes comply with all relevant restrictions while maximizing efficiency.

8.12.6 Communication and tracking

Efficient route optimization relies on real-time communication and tracking systems. This allows for continuous monitoring of vehicles' positions, progress, and any deviations from the planned routes. Communication and tracking enable better coordination, proactive response to disruptions, and improved operational control.

8.12.7 Performance analysis and continuous improvement

Route optimization is an ongoing process that requires continuous analysis and improvement. Monitoring and analyzing key performance indicators such as travel times, fuel consumption, delivery accuracy, and customer satisfaction help identify areas for optimization and guide decision-making to enhance overall route efficiency (Liu et al., 2019). By integrating these operations and processes, organizations can streamline operations, reduce costs, improve customer satisfaction, and optimize resource utilization through effective route optimization.

8.13 Vehicle routing problem and delivery optimization

The Vehicle Routing Problem (VRP) is a well-known combinatorial optimization problem in logistics and operations research. It involves determining the optimal routes for a fleet of vehicles to visit a set of customers or locations while satisfying various constraints and objectives. The VRP is commonly encountered in delivery, distribution, and transportation planning scenarios. The VRP involves several constraints that need to be considered during route planning. These constraints include vehicle capacity limits (to ensure that demands are met), time windows (defining specific time intervals in which customers can be visited), maximum travel distances or durations, and any other operational limitations (Mao et al., 2018). The primary objective of the VRP is to minimize the total cost or distance traveled by the vehicles. This can include minimizing fuel consumption, reducing vehicle idle time, minimizing the number of vehicles used, or minimizing total travel time. Other objectives may include maximizing customer satisfaction by meeting time window requirements or balancing workloads among vehicles. The VRP has numerous variants depending on additional complexities or specific requirements. Some common variants include the Capacitated VRP (CVRP), where each vehicle has a limited capacity. The Vehicle Routing Problem with Time Windows (VRPTW) considers time windows for customer visits, and the Vehicle Routing Problem with Pickup and Delivery (VRPPD) involves both pickups and deliveries. Solving the VRP is challenging due to its combinatorial nature (Yu et al., 2018). Various solution approaches are employed, including exact methods such as mathematical programming formulations (e.g., integer programming) and heuristic or metaheuristic techniques like genetic algorithms, simulated annealing, ant colony optimization, and tabu search. These methods aim to find near-optimal solutions within a reasonable computation time. The VRP has wide-ranging applications in logistics and transportation management. It is commonly used for delivery route planning, waste collection, courier services, public transportation, fleet management, and supply chain optimization. Effective VRP solutions can significantly improve operational efficiency, reduce costs, and enhance customer service. The Vehicle Routing Problem is a complex optimization problem with practical implications in transportation and logistics (Snoeck et al., 2020). Solving it efficiently can lead to significant benefits, enabling organizations to allocate resources better, optimize routes, reduce fuel consumption, and improve overall operational efficiency.

8.14 Optimization of delivery operations

Optimization of delivery operations streamlines the entire delivery workflow, from order processing to the final delivery, by minimizing costs, reducing delivery times, and maximizing resource utilization. Here are the critical aspects of optimizing delivery operations: (1) Route Optimization: One crucial aspect is optimizing delivery routes. This involves determining the most efficient sequence of stops and routes for delivery vehicles or drivers. Route optimization algorithms consider delivery locations, order volumes, time windows, traffic conditions, and vehicle capacities to minimize travel distance, reduce fuel consumption, and optimize time utilization. (2) Vehicle Allocation: Efficient vehicle allocation ensures that suitable vehicles are assigned to the corresponding delivery routes. This involves considering vehicle capacities, availability, and compatibility with the types of deliveries (Liu, 2021). By optimizing vehicle allocation, organizations can minimize underutilization or overloading of vehicles, leading to better resource utilization. (3) Load Balancing: Load balancing focuses on evenly distributing the workload among delivery vehicles or drivers. It ensures that the delivery capacity of vehicles is optimally utilized, reducing idle time and improving overall operational efficiency. Load balancing prevents the overloading of certain vehicles while others remain underutilized. (4) Time Window Management: Time window management involves scheduling deliveries within specific time intervals agreed upon with customers. Optimizing time windows ensures deliveries are made within the time frames, minimizing waiting times and improving customer satisfaction. Effective time window management helps avoid missed or late deliveries. (5) Real-time Tracking and Communication: Implementing real-time tracking and communication systems enables constant monitoring of delivery progress. It allows organizations to track delivery vehicles or drivers, receive real-time updates on order status, and proactively respond to delays or issues. Real-time communication enhances coordination and helps manage customer expectations. (6) Delivery Network Optimization: Optimization extends beyond individual deliveries to the entire delivery network. Organizations can reduce delivery distances, minimize transportation costs, and improve overall operational efficiency by strategically optimizing the location of distribution centers, warehouses, and fulfillment centers. (7) Analytics and Data-driven Insights: Leveraging analytics and data-driven insights enables organizations to understand their delivery operations better (Yu et al., 2018). By analyzing delivery data, organizations can identify bottlenecks, inefficiencies, and areas for improvement. Data-driven insights can inform decision-making, optimize routes, and identify trends to enhance delivery operations. (8) Continuous Improvement: Optimization of delivery operations is an ongoing process.

8.15 Last-mile delivery efficiency

Last-mile delivery refers to the final leg of the delivery process, where goods are transported from a transportation hub or distribution center to the end customer's location. It is often considered the most critical and challenging part of the delivery journey. Last-mile delivery plays a crucial role in customer satisfaction and can significantly impact the overall cost and efficiency of the delivery operation (Min et al., 2019). Last-mile delivery covers the shortest distance but can consume a significant amount of time due to traffic congestion, road conditions, and the density of delivery locations, as shown in Fig. 8.1.

Optimizing the route and scheduling becomes crucial to minimize travel time and ensure timely deliveries. These involve multiple delivery points within a single trip. Couriers or delivery personnel must efficiently plan their routes, considering factors like order volumes, delivery time windows, and the proximity of delivery locations. Optimizing the sequence of deliveries can enhance efficiency and reduce travel distance. It requires flexibility to accommodate specific customer preferences, such as delivery time slots or special instructions. Last-mile delivery in urban areas poses unique challenges, including limited parking, congested streets,

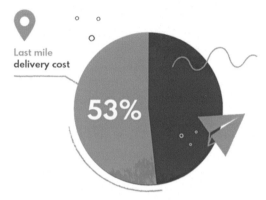

Figure 8.1 Last mile delivery cost.

and restricted access to certain areas. Delivery vehicles and personnel must navigate these challenges while complying with local regulations and restrictions. Integrating technology plays a significant role in optimizing last-mile delivery. Adopting sustainable delivery practices such as electric vehicles, route optimization algorithms, and delivery consolidation can minimize last-mile delivery's environmental impact. Efficient and well-executed last-mile delivery is crucial for customer satisfaction and overall supply chain effectiveness. By employing optimization strategies, leveraging technology, and adapting to the unique challenges of the last mile, organizations can improve delivery efficiency, reduce costs, and enhance the overall customer experience.

8.15.1 Application in a tangible real-world scenario

A real-time assessment of the last-mile delivery approach was conducted, considering the delivery pattern of an online shopping website, where products must be delivered between 9 am and 9 pm. The direct internal and direct external costs associated with the delivery process were meticulously calculated during this evaluation. The analysis revealed that the most efficient delivery alternative was using motorcycles, which managed to handle a greater volume of deliveries per route. Compared to other vehicle types, it became evident that the salary expenses for couriers using larger vehicles were substantially higher. Specifically, the internal cost for delivering a single parcel was the most economical when utilizing motorcycles, amounting to Rs. 203.55. In contrast, larger vehicles incurred an expense of Rs. 418.60 for the same task. The analysis included an estimation of the direct external costs associated with each of the examined vehicles. Additionally, when factoring in the supplementary impact of traffic, congestion costs were considered, reflecting the extra expenses incurred when adding another vehicle to the traffic flow. Notably, the motorcycle exhibited a minimal congestion cost of 2.1%, with the cost increase primarily attributed to congestion in urban areas.

Conversely, on-foot deliveries couldn't easily overcome challenges related to time and size constraints. Consequently, these findings carry significant implications for policymakers as they plan new policies and incentives to support the evolving landscape of urban deliveries. For transport and logistics operators, a crucial revelation lies in the assessment of costs per parcel. Despite its limited carrying capacity, the motorcycle remains the most cost-effective choice, capable of handling a substantial volume of deliveries.

8.16 Conclusion

By leveraging AI technologies, logistics firms can attain heightened supply chain transparency, refine route planning and resource allocation, facilitate predictive maintenance, enhance inventory management, automate warehousing tasks, and 40 risk management and security measures. An in-depth evaluation of the last-mile delivery approach was conducted in real-time, taking into account the delivery schedule of an online shopping platform, which mandates deliveries between 9 a.m. and 9 p.m.

This comprehensive assessment meticulously tallied the direct internal and external costs of the delivery process. The findings underscored the efficacy of motorcycles as the optimal choice for deliveries, given their ability to efficiently handle larger delivery volumes per route. Notably, the internal cost for delivering a single parcel was most economical when employing motorcycles, totaling Rs. 203.55, in stark contrast to the Rs. 418.60 incurred by larger vehicles for the same task.

The analysis also encompassed an estimation of the direct external costs associated with each of the evaluated vehicles, with motorcycles revealing minimal congestion costs of just 2.1%, primarily concentrated in urban areas. These advancements translate into cost savings, heightened operational efficiency, reduced downtime, improved customer satisfaction, and bolstered competitiveness in the global market. As AI continues to evolve, its transformative potential within the logistics sector will continue to expand, ushering in a more efficient, sustainable, and interconnected future.

References

Abduljabbar, R., Dia, H., Liyanage, S., & Bagloee, S. A. (2019). Applications of artificial intelligence in transport: An overview. *Sustainability*, *11*(1), 189.

Chen, Y. T., Sun, E. W., Chang, M. F., & Lin, Y. B. (2021). Pragmatic real-time logistics management with traffic IoT infrastructure: Big data predictive analytics of freight travel time for Logistics 4.0. *International Journal of Production Economics*, *238*, 108157.

Deng, L. (2018). Artificial intelligence in the rising wave of deep learning: The historical path and future outlook [perspectives]. *IEEE Signal Processing Magazine*, *35*(1), 180, -177.

Ding, Y., Jin, M., Li, S., & Feng, D. (2021). Smart logistics based on the internet of things technology: An overview. *International Journal of Logistics Research and Applications*, *24*(4), 323−345.

James, J. Q., Yu, W., & Gu, J. (2019). Online vehicle routing with neural combinatorial optimization and deep reinforcement learning. *IEEE Transactions on Intelligent Transportation Systems*, *20*(10), 3806−3817.

Janarden, M. (2017). *Moving the goalposts: AI and logistics*. Infosys Insights.

Jiang, J., & Ma, Y. (2020). Path planning strategies to optimize accuracy, quality, build time and material use in additive manufacturing: A review. *Micromachines*, *11*(7), 633.

Kuo, P. H., Li, T. H. S., Ho, Y. F., & Lin, C. J. (2015). Development of an automatic emotional music accompaniment system by fuzzy logic and adaptive partition evolutionary genetic algorithm. *IEEE Access*, *3*, 815−824.

Lee, S., Kim, Y., Kahng, H., Lee, S. K., Chung, S., Cheong, T., & Kim, S. B. (2020). Intelligent traffic control for autonomous vehicle systems based on machine learning. *Expert Systems with Applications*, *144*, 113074.

Liu, B. (2021). Logistics distribution route optimization model based on recursive fuzzy neural network algorithm. *Computational Intelligence and Neuroscience*, *2021*.

Liu, C., Feng, Y., Lin, D., Wu, L., & Guo, M. (2020). Iot based laundry services: An application of big data analytics, intelligent logistics management, and machine learning techniques. *International Journal of Production Research*, *58*(17), 5113−5131.

Liu, L., Wang, H., & Xing, S. (2019). Optimization of distribution planning for agricultural products in logistics based on degree of maturity. *Computers and Electronics in Agriculture*, *160*, 1−7.

Liu, T., Hu, X., Li, S. E., & Cao, D. (2017). Reinforcement learning optimized lookahead energy management of a parallel hybrid electric vehicle. *IEEE/ASME Transactions on Mechatronics*, *22*(4), 1497−1507.

Lu, H., Zhang, X., & Yang, S. (2020). A learning-based iterative method for solving vehicle routing problems. In *International conference on learning representations*.

Mammeri, Z. (2019). Reinforcement learning based routing in networks: Review and classification of approaches. *IEEE Access*, *7*, 55916−55950.

Mao, Q., Hu, F., & Hao, Q. (2018). Deep learning for intelligent wireless networks: A comprehensive survey. *IEEE Communications Surveys & Tutorials*, *20*(4), 2595−2621.

Min, Q., Lu, Y., Liu, Z., Su, C., & Wang, B. (2019). Machine learning based digital twin framework for production optimization in petrochemical industry. *International Journal of Information Management*, *49*, 502−519.

Radmanesh, M., Kumar, M., Guentert, P. H., & Sarim, M. (2018). Overview of path-planning and obstacle avoidance algorithms for UAVs: A comparative study. *Unmanned Systems*, *6*(02), 95−118.

Snoeck, A., Merchán, D., & Winkenbach, M. (2020). Route learning: A machine learning-based approach to infer constrained customers in delivery routes. *Transportation Research Procedia*, *46*, 229−236.

Stampa, G., Arias, M., Sánchez-Charles, D., Muntés-Mulero, V., & Cabellos, A. (2017). A deep-reinforcement learning approach for software-defined networking routing optimization. *arXiv preprint arXiv*, *1709*, 07080.

Sulova, S. (2021). Big data processing in the logistics industry. *Лкономика и компютърни науки Economics and computer science*, 6.

Tang, L., & Meng, Y. (2021). Data analytics and optimization for smart industry. *Frontiers of Engineering Management*, *8*(2), 157−171.

Valadarsky, A., Schapira, M., Shahaf, D., & Tamar, A. (2017). Learning to route, November *Proceedings of the 16th ACM workshop on hot topics in networks*, 185−191.

Waschneck, B., Reichstaller, A., Belzner, L., Altenmüller, T., Bauernhansl, T., Knapp, A., & Kyek, A. (2018). Optimization of global production scheduling with deep reinforcement learning. *Procedia Cirp*, *72*, 1264−1269.

Woschank, M., Rauch, E., & Zsifkovits, H. (2020). A review of further directions for artificial intelligence, machine learning, and deep learning in smart logistics. *Sustainability*, *12*(9), 3760.

Xie, J., Yu, F. R., Huang, T., Xie, R., Liu, J., Wang, C., & Liu, Y. (2018). A survey of machine learning techniques applied to software-defined networking (SDN): Research issues and challenges. *IEEE Communications Surveys & Tutorials*, *21*(1), 393−430.

Yu, C., Lan, J., Guo, Z., & Hu, Y. (2018). DROM: Optimizing the routing in software-defined networks with deep reinforcement learning. *IEEE Access*, *6*, 64533−64539.

Zantalis, F., Koulouras, G., Karabetsos, S., & Kandris, D. (2019). A review of machine learning and IoT in smart transportation. *Future Internet*, *11*(4), 94.

Zhao, L., Wang, J., Liu, J., & Kato, N. (2019). Routing for crowd management in smart cities: A deep reinforcement learning perspective. *IEEE Communications Magazine*, *57* (4), 88−89.

CHAPTER NINE

Revolutionizing sustainability: the role of robotics in supply chains

Pradeep Bedi[1,2], Sanjoy Das[3], S.B. Goyal[4], Anand Singh Rajawat[5] and Sandeep Kautish[6,7]

[1]Department of Computer Science, Regional Campus Manipur, Indira Gandhi National Tribal University, Amarkantak, Madhya Pradesh, India
[2]Department of Computer Science and Engineering, Galgotias University, Greater Noida, Uttar Pradesh, India
[3]Department of Computer Science, Indira Gandhi National Tribal University, Regional Campus Manipur, Makhan, India
[4]Faculty of Information Technology, City University, Petaling Jaya, Malaysia
[5]School of Computer Sciences & Engineering, Sandip University, Nashik, Maharashtra, India
[6]LBEF Campus (APU Malaysia) Kathmandu, Nepal
[7]Model Institute of Engineering and Technology, Jammu, Jammu and Kashmir, India

9.1 Introduction

9.1.1 Background and significance of supply chain automation using robots

Supply chains, strained by labor shortages and increased demand, turn to technology, particularly Robotics and Autonomous Systems (RAS) in the food industry. Duong et al. (2020) explores RAS applications across food quality, safety, waste, efficiency, and analysis, with a focus on factors like data availability, cybersecurity, skills, and finances. The use of autonomous robots has expanded across industries, enhancing efficiency, cleanliness, and safety. RAS streamlines operations, reduces errors, and facilitates rapid scaling to meet rising demand. This automation improves customer satisfaction and frees up staff for higher-value tasks while robots handle routine and hazardous work in warehouses and logistics centers. Modern robots driven by artificial intelligence (AI) are developed as per the needs of supply chains. They are adapted to dangerous areas and can serve as dependable employees, enhancing worker productivity. Workers require guidance to meet standards that were once taken for granted, especially when faced with a global health danger like COVID-19, with developing varieties spreading facilities on constant alert, as we have seen since the first wave of the pandemic, has been reported by Hofmann et al. (2019).

Computational Intelligence Techniques for Sustainable Supply Chain Management.
DOI: https://doi.org/10.1016/B978-0-443-18464-2.00007-8

Robotics is revolutionizing operations in the field of sustainable supply chains. In this sense, the term "robotics" refers to the multidisciplinary area that includes the design, development, maintenance, and programming of autonomous or supervised robots. In a variety of sectors, including industry, healthcare, agriculture, spatial exploration, and entertainment, these machines play a critical role in improving sustainable practices. In sustainable supply chain robotics, mechanical, electrical, and software engineering are skillfully combined throughout the design and manufacturing of robots. To achieve optimal operation, mechanical design elements carefully evaluate materials, techniques, and sensors. Electrical design, on the other hand, prioritizes effective control circuits and power systems that are in line with sustainable energy methods. Software design also includes algorithm creation and programming languages created expressly for robot use in the context of sustainable supply chains.

Robots in this discipline can be categorized according to their particular applications and degrees of autonomy. Industrial robots, which are frequently used in production operations, are skilled at carrying out repetitive activities like welding and assembly. Industrial robots take on these responsibilities, which expedite manufacturing procedures and reduce waste output, thereby fostering supply chain sustainability. Service robots, on the other hand, operate in non-industrial settings and carry out activities like cleaning and delivery. By maximizing resource use and minimizing human effort, these robots support sustainable practices. Finally, autonomous robots function self-sufficiently, completing tasks like exploration and surveillance while maintaining sustainable procedures. Classification of robots used in sustainable supply chains has been shown in Fig. 9.1.

In the supply chain, distribution center, and warehouse management communities, robots have recently attained prominence and are still playing a big part in warehouse automation. Fig. 9.1 illustrates various models of warehouse robots.

Industry 4.0 integrates transformative innovations like robotics, AI, and the Internet of things (IoT) into manufacturing to enhance efficiency, lower costs, and improve product quality. Robotics plays a pivotal role by automating operations, reducing downtime, and enabling real-time data analysis, ultimately benefitting both businesses and consumers (Duong et al., 2020, Flechsig et al., 2022). This chapter outlines the goals of dissecting robotics' contributions to Industry 4.0 and their impact on operational efficiency and customer value.

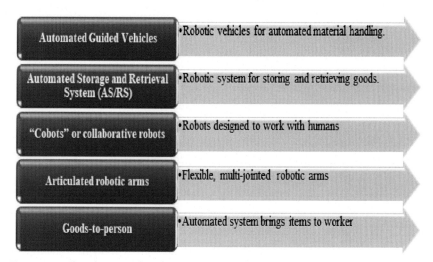

Figure 9.1 Classification of robots used in sustainable supply chain.

9.1.2 Objective of the chapter

The main objective of this chapter is to conduct a thorough analysis of supply chain automation utilizing robots, highlighting their purpose, benefits, challenges, and upcoming advances. This chapter also aims to provide real-world insights into the specific supply chain uses of robotic automation. Also, it seeks to present a theoretical framework for supply chain automation, which entails talking about ways to gather data, analyze it, and assess current supply chain procedures and performance indicators. Recognizing the challenges inherent in supply chain automation, this chapter will not only address these issues but also propose strategies to navigate and mitigate them.

9.1.3 Overview of the methodology and structure of the chapter

This chapter will provide a roadmap of the approaches taken to explore the integration of robotics within the supply chain, detailing how these technologies can streamline operations and improve efficiency. Integrating robotics in the supply chain involves incorporating robotic technologies to automate and optimize various processes within the supply chain. This includes assessing existing processes, identifying suitable applications for robotics, selecting appropriate robotic technologies, developing implementation plans, conducting pilot tests, training the workforce, and continuously monitoring and optimizing performance. The integration of robotics aims to improve efficiency, productivity, and cost-effectiveness in supply chain operations.

The introduction section provides an overview of supply chain automation using robots, emphasizing its significance and background, and discussing the application of supply chains in different sectors.

9.1.3.1 Methodological framework for supply chain automation

Introduces a framework for implementing supply chain automation, including data collection methods, analysis techniques, and identifying areas for improvement.

9.1.3.2 Challenges and future directions

Explores the challenges and limitations associated with supply chain automation and discusses emerging trends and future developments in robotics and automation.

9.1.3.3 Case studies

Presents specific case studies that demonstrate the application of robotics in supply chain optimization.

9.1.3.4 Conclusion

Summarizes the main findings and conclusions of the chapter, providing a concise wrap-up of the content discussed.

9.2 Methodological framework for supply chain automation

An organized approach to integrating automation technology and procedures into supply chain operations is provided by a methodical framework for supply chain automation which has been demonstrated in Fig. 9.2. It highlights the essential stages, factors to think about, and best practices for deploying automation successfully to maximize effectiveness, reduce operating expenses, and enhance supply chain performance in general. It entails evaluating the current situation, establishing automation goals, figuring out opportunities, choosing suitable technology, developing an implementation strategy, running pilot tests, and scaling up the

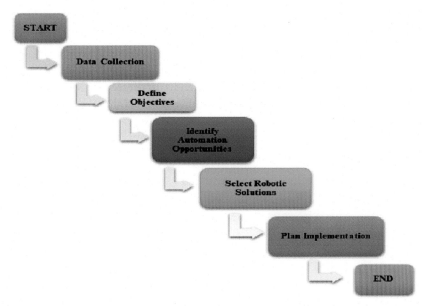

Figure 9.2 Methodological framework for supply chain automation.

automation. Throughout the process, emphasis is placed on personnel training, change management, and continuous improvement. The framework attempts to increase overall supply chain performance, minimize expenses, and maximize efficiency.

Utilizing many forms of data, including product, financial, sales, warehouse, inventory, and logistics data, big data analytics is an essential part of streamlining supply chain management procedures. Organizations must take a structured approach when implementing automation in the supply chain. This approach entails setting clear goals, determining feasibility, identifying opportunities for automation, choosing appropriate robotic solutions, planning execution, testing and validation, deployment and integration, monitoring and optimization, ongoing enhancement, and encouraging cooperation and knowledge sharing. By employing this strategy, businesses may boost efficiency, productivity, and sustainability while adjusting to changing customer demands and staying one step ahead of the competition. Let's get into detail about these points.

9.2.1 Data collection

By utilizing multiple forms of data, big data analytics in supply chain management (SCM) optimizes procedures. Product information provides connections between inventory products, improving inventory control. For resource optimization, financial data analyses production and logistical costs. Inventory levels and forecasting accuracy are improved by sales and demand data. Operations and resource allocation are streamlined using warehouse data. Stockouts are reduced and carrying costs are optimized via inventory data. Route planning, delivery scheduling, and customer service are all improved by logistics data.

9.2.2 Define objectives

Organizations define specific objectives and goals for supply chain automation in this step. Determine the precise areas of the supply chain where robotics can have a positive influence, such as increasing production, sustainability, and efficiency. For instance, companies might think about automating work in inventory management, order fulfillment, shipping, or quality control.

9.2.3 Assess feasibility

To determine whether integrating robotics into the supply chain is technically, operationally, and financially feasible, do a feasibility study. Examine elements including your infrastructure's preparedness to accommodate robotics, compatibility with current systems and procedures, cost-benefit analysis, and potential implementation concerns. This study assists in determining whether robot technology is practical and advantageous for the business's needs.

9.2.4 Identify automation opportunities

Determine which specific supply chain procedures and operations can be automated by robotics. To do this, it is necessary to analyze several areas, including warehousing, inventory control, order fulfillment, transportation, and quality control. Consider considerations like the difficulty of the task and the possible advantages automation may provide to rank these opportunities based on their potential impact and ease of execution.

9.2.5 Select robotic solutions

Analyze the many robotic products on the market that fit the outlined automation potential. Think about things like the robots' functionality (what they can do), scalability (how well they can manage growing quantities), ease of interface with current systems, flexibility to adapt to shifting requirements, and vendor support. Choose the most appropriate robotic technologies for each indicated automation area.

9.2.6 Plan implementation

Create a thorough implementation strategy that specifies the procedures, deadlines, and materials needed to implement robotics in the supply chain. During this phase of planning, consider many things, such as the training needs of staff who will work alongside robots, the infrastructure changes required for accommodating robotics technology, data integration to ensure smooth information flow, and change management tactics to support the adoption of automation. To achieve a seamless adoption process, cooperation with internal stakeholders, IT teams, and robotics providers is essential.

9.2.7 Test and validate

To guarantee that the chosen robotic solutions meet the specified objectives, conduct pilot tests and validation exercises. You will evaluate the effectiveness, dependability, and compatibility of the robotic systems with the current processes throughout this phase. By putting the robots to the test, you can learn more about how effectively they work and spot any areas that could use improvement. Before full-scale deployment, these insights enable the implementation plan to be improved and iterated.

9.2.8 Deployment and integration

After the robotic solutions have been tested, the company can apply them according to the approved implementation strategy throughout the supply chain. Connect robotics technology to current infrastructure, including data analytics platforms, enterprise resource planning (ERP) software, and warehouse management systems (WMS). To enable effective coordination and maximize the advantages of automation, make sure that the robotic systems and other elements of the supply chain ecosystem can exchange data and communicate effectively.

9.2.9 Monitor and optimize

Continuous monitoring of the robotic systems' performance while gathering pertinent data. Establish key performance indicators (KPIs) to gauge how automation is affecting important metrics like sustainability, productivity, and efficiency. Analyze the data collected to find potential areas for modification and further optimization. Monitor the robotic systems' performance regularly, and then use the information that is learned to improve and optimize how they work.

9.2.10 Continuous improvement

By frequently assessing the effectiveness of the robotic systems and locating areas for improvement, promote a culture of continuous improvement. Keep up on robot technology developments and look at new supply chain automation prospects. Continually investigate ways to improve the robotic systems' performance and efficiency and tailor them to changing business requirements.

9.2.11 Collaboration and knowledge sharing

This entails cooperating with stakeholders to share concepts, best practices, and lessons discovered. Through this partnership, fresh opportunities, problems, and creative solutions are all better understood. Sharing information keeps businesses informed of changes in robotics technology and market trends, encouraging continual progress and advancing the causes of social and environmental sustainability.

9.2.12 Data collection methods and analysis techniques

Techniques for gathering data and methods for analyzing it are essential for gaining insightful knowledge and improving supply chain management. Although the field of supply chain analytics has been around for more than a century, there have been substantial advancements in the mathematical models, data infrastructure, and applications that support it. With the advancement of statistical methods, predictive modeling, and machine learning, mathematical models have become better. With the advent of cloud computing, complex event processing (CEP), and the Internet of Things, data architecture has transformed. Fig. 9.3 shows the categories of collected data. Applications such as enterprise resource planning, warehouse management, logistics, and enterprise asset management have developed to provide information across conventional application boundaries.

Techniques for data gathering and analysis are crucial for supply chain management optimization. Observations/field studies, interviews/focus groups, surveys/questionnaires, and sensor-based data collecting are typical techniques. Surveys and questionnaires are used to collect structured data

Figure 9.3 Categories of collected data.

from stakeholders, whereas focus groups and interviews offer qualitative information. Field research and observations provide firsthand information on waste and inefficiencies in processes. Real-time monitoring and proactive decision-making are made possible by sensor-based data collecting. Data mining, supply chain performance measurements, simulation/modeling, and text/sentiment analysis are a few analysis methodologies. These methods reveal trends, enhance procedures, and offer perceptions for making choices. Organizations can enhance supply chain performance and achieve sustainable results by using these strategies and tactics.

9.2.13 Evaluating existing supply chain processes and performance metrics

It's necessary to evaluate existing procedures and performance measures to improve supply chain performance overall. Set up the critical performance measures by specifying them, for example, on-time delivery and customer satisfaction. Obtain appropriate data from multiple sources, then examine the metrics to spot inefficiencies and bottlenecks. Utilize the findings to motivate initiatives for ongoing development, such as process automation and optimization. Encourage open dialogue and cooperation among supply chain partners while monitoring and tracking progress. Organizations can improve the performance of their supply chains by routinely reviewing and making improvements.

Big data analytics were examined by Bag et al. (2020) as an operational excellence strategy to improve the performance of sustainable supply chains. Big data analytics, according to the authors, can be used to improve supply chain operations, and product quality, reduce waste and emissions, and increase customer happiness. According to the report, big data analytics may boost supply chain sustainability by giving useful insights into supply chain performance metrics, pointing out potential improvement areas, and assisting in data-driven decision-making. On the other hand, a review of quantitative techniques for supply chain resilience analysis was done by Hosseini et al. (2019). According to the authors, supply chain planning and operations can be made more resilient by including risk management measures. The study examined various quantitative techniques, such as simulation, optimization, game theory, and system dynamics, and provided details on their advantages and disadvantages.

Taking the usage of a cutting-edge neutrosophic methodology, Abdel-Baset et al. (2019) investigated green supply chain management practices. Based on the degree of uncertainty and consistency in the data, the authors created a framework to assess green supply chain practices. According to the study's outcomes, the proposed methodology can offer a more accurate assessment of green supply chain practices than conventional approaches. Furthermore, to choose supply chain sustainability measures, Abdel-Basset et al. (2019) developed a hybrid plithogenic decision-making strategy with quality function deployment. According to the authors, the suggested strategy can assist organizations in choosing the best sustainability indicators in accordance with stakeholder needs. The study found that the suggested strategy can strengthen decision-making and increase supply chain sustainability. In a similar study, Yadav et al. (2020) used an automotive scenario to establish a framework to address issues with sustainable supply chains utilizing Industry 4.0 and circular economy techniques. The framework, according to the authors, can aid businesses in implementing sustainable practices and enhancing the efficiency of their supply chains by utilizing industry 4.0 technology and circular economy principles. The investigation found that the proposed structure could enhance supply chains' sustainability and increase organizations' profitability.

These studies offer information on various strategies that can be employed to improve the sustainability and resilience of supply chain operations. The application of Industry 4.0, the circular economy, the neutrosophic approach, quantitative methodologies, big data analytics, and other emerging technologies can enhance supply chain performance and support businesses in implementing sustainable practices. These methods can help to enhance data-driven decision-making, highlight areas for development, and provide insightful information on supply chain performance metrics.

9.2.14 Identifying areas for improvement and automation opportunities

Enhancing organizational efficiency and maintaining competitiveness in today's ever-changing corporate environment requires the identification of areas for improvement and automation opportunities. Organisations must adopt a methodical strategy to find these areas which has been demonstrated in Fig. 9.4. To fully grasp the strengths, limitations, and pain spots of the current processes and workflows. Data analysis, stakeholder comments, and benchmarking against industry norms can all be used to perform this evaluation. Organizations can learn a lot about areas that need development and automation by analyzing pertinent data and talking to important stakeholders. Identifying potential for automation also requires remaining current with developing technologies and market trends. Organizations can pinpoint certain operations or processes that can be streamlined, automated, or improved to maximize efficiency, cut costs, and boost overall performance by combining these efforts. In the end, this proactive approach to pinpointing opportunities for automation and improvement guarantees that businesses are prepared to adjust to shifting consumer needs and achieve sustained growth.

To identify automation opportunities, these steps should be followed.

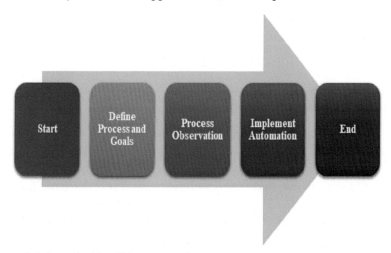

Figure 9.4 Steps for identifying automation opportunities.

9.2.14.1 Process and goal definition

Start by outlining the procedures followed by the company in detail. Automation can improve efficiency and effectiveness even if they are already good. To find areas where automation can be used, examine the data collection, transformation, and use procedures involved in these operations. Furthermore, describe the company's objectives and list the major players involved.

9.2.14.2 Process observation

Study the procedures in operation and think of creative ways to make them more useful. Look for processes that can be simplified or eliminated from the analysis. Engage those involved in the process and get their feedback on any problems or areas that may be made better. Make sure that any modifications adhere to the broad company objectives.

9.2.14.3 Implement automation

Establish the process modifications after finding possible automation enhancements, then communicate them with the appropriate stakeholders. Employ the automation solution and monitor as the workflow is improved, bottlenecks are removed, and efficiency is increased. If the automation is effective, companies can apply this strategy to other organizational procedures.

These procedures make it simple to locate automation opportunities, promote process enhancements, and boost supply chain operational effectiveness. It is essential to coordinate automation efforts with the objectives and routinely assess how automation affects the operations.

9.3 Supply chains in different sectors

A supply chain is a network of individuals and organizations in charge of creating a product, packaging it, and sending it to the consumer. The chain's first and last connections are the suppliers of raw materials, and its final link is the mode of transportation used to deliver the finished product to the client. The benefits of a streamlined supply chain, such as lower costs and higher production, can demonstrate the significance of supply chain management has been studied (Aday & Aday, 2020). To reduce costs and keep up with competition, businesses improve their supply chains. This includes each step taken to provide a client with a finished good or service, as analyzed by Chang et al. (2019). Fig. 9.5 illustrates the stages involved in acquiring raw materials, moving them to the production stage, and then delivering the finished goods to a warehouse or retail location where the consumer may pick them up.

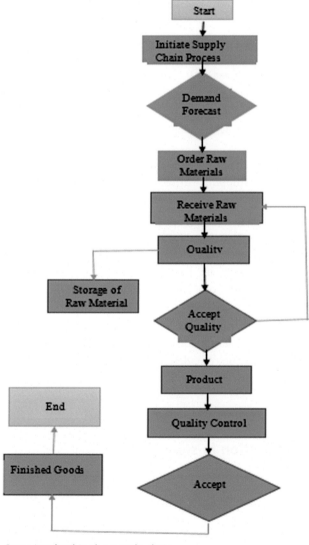

Figure 9.5 Steps involved in the supply chain.

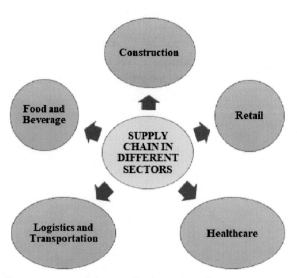

Figure 9.6 Different sectors where supply chain is used.

Organizations involved in the supply chain include manufacturers, suppliers, warehouses, shipping firms, distribution centers, and retailers (Belhadi et al., 2021). The visual presentation of various sectors where the supply chain is used has been demonstrated in Fig. 9.6. When a company receives a customer order, the supply chain starts operating. Product formation, promotional activities, processes, networks of distribution, financing, and customer service are thus among its core operations. Successful supply chain management can reduce the total expenditures of a business and increase profitability (Pettit et al., 2019). A broken link can have an expensive and negative impact on the chain as a whole. Several economic sectors, which include manufacturing, retail, healthcare, food and beverage, logistics and transportation, etc, depend on supply chain management.

9.4 Automation in supply chain

The rising levels of demand are beyond the capacity of conventional supply chain management. The implementation of automation technologies is the key to attaining a competitive edge. Tools for supply chain automation enable businesses new strategies for navigating unpredictability and maintaining the flow of goods and services. Automation of the supply

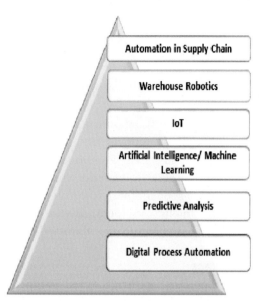

Figure 9.7 Types of technologies used in supply chain for automation.

chain is the process of handling supply chain jobs through technology rather than directly involving humans. Numerous sorts of technology are used in automation, which takes many different forms, that has been depicted in Fig. 9.7. Automation of various processes and duties within the scope of supply chain management is referred to as supply chain automation. It involves using tools like AI, machine learning, robotics, and data analytics to improve decision-making, streamline operations, and increase productivity (Dash et al., 2019).

9.4.1 Types of supply chain automation

9.4.1.1 Data capture automation

Data automation may significantly improve corporate operations at a time when companies need to save costs wherever they can. Due to the sheer volume of documents that must be processed every day, including invoices, inventory lists, purchase and sales orders, shipping notes, bills of lading, and others, data entry is one of the most time-consuming and error-prone operations in the whole supply chain. Furthermore, considering there is a speed restriction on how quickly employees can type data manually, as time goes on, their ability to remain focused and productive will certainly decline. Fortunately, data capture automation may use optical character recognition (OCR) and other comparable technologies to retrieve information from digital or scanned documents. This procedure simply takes a few seconds, and you may repeat it with all the documents.

9.4.1.2 Warehouse automation
Due to the various advantages warehouse robotics provide, like higher efficiency and order accuracy, businesses are investing in them (55% of them, to be exact). They also lessen workplace injuries brought on by mishaps or exhaustion because robots do the literal heavy lifting. Collaborative robots, intelligent forklifts, automated storage and retrieval systems (AS/RS), autonomous cars, automated sortation systems, and other devices are all included in automation for warehouses.

9.4.1.3 Internet of things
IoT devices can communicate data, track stored products, and report on their position and status. They are used by supply chain managers to reduce paperwork and track the movement of items. They are especially helpful in the food and retail industries for keeping track of products that are susceptible to changes in temperature or humidity. Wireless networks (WiFi, Bluetooth) and skilled staff are needed for IoT devices to function properly, but the potential ROI (maximizing stocks, limiting waste) is worthwhile the effort.

9.5 Predictive analytics

While modern supply chain management takes a proactive stance, traditional supply chain management often takes a reactive stance while anticipating client requirements. This is where predictive analytics comes in, by using data analysis to predict future trends. Demand forecasting, pricing strategy, & inventory management are just a few of the applications for this increasingly common automated technology (Mahlamäki et al., 2020).

A significant time and money saver in the supply chain is the automation of manual operations like document processing. Companies may improve their supply chain operations, get rid of bottlenecks, and establish a smoother workflow by automating these procedures. Employees can now concentrate on more worthwhile tasks like customer prospecting and relationship-building thanks to automation. Automation also makes departments and locations more visible and transparent, which lowers the possibility of errors in order processing and delivery. Automated shipping procedures increase transparency and customer satisfaction by enabling customers to follow their shipments in real-time (Omar et al., 2021). The production of accurate and beneficial data is one of the main benefits of automation. Companies can quickly obtain reliable data by dismantling data silos and developing connected systems. This makes it possible for leaders to make defensible decisions, especially when it comes to areas like demand planning, supported by reliable and current information. Additionally, automation makes it possible to produce dynamic reports that are simple for teams to share, encouraging teamwork and well-informed decision-making.

Automation is essential for delivering quicker and more effective customer support. Companies may guarantee quick delivery times, competitive pricing, and real-time order updates using automated processes. Meeting and exceeding client expectations is essential for market success at a time when consumers are becoming more and more demanding. Automation enables businesses to provide excellent customer experiences, boosting client loyalty and happiness (Dietrich et al., 2021).

9.6 Role of robots in supply chain automation

Automating processes is possible thanks to robotic process automation. The term "software automation" refers to a technology that operates within software and streamlines human operations. Between the company and its suppliers, it covers the flow of goods and services. Together, these words make up RPA in Supply Chain, which has made SCM efficient across sectors. On the other side, it is the brand-new lever that promotes process effectiveness and efficiency. It lessens our workload by automating simple and repetitive problems. It entails procedures for delivering a good or service to the customer. Vendors, merchants, producers, warehouses, transportation firms, and distribution centers are a few of the enterprises active in the sector. As demand rises, it enables firms to quickly scale up to meet their needs (van Hoek et al., 2022). Supply chain management is only one of several businesses that employ robotic process automation

(RPA) to automate repetitive, normal processes. It can be applied in the sector to automate a variety of tasks, including data entry, order processing, and inventory management.

9.6.1 Order processing and payments

Order processing is an appeal made by the client, and the provider then decides whether or not to comply. It avoids data entry errors by automatically obtaining sales order data across every one of the order types, including fax, email, mail, and EDI. To make order entry and fulfillment simpler, it maintains complicated business rules and automates key order-processing procedures.

9.6.2 Onboarding of partners

The process of integrating new products and services is known as onboarding. Intelligent bots were developed that synchronize and contribute to automating the procedure of onboarding partners via it.

9.6.3 Shipment scheduling and tracking

The procedure of time management and monitoring the delivery of goods is called shipment scheduling and tracking. Automation in shipment by electronically entering the data, scheduling it based on the data, and tracing it using a unique ID that is generated when the shipment is scheduled. Supply chain leaders with a focus on customer fulfillment can make use of robotic process automation as a "speed to value" solution today.

9.6.4 Invoicing

A document from the supplier to the customer that provides the details of the quantities and rates of the supplier's goods or services is called an invoice. The procedure for billing uses data entry, the extraction procedure of data, and computation through automation.

9.6.5 Procurement and inventory

The process of storing products and preparing them for sale, encompassing all final products that are available for sale, is known as procurement and inventory. A unique ID in the database triggers automatic data entry, which is automated by the Robotic Process Automation.

9.6.6 Supply and demand planning

It is a technique of predicting demand to produce and deliver the good or service more effectively and to ensure customer satisfaction. Utilize its program to automatically update the necessary goods' data and strategy for managing new product entries.

9.6.7 Customer services

Giving a customer prompt, responsive, and enthusiastic service is the traditional definition of good customer service. When clients request items or changes to goods, their requests are sent immediately to the organization's customer request portal. If there is a new need, the database will be expanded (Radke et al., 2020, Huang et al., 2022).

When linked with supply chain operations, robotic process automation (RPA) is a useful means of investment that provides several advantages and solutions. Firstly, it allows for ongoing inventory tracking and shipping management, sending out instant alerts when stock needs to be restocked. RPA collects and aggregates data from annual analyses through the automation of supply and demand planning, proactively making forecasting and decision-making more effective. Analyzing huge data sets and client purchasing patterns, also makes trend research easier and enables firms to spot trends and come to wise judgments. Based on established parameters including pricing, quantities, purchase volumes, and customer preferences, it may automatically produce purchase orders. By running tasks in an automated and organized manner, RPA brings efficiency to supply chain operations, with procurement managers focusing on higher-priority tasks and exceptions.

RPA eliminates the need for human involvement thanks to its round-the-clock functionality, allowing managers to concentrate on important activities that call for human judgment. Also, it simplifies coping with documents with vendors and suppliers and processing invoices while assuring accuracy and effectiveness. RPA expedites processes like data validation and verification, labor allocation optimization, and error reduction for jobs like extracting information from invoices. Optimizing logistics costs through effective pricing of procured materials is a significant area where RPA excels. Automated price lookup is a perfect candidate for RPA deployment because manual price lookups and comparisons across a variety of products can be time-consuming. It regularly does price searches and material comparisons, producing the best results for cost reduction. RPA also makes data entry and integration processes simpler, which minimizes errors and increases data accuracy. With the use of RPA, maintenance management is made more efficient, promising that equipment servicing responsibilities are completed. Helo and Hao (2022) explores the transformative effects of AI on global supply chain management (SCM), from the stages of planning and scheduling to optimization and transportation. It offers a concise review of AI's role in SCM and provides an analysis of emerging AI-driven business models through case studies. The research evaluates the practical AI solutions of various companies and their contributions to value creation within the supply chain, concluding with suggestions for designing AI-integrated business models for SCM applications. Jain (2019) addresses the challenge humans face in adapting to new technologies, such as AI and robotics, which are developing at an exponential rate. It underscores the importance of proactive discussion on the impact and future requirements of living and working with robots regularly. The paper emphasizes the need for education and dialogue regarding the societal roles of robots, their potential evolution, and their relationships with humans, including how these dynamics could alter human roles. It warns that without proper preparation, the rapid integration of robots could cause social upheaval and confusion. The paper also suggests that international regulations could shape robot development and utilization, preventing misuse. It discusses the current influence of robots on daily life and supply chains, their significant contributions to date, and forecasts their future implications.

9.6.7.1 Use cases

Modern supply chain automation is impossible without AI. It enables numerous automation technologies, including RPA, autonomous vehicles, warehouse robots, and digital workers, to effectively complete routine and error-prone operations. Automation of the supply chain using AI has been successful in a number of fields. Through smart automation and digital workers combining conversational AI and RPA, back-office automation processes like document processing may be optimized. The automatic and precise execution of tasks made possible by this integration increases productivity. Automation and AI can be used to optimize logistics throughout the supply chain. Significant investments are being made in transport automation technology, such as driverless trucks, by well-known businesses like Amazon, Tusimple, and Nuro, which is revolutionizing the effectiveness of logistics operations. Collaborative robots (cobots), which are AI-enabled technology, dramatically enhance warehouse management (Delgado et al., 2019). By automating warehouse chores, these cobots increase efficiency, productivity, and safety. Market prominent players like Ocado have significantly impacted the warehouse automation industry. Additionally, computer vision (CV) systems with AI capabilities are essential for automating product quality inspections. Manufacturers may improve productivity and accuracy in production lines while assuring greater quality assurance by utilizing tenacious AI-powered CV systems. The implementation of AI and computer vision in automated inventory management is yet another instance. Real-time scanning is one of the repetitive inventory chores that can be automated by bots that have computer vision and AI/ML capabilities. Such inventory scanning bots can be used by retail stores to simplify inventory management. To prevent future failures, it is necessary to carefully evaluate the economic feasibility and benefits over the long term of these solutions. Business leaders may make intelligent choices due to the forecasts provided by supply chain AI for specific regions. AI-powered forecasting systems offer precise demand insights relevant to each location by taking into account criteria like events, holidays, and trends that are distinct to that region. This personalization enables improved fulfillment procedures that satisfy particular geographical requirements, increasing operational effectiveness (Delgado et al., 2019).

9.7 Benefits of supply chain automation

Numerous advantages of supply chain automation can boost operational effectiveness and boost the effectiveness of the entire supply chain. However, it also presents some difficulties that businesses must overcome.

- Improved Operational Efficiency: Automation minimizes manual labor, streamlines procedures, and removes bottlenecks, resulting in accelerated order fulfillment.
- Cost Savings: Organisations can save a lot of money by automating operations since it lowers labor costs, reduces errors, improves inventory levels, and better utilizes resources (Dash et al., 2019).
- Improved Customer Satisfaction and Decreased Rework: Automation lowers the possibility of human error, assuring greater accuracy in tasks like data entry, order processing, and inventory management.

- Improved Visibility and Transparency: Automation enables improved management, surveillance, and control of inventories, orders, and shipments by enabling real-time visibility into supply chain activities. As a result, there will be more openness, fewer delays, and better customer service (Mastos et al., 2020).
- Better Decision-Making: Automation makes it easier to collect, analyze, and report data, which gives decision-makers useful information. It enables businesses to decide with confidence on demand forecasting, purchasing, inventory control, and logistics optimization (Tan & Sidhu, 2022).
- Scalability and Flexibility: Automated systems can quickly adapt to changing company requirements, allowing for flexibility and adaptability in managing growing volumes, a variety of product lines, and changing client expectations.

Supply chain automation has several advantages, including higher operational effectiveness, lower costs, more accuracy, better customer service, and quicker reaction times. It's crucial to keep in mind, though, that automation should be used cautiously, considering factors like cost-effectiveness, scalability, and the necessity of human knowledge and oversight in some situations. In today's fast-paced corporate environment, an intelligent automation strategy may revolutionize supply chain processes and provide a competitive advantage (Kopyto et al., 2020).

9.8 Challenges and future directions

9.8.1 Addressing challenges and limitations of supply chain automation

Businesses have several difficulties with supply chain management (SCM), including a lack of transparency, production schedules, and team synchronization. Intelligent data analytics are used by businesses in the manufacturing, retail, and Fast-Moving Consumer Goods (FMCG) sectors to address these issues. Supply chain analytics is the phrase for this. By doing so, businesses can determine consumer preferences and forecast future demand for their products or services.

- Integration with legacy systems: Integrating the automation technology with current legacy systems is one of the major implementation problems for RPA. Legacy systems may have intricate architectural designs and be incompatible with RPA tools, making smooth integration extremely difficult (Syed et al., 2020).

- Process identification and prioritization: One of the biggest challenges is figuring out which processes should be prioritized for automation. Businesses must carefully evaluate and appraise their processes to decide which ones are most suitable for automation, considering aspects like complexity, stability, and possible advantages (Santos et al., 2020).
- Change management and employee opposition: Implementing RPA frequently results in changes to job duties and responsibilities, which may cause employee resistance. Effective change management and clear staff communications on the advantages of RPA are essential for overcoming resistance and managing the change process (Santos et al., 2020).
- Data quality and variability: For automated activities to be completed, RPA depends on precise and reliable data inputs. Inconsistencies, errors, and variability in data, however, might make it difficult to deploy RPA and result in unsuccessful processes or inaccurate outputs (van Hoek et al., 2022).
- Scalability and maintenance: Scalability issues arise when RPA deployments expand. RPA scaling needs careful planning and coordination across numerous processes and systems. RPA solutions might be challenging to maintain and update to meet evolving company needs and technology improvements (Syed et al., 2020).
- Training and skill requirements: It might be difficult to develop and retain the skills required for managing and implementing RPA. To design, build, and manage RPA solutions, organizations require personnel with the necessary knowledge and skills, which calls for the right training and upskilling efforts (Javaid et al., 2021).

9.8.2 Emerging trends and future developments in robotics and automation

Numerous new tendencies in automation across several industries are shown by the study. These developments and opportunities are highlighted by industrial robotics, intelligent unmanned autonomous systems, AI, and automation in certain industries, such as the food industry and factory automation. Dzedzickis et al. (2021), for instance, describe advanced industrial robotics applications while highlighting fresh developments and opportunities. Cobots, or collaborative robots, that can work alongside people, transportable robots for a variety of jobs, and the fusion of robotics with cutting-edge technology like machine learning and virtual reality are a few examples. The development of intelligent unmanned autonomous systems is the main topic of Zhang et al. (2017). They highlight the development of robotic systems, drones, and autonomous vehicles for a range of uses in agriculture, transportation, and security. These systems make use of technologies including decision-making algorithms, sensor fusion, and computer vision. AI is examined by Gill et al. (2022) as a major force behind next-generation computers. Deep learning, reinforcement learning, and federated learning are some of the new trends in AI that they examine. In a variety of industries, including manufacturing, healthcare, and smart cities, these trends enable intelligent automation, predictive analytics, and decision-making skills. The investigation by Iqbal et al. (2017) primarily looks at the potential of robotics in the food sector. They emphasize the employment of robots in packaging, quality control, and food processing. New developments in robotic automation in precision agriculture, visual systems for food inspection, and specialized robotic arms are all examples of emerging trends. An overview of current technology and new trends in factory automation is given by Dotoli et al. (2019). Among the important topics they cover are the Industrial Internet of Things (IIoT), cyber-physical systems, cloud computing, and data analytics for smart manufacturing. Real-time monitoring, proactive maintenance, and adaptable production systems are made possible by these trends. These studies draw attention to the development of cooperative robots, intelligent unmanned systems, automation driven by AI, specialized robotics in certain industries, and the incorporation of developing technologies in automation. These themes are influencing the future of automation by providing fresh opportunities for improving effectiveness, productivity, and performance across numerous industries.

A number of researchers have offered insightful information about where automation in supply chain management may go in the future. The idea of Agri-Food 4.0, which entails integrating cutting-edge technologies like IoT, robots, and precision agriculture in the agricultural industry, is highlighted by Lezoche et al. (2020). The use of these technologies to improve agricultural productivity, supply chains, and sustainable practices is where the future is headed. The significance of Industry 5.0 and the Triple Bottom Line (TBL) strategy in supply chain management is emphasized by Varriale et al. (2023). To create supply chains that are sustainable and socially responsible, this method takes into account economic, environmental, and social factors. The next step is to combine cutting-edge technologies with human-centered strategies to build ethical and resilient supply chains. The use of AI in supply chain management is the main topic (Riahi et al., 2021). Developing AI-driven decision support systems, predictive analytics, and autonomous decision-making capabilities are some of the future research objectives. Big data analytics, blockchain, and IoT all hold enormous promise for enhancing supply chain efficiency and increasing overall operational effectiveness. These studies collectively shed insight into the future of automation in supply chain management, which includes the use of cutting-edge technologies, human-centered strategies, and the incorporation of AI. Organizations may improve the performance of their supply chains, advance sustainability, and promote stakeholder collaboration by adopting these future trends.

9.8.3 Potential research areas for further exploration

The supply chain improvement areas are shown in the diagram in many places. Time efficiency is the first area, and it has to do with how quickly orders are filled. The second issue is process complexity, which includes overseeing inventory visibility, logistics and shipping, and supplier cooperation. The third element is continuity, which also deals with reporting and data analytics, as well as returns management and reverse logistics. Potential research could delve deeper into each of these topics. For example, investigations into the field of time efficiency can concentrate on finding supply chain bottlenecks and applying process enhancements to boost throughput. Research could examine the use of technology, such as RFID, to enhance inventory visibility and supplier collaboration in the area of complexity. Research in the continuity field can concentrate on employing data analytics and reporting to understand supply chain performance and pinpoint development opportunities. Research could also examine techniques for handling reverse logistics and returns, such as creating effective return procedures or working with third-party logistics companies. Fig. 9.8 demonstrates the improvements required in automation.

Organizations may improve their operations, increase customer happiness, and gain a competitive edge in the market by tackling these areas.

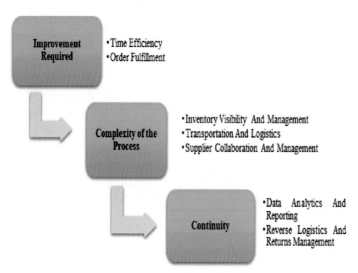

Figure 9.8 Improvement required in automation.

9.9 Sustainable supply chains

Clients, workers, shareholders, and governments have increased pressure on businesses in recent years to show better responsibility for society and the environment. This occurs as the business case for environmentally friendly operations becomes more compelling each year. Supply chains have gained attention from many firms as they consume a significant amount of resources and funds and frequently lead to unnecessary waste. Sustainability in the supply chain has thus become a major corporate objective (Jabbour et al., 2020). Businesses are now tracking the effects of their products and services on society and the environment from the starting point to the end point of the cycles.

The sustainability of the supply chain is the effort made by businesses to ensure that they take into consideration the effects that their operations have on both the environment and people. This includes the procurement of raw materials, production, storage, delivery, and every transportation connection in between (Bui et al., 2021). The objective is to have an advantageous effect on the people and communities that surround their operations while preventing environmental harm from components like energy usage, water usage, and generation of waste. These worries are in addition to the regular issues about revenue and profit in the corporate supply chain. The goal of sustainable supply chains also referred to as green or responsible supply chains, is to limit adverse effects on the environment and society while improving their financial performance (Kshetri, 2021). The idea of sustainability in supply chains incorporates several ideas, such as financial stability, responsibility for society, and concern for the environment, including ethical sourcing.

Companies all across the world have made efforts to reduce their carbon emissions, reduce waste, and enhance working conditions. The monitoring of complex programs that, for instance, prioritize renewable energy, recycle products and materials, or promote higher levels of social accountability among suppliers is made possible by the tracking of sustainability indicators in supply chain management (SCM) systems. In addition, businesses can take advantage of intelligence and pre-defined rules to make sure that goods aren't delivered needlessly. For instance, they can make sure that goods are dispatched from the nearest distribution center rather than one on the opposite end of the country. Fig. 9.9 provides insights into the various factors of sustainable supply chains.

Studies show that the supply chain is mostly responsible for a company's environmental impact. Supply chains usually involve energy-intensive manufacturing and distribution as goods are produced and

Figure 9.9 Key factors of sustainable supply chains.

shipped throughout the world. Therefore, organizations can frequently make the biggest influence by modifying their supplier chain rather than other corporate practices. The complexity of the many supplier relationships and border crossings makes it challenging to sustain the supply chain. Due to this intricacy, it could be challenging to see important operational aspects like the working conditions at a supplier's far-off factory.

9.9.1 Environmental sustainability

Countries are passing regulations and legislation that reduce carbon emissions including the impact of greenhouse gases in response to modifications in the environment across the world and the need to combat global warming. The urgency of resource depletion and climate change challenges is what motivates this. Resources are in short supply as a result of the expansion of economies like those of China and India and the rising world population. In light of this, customers, stakeholders, and companies are embracing sustainability and looking for eco-friendly products in greater numbers. As a result, companies are working to integrate sustainability tactics and environmentally sound decisions into their supply chains. By using environmentally friendly inputs and turning them into reusable outputs, this strategy, also known as "green and sustainable supply chain management," aims to build a sustainable supply chain. The goal is to lower expenditures while also helping the environment.

9.9.2 Social responsibility

As businesses are compelled to consider more about how their actions affect stakeholders, particularly the environment, they are paying more attention to corporate social responsibility (CSR) in supply chain management (SCM). SCM includes all processes that convert raw materials into completed goods, and it encompasses the efficient and economical management of the distribution of goods and services. Concerning the areas of human rights, occupational health and safety, sustainable production, and environmental practices as they relate to the supply chain, businesses need to be aware of their social obligations. These concerns include slavery, child labor, and working conditions.

9.9.3 Ethical sourcing and supply chain transparency

It implies that a company is aware of everything that is happening at every step of its supply chain and shares information about those activities both internally and externally, with facts to back it up. Companies' choices on supply chain transparency and the extent to which they will uphold ethical business practices can, however, be very diverse. Customers are more than four times more likely to believe in businesses that have a clear purpose and can explain how products are manufactured.

9.9.4 Collaboration and partnerships

To efficiently meet demand and ensure on-time, complete delivery, collaboration in the supply chain requires working with both internal and external partners to maintain an optimal flow through the supply chain. To make it simpler to identify and address issues, supply chain partners must develop real-time shared visibility and protocols. All parts of the supply chain's operations, including forecasting, capacity planning, and quality control, are collaborative. Given the scope and volume of data, transactions, and physical commodities traveling through today's global supply chains between suppliers, corporations, contract manufacturers, and logistics providers, it is only natural to expect that they must be highly connected systems (Jabbour et al., 2020, Bui et al., 2021, Kshetri, 2021, Abdel-Basset et al., 2020).

9.10 Case studies

9.10.1 Case studies on warehouse optimization through robotic automation

A design optimization method for redundantly actuated cable-driven parallel robots (CDPRs) utilized in automated warehouse systems was put forward by Zhang et al. (2020). The cable tensions and actuator placements for the CDPR were all optimized by the authors using a genetic algorithm. The suggested method produced a CDPR design with enhanced load-carrying capacity, workspace, and accuracy that was more effective and dependable. A dual-loop optimum control strategy for a robot manipulator utilized in warehouse automation was developed by Prakash et al. (2020). The tracking precision, disturbance rejection, and energy economy of the manipulator were all optimized by the authors using a feedback control technique. The suggested method led to better tracking precision, lower energy usage, and higher throughput.

Using dynamic demand, Li et al. (2020) suggested a scheduling approach for multi-robot intelligent warehousing systems. The real-time scheduling of robots and employment was optimized by the authors using an enhanced ant colony optimization technique. The suggested mechanism produced a reliable scheduling system that was productive and reduced the effects of dynamic demand variations. In warehouse contexts, Lee and Jeong (2021) proposed a mobile robot path optimization method based on reinforcement learning algorithms. The path of mobile robots in a warehouse was optimized by the authors using Q-learning and deep Q-learning algorithms. A more effective path planning system with high accuracy and reduced path times was produced as a result of the suggested methodology.

The Nimbro Picking system, a flexible part-handling system for warehouse automation, was introduced by Schwarz et al. (2017). The device is made up of a robot arm with a suction cup that can pick up and set objects of all sizes and shapes. The authors showed how well the system handled a variety of objects and produced high throughput rates. A human-robot co-working system for automation of warehouses was proposed by Rey et al. (2019), allowing human operators to collaborate with robots to choose and move things in a warehouse. They created a system prototype using ROS (Robot Operating System) and OpenCV (Open Source Computer Vision), which demonstrated enhanced effectiveness in terms of time and accuracy. A framework for creating automated order-picking robots for warehouses and retail businesses was proposed by Bormann et al. (2019). They identified the goods with the aid of computer vision algorithms and produced a picking plan, which was subsequently carried out by a robot arm. They picked goods with a high degree of efficiency and precision thanks to their strategy. A synchronization technique for robotic forward-reserve warehouses utilized for last-mile deliveries for e-commerce was put forth by Jiang and Huang (2022). They boosted order throughput and decreased order processing time by optimizing inventory allocation, order batching, and selecting paths using mathematical modeling and simulation. An algorithm for assigning tasks optimization in an automated warehouse system was presented by Baras et al. (2021). They tested their strategy using a simulation program they had created, and the results showed higher robot throughput and fewer wait times. As per the studies, the comparison of different parameters of robots in the automation process of warehouses was discussed in Table 9.1.

Table 9.1 Comparative study on warehouse optimization through robotics.

Study	Reach and payload	Flexibility	Speed	Space and footprint	Maintainability	Standardization
Zhang et al. (2020)	High	Yes	Fast and Efficient	Compact	Easy	Utilizes design optimization techniques specific to cable-driven parallel robots
Prakash et al. (2020)	Customizable	yes	Fast and Efficient	Compact	complex	Utilizes optimal control techniques tailored to robot manipulators
Li et al. (2020)	Customizable	yes	Efficient	Require space	complex	Provides a mechanism for scheduling in multi-robot warehouse systems
Lee and Jeong (2021)	Variable	yes	Fast and Efficient	Compact	Easy	Utilizes path optimization techniques based on reinforcement learning
Schwarz et al. (2017)	Variable	yes	Fast	Require space	complex	Provides versatile part-handling techniques for warehouse automation
Rey et al. (2019)	High	yes	Efficient	minimize space requirements	Easy	Utilizes ROS and OpenCV, industry-standard frameworks
Bornann et al. (2019)	Depending on the robot arm design	Yes	Fast	Compact	Easy	Utilizes computer vision techniques that can be standardized across systems
Jiang and Huang (2022)	customizable	yes	Efficient and fast	compact	Scalable and ease of maintenance	Mathematical modeling and simulation techniques can be standardized for similar warehouse setups
Baras et al. (2021)	customizable	Yes	Fast	Efficient space utilization	easy	Algorithmic approach can be standardized for similar automated warehouse systems

In this research, many issues of reach and payload, flexibility, speed, space and footprint, maintainability, and standardization are addressed in relation to improvements in warehouse automation systems. Each study offers distinctive insights and draws attention to particular methods or strategies that can be useful for developing and putting into practice automated warehouse systems.

9.10.2 Robotic sorting and packaging in e-commerce fulfilment

To increase production, accuracy, and efficiency in e-commerce warehouses, robotic systems have grown in significance.

Based on robotic arm technology, Wang et al. (2020) developed a products sorting robot system for e-commerce logistics warehouses. The system is built to sort products based on their size, weight, and final placement. The system's performance was assessed by the authors for speed and sorting accuracy, and they found encouraging results. The system's ability to efficiently and precisely sort products suggests that it has the potential to increase the effectiveness of e-commerce warehouse operations. For the parcel sorting procedure in vertical sorting systems in e-commerce warehouses, Tan et al. (2021) developed an optimization strategy. To assess the effects of various variables, including parcel size and sorting procedures, on the sorting effectiveness, the authors created a simulation model. They increased sorting efficacy with regard to sorting duration and throughput by using a genetic algorithm to optimize the process. The authors claimed that their method might improve sorting effectiveness and reduce costs for e-commerce warehouses.

An automated robot-picking system for use in warehouse applications for e-commerce was introduced by Liang et al. (2015). According to consumer orders, the system is built to pull things off shelves and pack them. The accuracy, speed, and flexibility of the system were all examined by the authors. They claimed that the system picked goods quickly and accurately and showed great flexibility in adapting to various product types and sizes. The approach, according to the authors, might aid e-commerce warehouses in increasing the effectiveness of order fulfillment. For e-commerce warehouses, Liu et al. (2022) created a quick sorting robot system. The system includes a multi-robot coordination mechanism to increase sorting speed and efficiency. It is designed to sort products according to their destination and delivery time. The system demonstrated considerable advantages over traditional sorting techniques, according to the authors' evaluation of the system's performance in terms of sorting time, throughput, and energy consumption. The technique, according to the investigators, might help e-commerce facilities improve sorting effectiveness and reduce operating costs. A study on a green intelligent logistics sorting system in a big data environment is presented by Gao and Liu (2020). The authors provide a method for optimizing the sorting procedure in logistics operations that combines clever algorithms and big data analytics. The issues of energy consumption and environmental impact in sorting systems are discussed in the study. The outcomes show that the suggested approach significantly improves energy efficiency and sustainability, lowering the logistics operations' carbon footprint.

A hybrid agent-based strategy is used by Barenji et al. (2019) to construct an intelligent e-commerce logistics platform. The authors suggest a platform for optimizing many facets of e-commerce logistics operations, such as order fulfillment, inventory management, and delivery scheduling. This platform makes use of AI, machine learning, and multi-agent systems. The study highlights how incorporating intelligent agents can strengthen decision-making procedures and boost operational effectiveness. The findings show that the suggested platform can manage dynamic and complicated logistical scenarios successfully, improving customer satisfaction and reducing costs. Several researchers have proposed automation in e-commerce through robotics, as discussed above. Depending on the working, capacity, and efficiency, a comparison of robots is presented in Table 9.2.

The table shows that a variety of robots and systems are employed in the logistics sector for a variety of actions. Each system has a different level of efficiency and capability; some robots are intended for particular jobs like picking or sorting, while others are more adaptable and can perform a variety of jobs. It is anticipated that as technology develops, even more advanced robots and systems will be created to handle a larger range of duties in the logistics sector.

9.10.3 Autonomous mobile robots for material handling in manufacturing

Numerous research investigates the use of mobile autonomous robots in industrial and material handling systems. The pickup and delivery problem, autonomous mobile robot improvement, multi-agent reinforcement learning structures, scheduling operations, and the usage of autonomous mobile manipulators are just a few of the several facets of this topic that

Table 9.2 Comparison of robots in Sorting and Packaging in E-commerce.

Study	Type of robot	Receiving	Picking	Packing	Shipping	Weight	Efficiency	Time
Wang et al. (2020)	Robotic arm technology	✓	✓	x	x	Up to 60 kg	97.8%	–
Tan et al. (2021)	Vertical sorting system	x		✓	x	–	99%	–
Liang et al. (2015)	Robots, conveyors	x	✓	✓	x	Up to 12 kg	98.2%	2.8 seconds per pick
Liu et al. (2022)	Rapid sorting robot	x	x	✓	x	–	high	–
Gao and Liu (2020)	Hybrid system	✓	x	x	✓	Up to 50 kg	high	–
Barenji et al. (2019)	Hybrid agent-based	✓	✓	✓	✓	–	high	–

are covered. A pickup and delivery problem (PDP) with recharge for material handling systems using AMRs was put by Jun et al. (2021). To optimize the routing and scheduling of AMRs for PDP tasks while taking the battery recharging limits into account, the authors created a mixed-integer linear programming model. The outcomes demonstrated that the suggested methodology is capable of successfully optimizing the PDP problem with recharge. An autonomous mobile robot with a manipulator for use in manufacturing contexts was created by Datta et al. (2008). The material dealing with, installation, and inspection operations were all part of the robot's design. The authors showed how to successfully use the robot in an industrial application by using a supervisory control system to coordinate the robot's actions.

A multi-agent reinforcement learning (MARL) structure for intelligent manufacturing with AMRs was presented by Agrawal et al. (2021). The authors improved the scheduling and routing of AMRs in a manufacturing plant using a MARL algorithm. The findings demonstrated that the suggested framework can successfully increase the productivity of the production processes. The use of mobile robots to schedule intelligent intra-factory material delivery activities was studied by Kousi et al. (2019). To improve the scheduling and routing of AMRs in a manufacturing facility, the authors created a scheduling model. The outcomes demonstrated that the suggested model may successfully shorten the cycle time of material handling procedures and boost manufacturing process effectiveness.

Omnivil, an autonomous mobile manipulator for flexible production, was introduced by Engemann et al. (2020). The robot's functions in a production environment include material handling, assembly, and inspection. The authors showed how to successfully use the robot in an industrial setting by using a hierarchical control architecture to coordinate the robot's actions. AMRs provide many benefits over conventional material handling systems, including more productivity, efficiency, and flexibility. However, further study is required to address issues like battery life, system integration, and affordability to enable AMRs to be more widely used in manufacturing and logistics. Table 9.3, discusses the performance of the robot studies above in the manufacturing industry.

9.10.4 Last-mile delivery optimization with drone technology

The integration of advanced technologies such as drones and autonomous robots has led to substantial breakthroughs in last-mile delivery, the last

Table 9.3 Performance comparison of robots in manufacturing sector.

Study	Sensing	Perceiving	Decision making	Action taking	Type of material handling
Jun et al. (2021)	–	–	✓	✓	Pickup and delivery
Datta et al. (2008)	✓	✓	✓	✓	Manufacturing
Agrawal et al. (2021)	–	–	✓	✓	Intelligent manufacturing
Kousi et al. (2019)	–	–	✓	–	Intra-factory material supply
Engemann et al. (2020)	–	–	✓	✓	Flexible production
Ammar et al. (2021)	–	–	✓	–	Material quality management
Dörfler et al. (2019)	✓	✓	✓	✓	Robotic fabrication
Xiao et al. (2020)	–	–	✓	–	Advanced manufacturing
Dang et al. (2019)	–	–	✓	–	Transportation and manufacturing

link in the supply chain that transports goods to customers. Recently, the idea of last-mile delivery using drones has drawn a lot of attention. A thorough analysis of the market viability and citizen involvement for last-mile drone delivery across European cities was carried out by Aurambout et al. (2019). They projected the market's prospective demand for delivery services by drones and examined the factors influencing this technology's implementation. The effects and challenges of using intelligent technologies for modernizing last-mile delivery were studied by Sorooshian et al. (2022). They talked about the advantages and difficulties of incorporating technologies like drones, self-driving cars, and smart lockers, emphasizing the necessity for infrastructural development and legal frameworks.

The optimization and evaluation of a robot-assisted last-mile delivery system were the main topics of the study by Simoni et al. (2020). To optimize the routing and programming of autonomous robots, they created algorithms and mathematical frameworks that incorporated delivery time, vehicle capacity, and consumer preferences into account. For last-mile logistics, Lemardelé et al. (2021) investigated the possibilities of both drones and ground autonomous delivery systems. They discussed issues including energy efficiency, safety, and integration with current distribution networks as they explored the advantages, difficulties, and deployment possibilities of these technologies. An application and concept for a multimodal autonomous last-mile delivery system were suggested by Samouh et al. (2020). To optimize the delivery process, they provided a framework that incorporates drones, ground robots, and conventional vehicles while considering variables like real-time traffic circumstances, package size, and delivery urgency.

A numerical case study was carried out in Milan, Italy, by Borghetti et al. (2022) to investigate the real-world use of drones for last-mile delivery. Taking into account variables like product weight, delivery distance, and airspace limitations, they evaluated the viability and effectiveness of drone delivery in an urban environment. Chen, Leon, and Ractham (2022), concentrated on the emerging market economy, looked at customer acceptance of last-mile drone delivery services. They found elements impacting consumer acceptance of drone delivery, such as perceived benefits, trust, and regulatory issues, through an investigation of consumer behavior and preferences. Skoufi et al. (2021) developed a techno-economic strategy regarding last-mile drone delivery. They examined the advantages and disadvantages of drone delivery systems while taking into account variables including operational costs, energy use, and environmental effects. Insights into the financial sustainability of drone-based last-mile delivery services were the goal of their investigation. Chen and Demir (2022) investigated various drone and delivery robot models and last-mile delivery capabilities. They talked about how these technologies may be used to optimize delivery routes, boost productivity, and cut costs. The research provided many operational examples and mathematical models that can help with decision-making for the logistics of last-mile delivery.

AGVs and drones for last-mile delivery were evaluated by Fehling and Saraceni (2022), with an emphasis on the lessons learned from Germany. They explored the advantages and difficulties of integrating drones and AGVs while taking things like infrastructure needs, rules, and public acceptance into consideration.

The comparative study given in Table 9.4, shows collectively contributes to the understanding of the use of drones for last-mile delivery. They address various aspects, including practical implementation, customer acceptance, techno-economic analysis, operational models, and feasibility considerations. While drones hold significant potential for transforming last-mile logistics, further research is needed to address

Table 9.4 Comparative studies based on performance of drones for last-mile delivery.

Study	Practical implementation	Customer acceptance	Techno-economic analysis	Operational models	Feasibility considerations
Aurambout et al. (2019)	High	Moderate	Moderate	N/A	High
Sorooshian et al. (2022)	Moderate	High	Low	N/A	Moderate
Simoni et al. (2020)	Moderate	Low	High	High	High
Lemardelé et al. (2021)	Moderate	Moderate	Moderate	High	High
Samouh et al. (2020)	High	Low	Low	High	Moderate
Borghetti et al. (2022)	High	Moderate	Moderate	N/A	High
Chen, Leon, and Ractham (2022)	Moderate	High	Low	N/A	Moderate
Skoufi et al. (2021)	Moderate	Low	High	N/A	High
Chen and Demir (2022)	High	Low	High	High	High
Fehling and Saraceni (2022)	Moderate	Moderate	Moderate	N/A	High

technical, operational, regulatory, and societal challenges for their widespread adoption in real-world delivery systems.

9.10.5 Collaborative robots (cobots) in assembly line operations

Due to the potential advantages collaborative robots (cobots) could provide in terms of enhancing productivity, ergonomics, and cost-effectiveness, cobot integration into assembly lines has attracted a lot of attention. Studies on a variety of topics, including ergonomics, economics, decision support systems, risk assessment, and modeling, have been done on the integration of cobots into assembly lines. A multi-objective migrating birds optimization technique was developed by Li et al. (2021) to solve the collaborative robotics and cost-oriented assembly line balance problem. Their research centered on how to best distribute tasks among humans and robots operating in a production line. They sought to develop effective and economical assembly line balancing by taking cost objectives into account and utilizing collaborative robots.

To balance ergonomic risk and cycle time in collaborative human-robot assembly lines, Stecke and Mokhtarzadeh (2022) tackled the issue. To concurrently optimize cycle time and ergonomic risk factors, they suggested a mathematical model and optimization strategy. Their research emphasized the significance of taking ergonomic considerations into account when designing assembly lines and underlined the potential advantages of cooperative human-robot systems. An examination of assembly-time performance (ATP) in manufacturing operations with collaborative robots was done by Chen, Leon, and Ractham (2022). To assess how several variables, such as robot utilization, task distribution, and system configuration, affected the overall ATP, they used a systems approach. Their research offered information on how to use collaborative robots effectively to increase worker productivity and effectiveness in manufacturing operations.

Weckenborg and Spengler (2019) suggested a cost-focused method for balancing manufacturing lines with collaborative robots while considering ergonomics. To reduce the risk of musculoskeletal illnesses in human employees, their study included ergonomic risk assessment in the assembly line balancing process. They emphasized the significance of including ergonomics in the planning of cooperative manufacturing lines involving humans and robots. The Benders' decomposition algorithm for balancing assembly lines with collaborating robots was compared by Sikora and Weckenborg (2022). Their research centered on assessing the effectiveness of various decomposition tactics and determining the best method for using collaborative robots to address assembly line balancing issues. They sought to shed light on effective optimization methods for balance on the manufacturing line.

By utilizing cobots and exoskeletons, Weckenborg et al. (2022) suggest a way to balance ergonomics and costs in assembly lines. To examine the effectiveness of various assembly line setups with and without cobots and exoskeletons, the scientists created a simulation model. Their findings imply that integrating cobots and exoskeletons can increase productivity significantly while decreasing the risk of musculoskeletal illnesses.

On an assembly line with numerous cobots, Gjeldum et al. (2022) address the issue of task distribution. They provide a decision support system that optimizes work allocation by taking into account many aspects, such as task complexity, cobot capabilities, and assembly line layout. By using a case study of an automobile assembly line, the authors were able to demonstrate the validity of their methodology and demonstrate how the suggested system may boost output and cut down on cycle time. The human-robot interaction component of cobot integration in assembly lines is the main priority (Oberc et al., 2019). They investigated a manual assembly line where employees had received cobot training. To teach employees how to operate with cobots safely and successfully, the authors created a training program that consists of both theoretical and practical modules. According to their findings, the training program enhanced the employees' understanding of and comfort using robots.

The availability of cobots, equipment, and setup time are all factors that are considered in the study of Maruf (2022) model for human-robot collaborative assembly lines. The author created a mathematical model to optimize the layout of the assembly line and task distribution while taking into account many limitations, including the number of cobots, the duration of the assembly process, and the compatibility of the workpieces. The case study's outcomes demonstrated that the suggested model can boost output while decreasing setup time. According to the degree of automation, Stone et al. (2021) suggest a method for evaluating the safety of interactions between humans and robots. The Cobot and Robot Risk Assessment (CARRA) technique was created by the authors as a risk assessment tool that considers potential hazards related to various levels of automation. Through the use of an assembly line case study, the authors validated their strategy, demonstrating how the suggested technique might increase safety and lower the likelihood of accidents.

In Table 9.5, the term "ergonomics" is used to indicate how the study addresses worker comfort and safety throughout assembly line operations. The study's analysis of economics examines the effectiveness and cost-effectiveness of using collaborative robots in assembly line tasks. In relation to assembly line balancing and robot allocation, the use of tools or systems known as "decision support systems" is referred to. The study's method of addressing cooperation and communication in assembly line tasks between humans and robots is known as "human-robot interaction." Risk assessment refers to the study's strategy for resolving potential risks and safety problems associated with the usage of collaborative robots in assembly line activities.

Table 9.5 Performance metrics of cobots in assembly line operations.

Study	Ergonomics	Economics	Decision support systems	Human-robot interaction	Risk assessment
Li et al. (2021)	Moderate	High	No	Moderate	Low
Stecke and Mokhtarzadeh (2022)	High	Moderate	No	High	High
Chen, Huang, et al. (2022)	Moderate	Moderate	No	Moderate	Low
Weckenborg and Spengler (2019)	High	High	No	Moderate	Low
Sikora and Weckenborg (2022)	Moderate	Moderate	Yes	Moderate	Low
Weckenborg et al. (2022)	High	High	No	High	Low
Gjeldum et al. (2022)	Moderate	Moderate	Yes	Moderate	Low
Oberc et al. (2019)	Low	Low	No	High	High
Maruf (2022)	High	Moderate	No	High	Low
Stone et al. (2021)	Low	Low	No	Low	High

Collectively, these case studies add to the body of research that is expanding in the fields of warehouse optimization, e-commerce fulfillment, production processing materials, and last-mile delivery. They demonstrate the developments in robotic automation, autonomous mobile robots, and drone technology as well as the difficulties they have faced and the promise they have to improve different elements of logistics and supply chain operations. Further improvements and developments in the industry will surely result from continued research and development in these areas, ultimately improving the effectiveness, accuracy, and productivity of logistics and supply chain systems.

9.11 Emerging trends and research gaps

Automation in the current supply chain is mostly driven by AI. Due to its integration across numerous domains, the supply chain sector can benefit from streamlined operations, increased efficiency, greater quality control, and region-specific decision-making, which promotes overall optimization and success. There are several new trends in robotics usage in sustainable supply chains. These patterns show that robot technology is increasingly being adopted and integrated into supply chain activities to improve sustainability practices. There have been several studies on the use of robotics in sustainable supply chain management, its effects on circular economies, financial matters, quality-driven post-harvest supply chains, and the use of agricultural drones and IoT to comprehend the food supply chain in the post-COVID-19 era.

In a used-car resale business, Sathiya et al. (2021) investigated the employment of mobile robots and evolutionary optimization algorithms for supply chain management. The study focuses on how these innovations can improve efficiency in operations, lower carbon emissions, and improve resource use, all of which support sustainable business practices in the automotive sector. The importance of environmental robotics in fostering sustainability in circular economies is discussed by Grau Ruiz and O'Brolchain (2022). The study demonstrates how robotic technology can aid in the recovery of resources, closed-loop systems, and reducing waste, ultimately helping in the shift towards a more environmentally friendly and regenerative future. E-Fatima et al. (2023) investigate the use of robotic process automation (RPA) in beef supply chains while considering its financial aspects and sustainability effects. The study focuses on how RPA can boost sustainability metrics in the beef business while lowering costs and increasing operational efficiency. To build a sustainable chain system, Bechtsis et al. (2018) present a methodology for incorporating Intelligent Autonomous Vehicles (IAVs) in digital supply chains. The report highlights how IAVs can improve supply chain operations by lowering emissions, increasing sustainability, and improving operational efficiency.

With an emphasis on quality-driven operations, Chauhan et al. (2022) concentrate on the use of robotics in post-harvest supply chains. To create a sustainable and high-quality supply chain, the project investigates how robotics can boost product quality, minimize waste, and increase efficiency in post-harvest processes. The deployment of agricultural drones and the Internet of Things (IoT) to comprehend the food supply chain in the post-COVID-19 era is explored by Dutta and Mitra (2021). The utilization of these advancements and their effects on traceability, monitoring, and general efficiency in the food supply chain are examined in the study.

To better understand the long-term impacts of integrating these technologies on the environment, society, and economy, further research is required in the field of robotics within sustainable supply chains. There is a shortage of studies examining the wider sustainability effects of robotics, including the lowering of greenhouse gas emissions, the impact on labor dynamics, along the overall resilience of the supply chain, despite some research highlighting some of the possible advantages of robotics, such as enhanced efficiency and cost limitation. Additionally, it is essential to carry out more studies to analyze the adaptability and usefulness of robotics across diverse industries and supply chain contexts, considering aspects like product variance, supply chain complexity, and regulatory constraints. By filling in these knowledge gaps, it will be possible to gain an expanded awareness of how robotics might improve supply chain operations' sustainability.

9.12 Conclusion and future scope

In conclusion, the methodology framework for supply chain automation is presented, which offers a structured technique for integrating automation processes and technology into supply chain operations. Organizations may maximize productivity, cut costs, and enhance supply chain performance by using this approach. The framework establishes a focus on the significance of data gathering and analysis, clearly defining objectives, evaluating viability, identifying automation opportunities, choosing appropriate robotic solutions, planning execution, evaluating and confirming implementation and integration, tracking and optimization, ongoing enhancement, and encouraging cooperation and knowledge sharing. The chapter also emphasizes the importance of analyzing current procedures and performance indicators, as well as pinpointing potential areas for development and automation. To successfully deploy supply chain automation, it is necessary to solve difficulties with integration with legacy systems, process identification and prioritization, change management, data quality, scalability, and talent needs. Future supply chain operations hold great potential due to growing developments in robotics, AI, automation, and intelligent unmanned autonomous systems. Approximately 32% of improvement is shown by using robotics in sustainable supply chains. For organizations looking to use automation for sustainable growth and competitive advantage in the dynamic business landscape, the methodological approach and emerging trends covered in this chapter offer helpful insights and advice.

The future potential of robotics in supply chains is immense, sustainable, circular economy by improving resource efficiency and reducing waste. With advancements in technology, robots are poised to revolutionize the way goods are handled and transported. They can work alongside humans, automating repetitive tasks and increasing efficiency. Robotic systems like Automated Guided Vehicles (AGVs) and Automated Storage and Retrieval Systems (AS/RS) are becoming more sophisticated, enabling seamless and precise movement of goods. Additionally, drones and autonomous delivery vehicles are emerging as viable options for last-mile delivery. These innovations promise to not only improve speed and accuracy in logistics but also reduce costs and environmental impact. As robotics continues to evolve, it is likely to play a pivotal role in shaping the future of supply chains worldwide. Advances in robotics will lead to better, more personalized consumer services, while simultaneously reducing environmental and social risks by lowering emissions and improving worker safety. Robotics will also drive economic innovation, creating new business opportunities and jobs in technology-related fields. Moreover, strategic collaborations and international standards are anticipated to evolve, fostering responsible and balanced growth in robotic integration into supply chains.

References

Abdel-Baset, M., Chang, V., & Gamal, A. (2019). Evaluation of the green supply chain management practices: A novel neutrosophic approach. *Computers in Industry*, *108*(1), 210–220.

Abdel-Basset, M., Mohamed, R., Sallam, K., & Elhoseny, M. (2020). A novel decision-making model for sustainable supply chain finance under uncertainty environment. *Journal of Cleaner Production*, *269*, 122324.

Abdel-Basset, M., Mohamed, R., Zaied, A. E. N. H., & Smarandache, F. (2019). A hybrid plithogenic decision-making approach with quality function deployment for selecting supply chain sustainability metrics. *Symmetry*, *11*(7), 903.

Aday, S., & Aday, M. S. (2020). Impact of COVID-19 on the food supply chain. *Food Quality and Safety*, *4*(4), 167–180.

Agrawal, A., Won, S. J., Sharma, T., Deshpande, M., & McComb, C. (2021). A multi-agent reinforcement learning framework for intelligent manufacturing with autonomous mobile robots. *Proceedings of the Design Society*, *1*, 161–170.

Ammar, M., Haleem, A., Javaid, M., Walia, R., & Bahl, S. (2021). Improving material quality management and manufacturing organizations system through Industry 4.0 technologies. *Materials Today: Proceedings*, *45*, 5089–5096.

Aurambout, J. P., Gkoumas, K., & Ciuffo, B. (2019). Last mile delivery by drones: An estimation of viable market potential and access to citizens across European cities. *European Transport Research Review, 11*(1), 1—21.

Bag, S., Wood, L. C., Xu, L., Dhamija, P., & Kayikci, Y. (2020). Big data analytics as an operational excellence approach to enhance sustainable supply chain performance. *Resources, Conservation and Recycling, 153*, 104559.

Baras, N., Chatzisavvas, A., Ziouzios, D., & Dasygenis, M. (2021). Improving automatic warehouse throughput by optimizing task allocation and validating the algorithm in a developed simulation tool. *Automation, 2*(3), 116—126.

Barenji, A. V., Wang, W. M., Li, Z., & Guerra-Zubiaga, D. A. (2019). Intelligent E-commerce logistics platform using hybrid agent based approach. *Transportation Research Part E: Logistics and Transportation Review, 126*, 15—31.

Bechtsis, D., Tsolakis, N., Vlachos, D., & Srai, J. S. (2018). Intelligent autonomous vehicles in digital supply chains: A framework for integrating innovations towards sustainable value networks. *Journal of Cleaner Production, 181*, 60—71.

Belhadi, A., Kamble, S., Jabbour, C. J. C., Gunasekaran, A., Ndubisi, N. O., & Venkatesh, M. (2021). Manufacturing and service supply chain resilience to the COVID-19 outbreak: Lessons learned from the automobile and airline industries. *Technological Forecasting and Social Change, 163*, 120447.

Borghetti, F., Caballini, C., Carboni, A., Grossato, G., Maja, R., & Barabino, B. (2022). The use of drones for last-mile delivery: A numerical case study in Milan, Italy. *Sustainability, 14*(3), 1766.

Bormann, R., de Brito, B.F., Lindermayr, J., Omainska, M., & Patel, M. (2019). Towards automated order picking robots for warehouses and retail. In *Computer Vision Systems: 12th International Conference, ICVS 2019*, Thessaloniki, Greece, September 23—25, 2019, Proceedings 12 (pp. 185—198). Springer International Publishing.

Bui, T. D., Tsai, F. M., Tseng, M. L., Tan, R. R., Yu, K. D. S., & Lim, M. K. (2021). Sustainable supply chain management towards disruption and organizational ambidexterity: A data driven analysis. *Sustainable Production and Consumption, 26*, 373—410.

Chang, S. E., Chen, Y. C., & Lu, M. F. (2019). Supply chain re-engineering using blockchain technology: A case of smart contract based tracking process. *Technological Forecasting and Social Change, 144*, 1—11.

Chauhan, A., Brouwer, B., & Westra, E. (2022). Robotics for a quality-driven post-harvest supply chain. *Current Robotics Reports, 3*(2), 39—48.

Chen, C., & Demir, E. (2022). *Drones and delivery robots: Models and applications to last mile delivery. The palgrave handbook of operations research* (pp. 859—882). Cham: Springer International Publishing.

Chen, C., Leon, S., & Ractham, P. (2022). Will customers adopt last-mile drone delivery services? An analysis of drone delivery in the emerging market economy. *Cogent Business & Management, 9*(1), 2074340.

Chen, N., Huang, N., Radwin, R., & Li, J. (2022). Analysis of assembly-time performance (ATP) in manufacturing operations with collaborative robots: A systems approach. *International Journal of Production Research, 60*(1), 277—296.

Dang, Q. V., Nguyen, C. T., & Rudová, H. (2019). Scheduling of mobile robots for transportation and manufacturing tasks. *Journal of Heuristics, 25*, 175—213.

Dash, R., McMurtrey, M., Rebman, C., & Kar, U. K. (2019). Application of artificial intelligence in automation of supply chain management. *Journal of Strategic Innovation and Sustainability, 14*(3), 43—53.

Datta, S., Ray, R., & Banerji, D. (2008). Development of autonomous mobile robot with manipulator for manufacturing environment. *The International Journal of Advanced Manufacturing Technology, 38*, 536—542.

Delgado, J. M. D., Oyedele, L., Ajayi, A., Akanbi, L., Akinade, O., Bilal, M., & Owolabi, H. (2019). Robotics and automated systems in construction: Understanding industry-specific challenges for adoption. *Journal of Building Engineering, 26*, 100868.

Dietrich, F., Ge, Y., Turgut, A., Louw, L., & Palm, D. (2021). Review and analysis of blockchain projects in supply chain management. *Procedia Computer Science, 180*, 724–733.

Dotoli, M., Fay, A., Miśkowicz, M., & Seatzu, C. (2019). An overview of current technologies and emerging trends in factory automation. *International Journal of Production Research, 57*(15–16), 5047–5067.

Duong, L. N., Al-Fadhli, M., Jagtap, S., Bader, F., Martindale, W., Swainson, M., & Paoli, A. (2020). A review of robotics and autonomous systems in the food industry: From the supply chains perspective. *Trends in Food Science & Technology, 106*, 355–364.

Dutta, P. K., & Mitra, S. (2021). Application of agricultural drones and IoT to understand food supply chain during post COVID-19. *Agricultural Informatics: Automation Using the IoT and Machine Learning, 67–87*.

Dzedzickis, A., Subačiūtė-Žemaitienė, J., Šutinys, E., Samukaitė-Bubnienė, U., & Bučinskas, V. (2021). Advanced applications of industrial robotics: New trends and possibilities. *Applied Sciences, 12*(1), 135.

Dörfler, K., Hack, N., Sandy, T., Giftthaler, M., Lussi, M., Walzer, A. N., & Kohler, M. (2019). Mobile robotic fabrication beyond factory conditions: Case study Mesh Mould wall of the DFAB HOUSE. *Construction Robotics, 3*, 53–67.

E-Fatima, K., Khandan, R., Hosseinian-Far, A., & Sarwar, D. (2023). The adoption of robotic process automation considering financial aspects in beef supply chains: An approach towards sustainability. *Sustainability, 15*(9), 7236.

Engemann, H., Du, S., Kallweit, S., Cönen, P., & Dawar, H. (2020). Omnivil—an autonomous mobile manipulator for flexible production. *Sensors, 20*(24), 7249.

Fehling, C., &Saraceni, A. (2022). Feasibility of Drones & Agvs in the Last Mile Delivery: Lessons from Germany. Available at SSRN 4065011.

Flechsig, C., Anslinger, F., & Lasch, R. (2022). Robotic process automation in purchasing and supply management: A multiple case study on potentials, barriers, and implementation. *Journal of Purchasing and Supply Management, 28*(1), 100718.

Gao, X., & Liu, L. (2020). Green intelligent logistics sorting system in big data environment. In *IOP Conference Series: Materials Science and Engineering* (Vol. 711, No. 1, p. 012040). IOP Publishing.

Gill, S. S., Xu, M., Ottaviani, C., Patros, P., Bahsoon, R., Shaghaghi, A., & Uhlig, S. (2022). AI for next generation computing: Emerging trends and future directions. *Internet of Things, 19*, 100514.

Gjeldum, N., Aljinovic, A., CrnjacZizic, M., & Mladineo, M. (2022). Collaborative robot task allocation on an assembly line using the decision support system. *International Journal of Computer Integrated Manufacturing, 35*(4–5), 510–526.

Grau Ruiz, M. A., & O'Brolchain, F. (2022). Environmental robotics for a sustainable future in circular economies. *Nature Machine Intelligence, 4*(1), 3–4.

Helo, Petri, & Hao, Yuqiuge (2022). Artificial intelligence in operations management and supply chain management: An exploratory case study. *Production Planning & Control, 33* (16), 1573–1590. Available from https://doi.org/10.1080/09537287.2021.1882690.

Hofmann, E., Sternberg, H., Chen, H., Pflaum, A., & Prockl, G. (2019). Supply chain management and Industry 4.0: Conducting research in the digital age. *International Journal of Physical Distribution & Logistics Management, 49*(10), 945–955.

Hosseini, S., Ivanov, D., & Dolgui, A. (2019). Review of quantitative methods for supply chain resilience analysis. *Transportation Research Part E: Logistics and Transportation Review, 125*, 285–307.

Huang, Z., Mao, C., Wang, J., & Sadick, A. M. (2022). Understanding the key takeaway of construction robots towards construction automation. *Engineering, Construction and Architectural Management, 29*(9), 3664−3688.

Iqbal, J., Khan, Z. H., & Khalid, A. (2017). Prospects of robotics in food industry. *Food Science and Technology, 37*, 159−165.

Jabbour, C. J. C., Fiorini, P. D. C., Ndubisi, N. O., Queiroz, M. M., & Piato, É. L. (2020). Digitally-enabled sustainable supply chains in the 21st century: A review and a research agenda. *Science of the Total Environment, 725*, 138177.

Jain, V. N. (2019). Robotics for supply chain and manufacturing industries and future it holds. *International Journal of Engineering Research & Technology, 8*, 66−79.

Javaid, M., Haleem, A., Singh, R. P., & Suman, R. (2021). Substantial capabilities of robotics in enhancing industry 4.0 implementation. *Cognitive Robotics, 1*, 58−75.

Jiang, M., & Huang, G. Q. (2022). Intralogistics synchronization in robotic forward-reserve warehouses for e-commerce last-mile delivery. *Transportation Research Part E: Logistics and Transportation Review, 158*, 102619.

Jun, S., Lee, S., & Yih, Y. (2021). Pickup and delivery problem with recharging for material handling systems utilising autonomous mobile robots. *European Journal of Operational Research, 289*(3), 1153−1168.

Kopyto, M., Lechler, S., von der Gracht, H. A., & Hartmann, E. (2020). Potentials of blockchain technology in supply chain management: Long-term judgments of an international expert panel. *Technological Forecasting and Social Change, 161*, 120330.

Kousi, N., Koukas, S., Michalos, G., & Makris, S. (2019). Scheduling of smart intra−factory material supply operations using mobile robots. *International Journal of Production Research, 57*(3), 801−814.

Kshetri, N. (2021). Blockchain and sustainable supply chain management in developing countries. *International Journal of Information Management, 60*, 102376.

Lee, H., & Jeong, J. (2021). Mobile robot path optimization technique based on reinforcement learning algorithm in warehouse environment. *Applied sciences, 11*(3), 1209.

Lemardelé, C., Estrada, M., Pagès, L., & Bachofner, M. (2021). Potentialities of drones and ground autonomous delivery devices for last-mile logistics. *Transportation Research Part E: Logistics and Transportation Review, 149*, 102325.

Lezoche, M., Hernandez, J. E., Díaz, M. D. M. E. A., Panetto, H., & Kacprzyk, J. (2020). Agri-food 4.0: A survey of the supply chains and technologies for the future agriculture. *Computers in Industry, 117*, 103187.

Li, Z., Barenji, A. V., Jiang, J., Zhong, R. Y., & Xu, G. (2020). A mechanism for scheduling multi robot intelligent warehouse system face with dynamic demand. *Journal of Intelligent Manufacturing, 31*, 469−480.

Li, Z., Janardhanan, M. N., & Tang, Q. (2021). Multi-objective migrating bird optimization algorithm for cost-oriented assembly line balancing problem with collaborative robots. *Neural Computing and Applications, 33*, 8575−8596.

Liang, C., Chee, K.J., Zou, Y., Zhu, H., Causo, A., Vidas, S., & Cheah, C.C. (2015, October). Automated robot picking system for e-commerce fulfillment warehouse application. In The 14th IFToMM World Congress.

Liu, C., Xu, Y., Tang, C., & Chen, D. (2022, December). Rapid Sorting Robot System for E-commerce Warehouse. In 2022 IEEE International Conference on Robotics and Biomimetics (ROBIO) (pp. 1521−1525). IEEE.

Mahlamäki, T., Storbacka, K., Pylkkönen, S., & Ojala, M. (2020). Adoption of digital sales force automation tools in supply chain: Customers' acceptance of sales configurators. *Industrial Marketing Management, 91*, 162−173.

Maruf, A. (2022). The development of human-robot collaborative assembly line model by considering availability of robots, tools, and setup time. *JurnalIlmiahTeknikIndustri, 21*(2), 321−329.

Mastos, T. D., Nizamis, A., Vafeiadis, T., Alexopoulos, N., Ntinas, C., Gkortzis, D., & Tzovaras, D. (2020). Industry 4.0 sustainable supply chains: An application of an IoT enabled scrap metal management solution. *Journal of Cleaner Production, 269*, 122377.

Oberc, H., Prinz, C., Glogowski, P., Lemmerz, K., & Kuhlenkötter, B. (2019). Human robot interaction—learning how to integrate collaborative robots into manual assembly lines. *Procedia Manufacturing, 31*, 26—31.

Omar, I. A., Jayaraman, R., Debe, M. S., Salah, K., Yaqoob, I., & Omar, M. (2021). Automating procurement contracts in the healthcare supply chain using blockchain smart contracts. *IEEE Access, 9*, 37397—37409.

Pettit, T. J., Croxton, K. L., & Fiksel, J. (2019). The evolution of resilience in supply chain management: A retrospective on ensuring supply chain resilience. *Journal of Business Logistics, 40*(1), 56—65.

Prakash, R., Behera, L., Mohan, S., & Jagannathan, S. (2020). Dual-loop optimal control of a robot manipulator and its application in warehouse automation. *IEEE Transactions on Automation Science and Engineering, 19*(1), 262—279.

Radke, A. M., Dang, M. T., & Tan, A. (2020). Using robotic process automation (RPA) to enhance item master data maintenance process. *LogForum, 16*(1).

Rey, R., Corzetto, M., Cobano, J.A., Merino, L., & Caballero, F. (2019, September). Human-robot co-working system for warehouse automation. In 2019 24th IEEE International Conference on Emerging Technologies and Factory Automation (ETFA) (pp. 578—585). IEEE.

Riahi, Y., Saikouk, T., Gunasekaran, A., & Badraoui, I. (2021). Artificial intelligence applications in supply chain: A descriptive bibliometric analysis and future research directions. *Expert Systems with Applications, 173*, 114702.

Samouh, F., Gluza, V., Djavadian, S., Meshkani, S., & Farooq, B. (2020, September). Multimodal autonomous last-mile delivery system design and application. In 2020 IEEE International Smart Cities Conference (ISC2) (pp. 1—7). IEEE.

Santos, F., Pereira, R., & Vasconcelos, J. B. (2020). Toward robotic process automation implementation: An end-to-end perspective. *Business Process Management Journal, 26*(2), 405—420.

Sathiya, V., Chinnadurai, M., Ramabalan, S., & Appolloni, A. (2021). Mobile robots and evolutionary optimization algorithms for green supply chain management in a used-car resale company. *Environment, Development and Sustainability, 23*, 9110—9138.

Schwarz, M., Milan, A., Lenz, C., Munoz, A., Periyasamy, A.S., Schreiber, M., & Behnke, S. (2017, May). Nimbro picking: Versatile part handling for warehouse automation. In 2017 IEEE International Conference on Robotics and Automation (ICRA) (pp. 3032—3039). IEEE.

Sikora, C. G. S., & Weckenborg, C. (2022). Balancing of assembly lines with collaborative robots: Comparing approaches of the Benders' decomposition algorithm. *International Journal of Production Research*, 1—17.

Simoni, M. D., Kutanoglu, E., & Claudel, C. G. (2020). Optimization and analysis of a robot-assisted last mile delivery system. *Transportation Research Part E: Logistics and Transportation Review, 142*, 102049.

Skoufi, E., Filiopoulou, E., Skoufis, A., & Michalakelis, C. (2021). Last Mile Delivery by Drone: A Technoeconomic Approach. In Economics of Grids, Clouds, Systems, and Services: 18th International Conference, GECON 2021, Virtual Event, September 21—23, 2021, Proceedings 18 (pp. 27—35). Springer International Publishing.

Sorooshian, S., KhademiSharifabad, S., Parsaee, M., & Afshari, A. R. (2022). Toward a modern last-mile delivery: Consequences and obstacles of intelligent technology. *Applied System Innovation, 5*(4), 82.

Stecke, K. E., & Mokhtarzadeh, M. (2022). Balancing collaborative human—robot assembly lines to optimise cycle time and ergonomic risk. *International Journal of Production Research, 60*(1), 25—47.

Stone, R.T., Pujari, S., Mumani, A., Faies, C., & Ameen, M. (2021, September). Cobot and robot risk assessment (CARRA) method: An automation level-based safety assessment tool to improve fluency in safe human cobot/robot interaction. In Proceedings of the Human Factors and Ergonomics Society Annual Meeting (Vol. 65, No. 1, pp. 737–741). Sage CA: Los Angeles, CA: SAGE Publications.

Syed, R., Suriadi, S., Adams, M., Bandara, W., Leemans, S. J., Ouyang, C., & Reijers, H. A. (2020). Robotic process automation: Contemporary themes and challenges. *Computers in Industry, 115*, 103162.

Tan, W. C., & Sidhu, M. S. (2022). Review of RFID and IoT integration in supply chain management. *Operations Research Perspectives, 9*, 100229.

Tan, Z., Li, H., & He, X. (2021). Optimizing parcel sorting process of vertical sorting system in e-commerce warehouse. *Advanced Engineering Informatics, 48*, 101279.

van Hoek, R., Gorm Larsen, J., & Lacity, M. (2022). Robotic process automation in Maersk procurement—applicability of action principles and research opportunities. *International Journal of Physical Distribution & Logistics Management, 52*(3), 285–298.

Varriale, V., Cammarano, A., Michelino, F., & Caputo, M. (2023). Industry 5.0 and triple bottom line approach in supply chain management: The state-of-the-art. *Sustainability, 15*(7), 5712.

Wang, Y., Liang, Y., Chen, D., Liu, Y., & Wang, M. (2020, September). A goods sorting robot system for e-commerce logistics warehouse based on robotic arm technology. In 2020 IEEE International Conference on Real-time Computing and Robotics (RCAR) (pp. 310–314). IEEE.

Weckenborg, C., & Spengler, T. S. (2019). Assembly line balancing with collaborative robots under consideration of ergonomics: A cost-oriented approach. *IFAC-PapersOnLine, 52*(13), 1860–1865.

Weckenborg, C., Thies, C., & Spengler, T. S. (2022). Harmonizing ergonomics and economics of assembly lines using collaborative robots and exoskeletons. *Journal of Manufacturing Systems, 62*, 681–702.

Xiao, H., Muthu, B., & Kadry, S. N. (2020). Artificial intelligence with robotics for advanced manufacturing industry using robot-assisted mixed-integer programming model. *Intelligent Service Robotics*, 1–10.

Yadav, G., Luthra, S., Jakhar, S. K., Mangla, S. K., & Rai, D. P. (2020). A framework to overcome sustainable supply chain challenges through solution measures of industry 4.0 and circular economy: An automotive case. *Journal of Cleaner Production, 254*, 120112.

Zhang, F., Shang, W., Zhang, B., & Cong, S. (2020). Design optimization of redundantly actuated cable-driven parallel robots for automated warehouse system. *IEEE Access, 8*, 56867–56879.

Zhang, T., Li, Q., Zhang, C. S., Liang, H. W., Li, P., Wang, T. M., & Wu, C. (2017). Current trends in the development of intelligent unmanned autonomous systems. *Frontiers of Information Technology & Electronic Engineering, 18*, 68–85.

CHAPTER TEN

Computational techniques for sustainable green procurement and production

Bhakti Parashar[1], Sandeep Kautish[2,3] and Amrita Chaurasia[4]
[1]VIT-Bhopal University, Bhopal, Madhya Pradesh, India
[2]LBEF Campus (APU Malaysia) Kathmandu, Nepal
[3]Model Institute of Engineering and Technology, Jammu, Jammu and Kashmir, India
[4]School of Commerce, Finance and Accountancy Christ (Deemed to be University), Ghaziabad, Uttar Pradesh, India

10.1 Introduction

Information and communication technologies have altered how we live, work, learn, and have fun, but it has also altered our surroundings. As a result, new job opportunities are emerging all over the world as computer literacy becomes a requirement for survival in almost every field. Because of the capacity of computers to instantly and inexpensively save, retrieve, and update massive amounts of data, it is extensively utilized in many clerical, bookkeeping, and support record-keeping functions in businesses (Raza et al., 2012). However, environmental concerns arise at every computer's life cycle stage, from manufacturing to use and disposal. Several experts and authors have shared their findings from studies on the environmental impact of ICT. The argument over the effectiveness of green technology for environmentally conscious and achievable IT continues. Sustainability and environmental friendliness are major concerns for all industries around the world. Several businesses are also working to protect the environment by implementing novel procedures, developing fresh approaches, and getting internationally recognized credentials such as ISO 14000 (Raza et al., 2012; Song & Xue-Xian, 2018).

In today's competitive global economy, IT has mandated the implementation of computers, the internet, centers for data storage, servers, and other machinery in almost every sector to assist businesses in turning smarter by improving revenue and managerial productivity. On the other

Computational Intelligence Techniques for Sustainable Supply Chain Management.
DOI: https://doi.org/10.1016/B978-0-443-18464-2.00004-2

hand, greater utilization of technology raises the possibility of increasing consumption of energy, as well as being responsible for a growing carbon footprint and greenhouse gas (GHGs) carbon dioxide emissions that harm the environment and cause climate change. As a result, major global gatherings such as the World Economic Forum, the Asia-Pacific Economic Cooperation Summit, the G7/G8 summit, and the 2015 United Nations Climate Change Conference Global Gender Gap Report, 23 (2023), have prioritized these issues on their agendas. According to the World Economic Forum, our ability to survive the next global disaster is dwindling due to strained public-sector spending and competing security agendas Over the next 10 years, fewer countries will have the economic freedom to invest in potential expansion, environmentally friendly technologies, higher education, care, and medical systems.

The gradual deterioration of public infrastructure and services in both developing and developed countries may be imperceptible at first, but the cumulative effects will be devastating to human capital and development — a critical mitigant to other global challenges. Making environmentally responsible product purchasing and manufacturing decisions is part of green procurement and manufacturing. Vachon (2015), emphasized the significance of investigating the impact of supply chain collaboration, such as joint environmental regulations and pollution control programs. Supplier collaboration and a green mindset lower procurement costs while assuring social and economic benefits to businesses, and computational approaches, which give insights into the environmental consequences of various phases of the product life cycle, may be utilized to assist in accomplishing these goals.

By enabling a circular economic model that incorporates material monitoring via global positioning systems (GPS), computational technology promotes sustainability measures like recycling, composting, green procurement, and remanufacturing (Büchi et al., 2020; Chang et al., 2019; Kouhizadeh et al., 2019; Mani et al., 2020). Also, green computing benefits everyone, not just the client, business, or country. It contributes to lower energy usage, waste, costs, and how we use technology, all of which have a positive impact on the environment (Raza et al., 2012).

Many concepts, such as the lean manufacturing method, innovative manufacturing, green production, kaizen, and lean-six sigma, are currently in use (Li, 2019; Tripathi et al., 2021). This concept is used to increase productivity by reducing staff, waste energy, and products resulting from work that does not add value to the workshop (Tripathi et al., 2021; Da Costa et al., 2020). Excessive waste always hurts an industry as it causes financial loss (Ali et al., 2020; Tjahjadi et al., 2020)

10.1.1 Purpose of the study

To address the aforementioned issue, it is critical to eliminate all types of waste in the sector through the use of acceptable alternative methods. Lean manufacturing is a popular method in today's world for reducing various types of waste (Siegel et al., 2019). Green manufacturing seeks to minimize the adverse effects of environmental problems on the production line, such as wasted materials and energy (Shokri & Li, 2020). Green technologies enable businesses to increase long-term output while using fewer resources (Prasad et al., 2016; Sagnak & Kazancoglu, 2016).

The primary goal of the idea of green is to identify methods to remove impediments to management system effectiveness (Oliveira et al., 2018). Worker inefficiency, resource depletion, increasing inventory, hazardous working conditions, and disorganized machinery are some of the issues that arise as a result of the decline in workplace conditions and the deterioration of the environment (Cherrafi et al., 2019; Choudhary et al., 2019; Inman & Green, 2018; Zhu, 2020; Kurdve et al., 2014). As a result, the goal of this chapter is to raise awareness about green computing while also providing an overview of key areas where IT organizations can save money and energy. Furthermore, we look into a formal approach to green computing, including standards and conformance, as well as some of the challenges it faces.

10.2 Real-life examples

Despite a lot of challenges, there are many real-life practical examples of computational methods for environmentally friendly purchasing and production:

- The Coca-Cola Company: Coca-Cola employs a range of computational techniques to support its efforts in green production and purchasing. The company, for instance, uses SERS to rate its suppliers' environmental performance and LCA to evaluate the environmental impact of its products. Coca-Cola additionally employs data analytics to monitor and manage its environmental performance as well as simulation modeling to simulate the environmental effects of various scenarios for procurement and production (Cocacola company, n.d.)

- IKEA: IKEA is another business that supports its environmentally friendly production and purchasing initiatives with computational techniques. For instance, IKEA uses EMS to manage its environmental performance and EPDs to provide information on the environmental performance of its products. IKEA additionally employs optimization algorithms to resolve issues with green procurement, such as the issue of reducing the environmental impact of its product line (IKEA and sustainability efforts. n.d.).

- Apple: Apple is a well-known company for its dedication to sustainability. Apple supports its eco-friendly production and purchasing efforts with a variety of computational techniques. For instance, Apple uses SERS to rate its suppliers' environmental performance and LCA to evaluate the environmental impact of its products. Additionally, Apple uses simulation modeling to simulate the environmental effects of various scenarios for procurement and production and data analytics to monitor and manage its environmental performance (Apple, n.d.).

These are just a few examples of how computational methods are being applied to support green purchasing and production. We can anticipate even more creative applications of these techniques in the future as the market for sustainable goods and services expands.

10.3 Meaning and background

Engineers have been attempting to tackle difficulties relating to the operation and maintenance of machinery since the beginning of the first industrial revolution. They also aimed to improve the overall efficiency of manufacturing operations, as well as production organization and other relevant areas. An effective method is developed to investigate work-related problems and solutions are presented as planned. However, the emergence of technology and the development of computational techniques heralded a new era in company issue resolution, because of the development of computing, the answer is predictable but very accurate. Optimization techniques may be used to improve manufacturing process efficiency by identifying the ideal process parameters, as well as to tackle tough production-related difficulties like scheduling (Belhadi et al., 2020).

Industry 4.0 emphasizes the development of a new approach to addressing environmental issues and manufacturing waste issues through the use of an appropriate process optimization technique. By observing end-to-end management systems, process optimization is used to maximize efficiency under restricted limitations. Smart lean-green is an open innovation method critical to addressing shop-floor productivity challenges in the current Industry 4.0 environment. This method aids in the elimination of numerous inefficiencies at the manufacturing ground level (Apple, n.d.). Lean-green is a method of reducing waste in manufacturing processes and workplaces that combines two methods (Karkalos et al., 2019).

Social and economic development now primarily consists of digitization and going green. Academic and business communities have given digital technology a lot of thought and recognition as a crucial enabler of green supply chain management. Research in this developing area of technology is growing but has not yet reached saturation due to the advent of the Industry 4.0 era and the rapid development of digital technology (Duarte & Machado, 2017).

- Computational techniques and Industry 4.0: Computation methods are a quick, simple, dependable, and effective approach to using computers to solve computational, scientific, engineering, geometrical, geographical, and statistical problems. As a result, computational problem-solving methods are typically time-step-based. The iterative, looping, stereotyped, or modified processes used in the step-by-step technique are far less stressful than addressing problems manually. Computational approaches may also concentrate on overcoming computational challenges or issues through the use of algorithms, scripts, or command-line interfaces. A computational technique may include many factors or variables that characterize the system or model being investigated. The system simulates or animatedly tests the variables' interdependence to see how changes in one or more parameters affect the outputs (Wang et al., 2023).
- Industry 4.0 and green procurement and production: Because of the global industry's intense competition, industry professionals need to establish a transparent, creative strategy for managing production that can be implemented in virtually any kind of manufacturing process in Industry 4.0. A variety of process optimization approaches were combined and labeled as hybrid approaches to achieve this (Emetere, 2019). Among the techniques used were lean-kaizen, lean-green,

lean-smart, and lean–six Sigma. Of all of these approaches, the lean-green approach was found to be the most effective in eliminating environmental waste, including shop-floor waste. The lean-green approach improved employee morale and quality.

With the development of Industry 4.0, the Internet of Things, and Smart Manufacturing conceptual frameworks, there are numerous alternatives for reducing energy consumption and GHG emissions, such as a cyber-physical system that includes a manufacturing services layer, a fog manufacturing services layer, a cloud manufacturing services layer, and a blockchain-based service-oriented middleware (Tripathi et al., 2022). Green computing aims to reduce energy consumption and GHG emissions while also recycling and reusing energy beneficially and efficiently. Cloud data centers, on the other hand, consume a significant amount of energy and emit a significant amount of CO_2, which must be addressed by businesses that implement greener manufacturing processes (Hawking, 2018).

• Computational techniques and green procurement and production: Researchers, business processes, and companies are currently monitoring computational techniques in green procurement and production, also known as "Green IT," to operate a power-efficient system. Originally, the key objective of computers was to solve complex problems in less time and with greater precision. Green computing, on the other hand, has recently gained prominence in organizations as a means for enhancing energy efficiency and lowering device usage of electricity (Ahuja & Muthiah, 2021).

The primary goal of computational techniques for green computing is to study and practice resource processing in an efficient and environmentally responsible manner. Hazardous materials such as mercury, cadmium, and other harmful metals are present at several workstations. We will pollute the atmosphere and have a significant impact on the ecosystem as we dispose of these computers (Fig. 10.1).

The given figure depicts the significance of computational techniques in GPP:

• Industrial resilience: Sustainability has been an important issue for green procurement and purchasing in sectors and businesses all around the world (Lozano, 2012). Dawson and Probert (2007) provided concrete approaches for sustainable procurement. Klassen and Vachon (2003) and Vachon and Mao (2008) stressed the need to investigate the impact of supply chain collaboration, such as joint environmental

Figure 10.1 Importance of computational techniques in green procurement and production.

regulations and pollution-reduction efforts. Partnership and supplier attitudes towards green practices lower procurement costs while providing social and economic benefits to the industry.

- Optimum utilization of resources: Green computing is an efficient, economical, and environmentally friendly method of computer usage, disposal, design, maintenance, and production, as well as its negative impact on a variety of sectors. Because economic and environmental performance are triadic, the degree of economic and societal value of IT is defined by its eco-sustainability (Molla et al., 2011).
- Cognitive production mechanism: Green computing (GIT) is a term used to describe applications of digital technology that are beneficial to the environment Melville (2012). Murugesan (2008) defines GIT as "the promotion of the creation, production, and supervision of computer hardware while minimizing the use of energy throughout its life process." The concept of "green" also includes the effective utilization

of technology resources as well as its use as a useful tool for businesses wishing to make significant efforts to minimize the adverse environmental effects of their creative manufacturing mechanisms (Harmon & Auseklis, 2009).

- Quality product with limited resources: Computational technology in green procurement and manufacturing is a set of practical approaches for ensuring that information technology is produced, disseminated, and used in an ecologically friendly, sustainable, and energy-efficient manner. This strategy is used to ensure effective energy consumption and to promote waste product recycling. Most organizations have addressed their environmental concerns by deploying Green IT to reduce electricity consumption and, as a result, costs (Griffins, 2005).
- Improvement in productive efficiency: The computational method is used in green procurement and manufacturing to create, produce, and discard computers, servers, screens, printing devices, and other storage devices, among other things. Exploring this area of technology necessitates focusing on critical issues such as computer energy efficiency and the development of algorithms that can be used to manage computer technology effectively. It has been discovered that the IT department of a company consumes approximately 60% of total energy consumption. Organizations that employ green computing are expected to give value to both their customers and their companies (Leong et al., 2020).

Buying products and services with a lower impact on the environment is known as "green procurement." It is an essential part of sustainable supply chain management and can assist businesses in lowering their environmental impact and boosting their bottom line. Numerous computational methods can be applied to support green procurement. These methods can be used to evaluate how products and services affect the environment, find and choose suppliers who adhere to environmental standards, and monitor and control environmental performance.

The most popular computational methods for green procurement are as follows:

- LCA is a technique for evaluating a product or service's environmental impact throughout its entire life cycle, from the extraction of raw materials to disposal.
- Environmental product declarations (EPDs) are standardized documents that offer details on a good or service's impact on the environment.

- Supply chain environmental rating systems (SERS): Tools called SERS are used to evaluate suppliers' environmental performance.
- Environmental management systems (EMS) are frameworks that assist businesses in controlling their environmental performance.

10.4 Literature review

To effectively increase performance, the improved flexible lean and green strategy has incorporated elements of Industry 4.0 to carry out process optimization by focusing on particular efficiency warning signs. Based on data, programmers are becoming a powerful tool for enhancing operations in the context of Industry 4.0 (Bravo et al., 2022). Ecological waste is the illicit utilization of resources from the earth, which could potentially be harmful to humanity and the environment. Investigators have concentrated on removing operations that generate hazardous waste because they have a direct impact on business efficiency, fiscal circumstances, employee security, wellness, and preparedness. A few studies have been published to promote the energizing of business information centers.

The goal of employing computational techniques in green procurement and manufacturing is to minimize the consumption of potentially hazardous substances, optimize the utilization of energy throughout the lifespan of the good, and encourage the potential for recycling or biological degradation of obsolete goods and scrap from factories. For a sustainable future, green products are essential. Lack of knowledge about consumer intentions towards green products and research gaps makes it difficult to translate the availability of green products into actual consumer and market acceptance (Sun et al., 2022). As economies and industries around the world grew and developed, consumption rose quickly (Akhtar et al., 2021). The negative effects that this phenomenon has had on the environment are a major source of concern (Perez-Castillo & Vera-Martinez, 2021; Wiprächtiger et al., 2020). The fact that waste is produced during the development and production of every product, as well as during its manufacture, distribution, consumption, and final disposal, is the reason for this concern. Computational techniques in green production and procurement also strive for financial viability, streamlined system efficiency, and ease of use, all while adhering to our societal and ethical responsibilities. It has numerous advantages, and everyone can benefit from it, beyond the consumer, business, or country. It contributes to lower energy demands, waste, and costs, as well as how we use technology, all of which benefit the environment.

The circular economy is being advocated more and more as a means of separating environmental consequences from economic growth (Marrucci et al., 2019). Circularity is made possible by low-carbon use and production mechanisms, which also support the achievement of sustainable development goals (Barquet et al., 2020). The circular economy aims to balance economic growth, resource sustainability, and environmental protection (Chen et al., 2020). Without fundamentally altering processes, GPP can offer opportunities for quick fixes in circles. To maximize value, sustainable procurement of cutting-edge products and services can be used more frequently (Cherrafi et al., 2017).

Majid and Romli's research study describes the use of green computing that assists IT-based industry data centers in terms of cost reductions in expenditure, conserving energy, environmentally friendly disposal of waste, biodiversity preservation, and lowering carbon dioxide emissions in pursuing the goal of environmental sustainability. Individual metrics are also included in the presented life cycle analysis techniques to classify data center practice into quantifiable units based on lowering expenses, energy utilization, sustainable management of waste, natural resource conservation, and carbon dioxide (CO_2) diminution (Anthony et al., 2018). GIT is located at the crossroads of energy-related research, engineering and cleaner production, and environmental science. The present research discovered scientific and organizational practices related to both environmental quality assurance and green computing, in addition to the development of such practices over the years. Converting technological issue-related manufacturing adds to the academic literature. It also determines GIT reference-related structures and associations, permitting researchers to recognize discrepancies in the literature and identify areas where the GIT subject is insufficient. The following section discusses future research directions: On a practical level, the present study's outcomes are of interest to academics and professionals who plan to dedicate time and effort to GIT-making decisions (pre-adoption), managerial practice expansion (adoption), and the societal advantages and measurements of GIT.

Pawlish and Varde (2010) proposed an environmentally conscious campus data center decision-making system that collects current campus data and depends on decision trees and case-based logic. Uddin et al. (2012) used virtualization in conjunction with ecological information technology, a predetermined structure for increasing data center efficiency in terms of energy. Green IT was used by the researchers to split data center components into multiple groups of resources based on factors like energy utilization ratio, workload consumption ratio, which is also CO_2 emissions, and so on. To reduce the energy consumption ratio of deployed systems, the current structure is based on the use of cloud computing and the process of virtualization.

To clearly emphasize green technologies, our environment needs fast pollution recovery. This is possible with the use of this technology, which may minimize pollutants while also improving cleanliness. Currently, both developed and emerging countries are turning to green technology to protect the environment from unwanted effects. Technology indicates that environmental degradation is increasing due to human intervention, emphasizing the importance of slowing down and adopting healthier lifestyles. As a result, using green technology is going to be not only necessary but also mandatory in years to come. As the earth's energy reserves deplete, we will need to depend on other sources of energy. Green technology opens up new possibilities while also encouraging cleanliness and freshness. As soon as possible we recognize the importance of green technology, better simplified and improved computing while also making the world healthier and more enjoyable (Kumar et al., 2016).

10.5 Challenges

Green procurement is still in its early stages, and it confronts several problems. Those working in the public sector have a poor degree of comprehension and expertise. The client and stakeholder organizations are impeding the adoption of green procurement. Moreover, the government's regulation of green procurement remains weak, causing customers and developers to be hesitant to embrace green practices (Parikka-Alhola, 2008). Green procurement is not a legal obligation in Azadnia et al. (2015) and the governing body merely provides macroeconomic oversight of building efforts to reduce negative environmental effects.

Additionally, there exist legislative and regulatory loopholes to assist green purchasing. Additional actions are needed to promote improved sustainability to address this issue (Kirkire et al., 2018) Current methods for evaluating sustainability criteria are inadequate (Sarkis et al., 2011).

As a result, more comprehensive tools for situations in which sustainability is a main concern are needed. A lack of consumer demand for reusable building components, as well as a scarcity of ecologically sound items are another barrier to green procurement implementation. When it comes to making better use of technology, fields where innovative green computing principles must be advanced and improved include but aren't restricted to Fig. 10.2.

- Data storage: Energy is widely consumed in data storage facilities. According to the US Department of Energy, data center facilities consume up to 100−200 times more power than standard workplaces. Electricity is widely used in data center facilities. According to the US Department of Energy, data centers consume hundred to two hundred times the energy of typical office buildings (Adservio Tam, 2022).

- Product lifec ycle: Product life cycle assessment is a method of investigating the environmental impact of all stages of an industrial good, process, or service's life cycle. For instance, the environmental impacts of a product that is produced are looked into from the extraction of raw materials and their processing to manufacturing, shipment, and consumption, including the recycling or ultimate disposal of the resources used in its production. A life cycle assessment (LCA) study entails meticulously recording all of the resources and components needed throughout every stage of the value chain for a product, operation, or service, as well as calculating the associated environmental emissions. LCA assesses the likelihood of cumulative adverse environmental

Figure 10.2 Challenges with green computing.

impacts as a result. The product's overall environmental perception will be recorded and modified (Adservio Tam, 2022).

- Optimization of technology: An average Internet search emits 7 g of CO_2, according to a Harvard physicist study published in 2009. Algorithm efficiency influences the total amount of resources required by computers for any given computing operation. There are several efficiency compromises to consider when developing programs, such as switching from a slow (e.g., linear) search strategy to a fast (e.g., hashed or indexed) one (Adservio Tam, 2022).

- Power monitoring: When not in use, certain electrical devices, such as photocopiers, laptops with central processing units, GPUs, screens, and printing devices, turn off or shift to a low-power mode. This is known as PC power control in computers, and it depends on the ACPI standard, which substitutes APM. In addition to areas that require improvements, others include more effective and faster storage, displays made with advanced technologies that use significantly less energy, such as reflecting displays, and others. The cryptocurrency mining process, which employs graphics processing units (GPUs), consumes a lot of energy and leaves a carbon footprint. As a result, green computing strategies emphasize the importance of switching off computer systems when they are not in use for an extended period. Green computing techniques Plan IT resources whenever the system runs low on power and is idle. If the execution power status falls below a certain threshold, the standby execution mode is used to save power. Another significant challenge in green computing is managing aging IT resources. Older gear consumes more power and must be replaced or discarded. As a result, the recycling method for aging IT resources must be implemented. Similarly, business policies banning the usage of paper printouts should be enforced (Adservio Tam, 2022).

10.6 Proposed solution

Data center emissions of carbon dioxide, prices, illegal waste disposal, depletion of natural resources, and overuse of energy have all increased as processing and computing power requirements continue to rise. This provides a problem for firms that rely on information technology to achieve sustainability in data centers (IT). By utilizing data centers for rendering solutions for workers, professionals, and potential customers, green computing strategies can assist IT-based businesses in achieving sustainability. However, it is well understood that business servers consume an enormous quantity of energy and experience additional expenses in cooling operations, making it difficult to fulfill the needs of accuracy and efficiency in data centers while also supporting more sustainable implementation practices and cost reduction. Some facts about computing are as follows: -

- Screen savers do not conserve electricity.
- Running a computer requires power and raises computing costs.
- A standard workstation with a 17-inch flat panel LCD uses approximately 100 watts, with the computer consuming 65 watts and the display consuming 35 watts. If left on 24 hours a day, it will consume 874-kW hours of power and emit nearly a tonne of CO_2 (Bolla et al., 2010).
- In information-intensive organizations, data centers usually contribute to 25% of total company IT costs and can account for more than 50% of the total commercial carbon footprint (Deng et al., 2014), Since 1996, the cost of electricity and cooling in data centers has increased by 800% (Sinha et al., 2016).
- Most data centers in the United States are predicted to spend as much on electricity as they do on hardware during the next 5 years, and twice as much on server maintenance and administration (Ong et al., 2012; Sinha et al., 2016).

While referring to issues like waste disposal, GHG emissions, energy conservation, packaging, and modes of transport, green purchasing encourages optimal performance and minimal environmental impact. Suppliers will have the greatest environmental impact on the vast majority of businesses involved in assembly or similar processes (Kazandjieva et al., 2012). Turning off the computer's display when not in use or using power-saving screens such as LED, or LCD as a replacement for traditional CRT screens, folder expressing practices, computer virtualization, and using more energy-efficient and less disruptive cooling techniques are among the various green computing practices (Nuttapon & Gabriel, 2012; Saha, 2014). The following are conceptual and operational summaries of the life cycle methods to be used in Green computing practices in IT-based commercial data centers (Fig. 10.3).

- Green architecture: Green design aims to efficiently evaluate, create, and synthesize environmentally beneficial things. As a result, hazardous chemicals were disposed of in IT-based firms without regard for environmental hazards (Saha, 2014). As a result, in the field of information

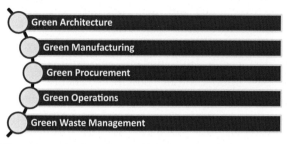

Figure 10.3 Life cycle methods to be used in green computing practices.

technology (IT), sustainable design must include energy-efficiency measures as well as the use of green computer systems, condensation equipment, and hardware for computers such as Light light-emitting diode (LED) displays (Agarwal & Nath, 2011; Brown et al., 2007).

- Green manufacturing: This stage of information center development in IT-based industries includes efforts to create a greener future by shifting data center tasks to more capable operations and increasing the recycling and reusability capacity of existing data center-built environments (Harmon & Auseklis, 2009). Under Green manufacturing, every activity associated with operating the data center's electrical parts, processors, and other corresponding components should have little or no environmental impact. Saha (2014). agreed, stating that in Green production, only electronic equipment with a low or no environmental impact should be used.

- Green procurement: IT-based businesses may also use green purchasing techniques including reducing, recycling, and reusing IT infrastructures in their purchase activities (Nuttapon & Gabriel, 2012). Furthermore, green procurement emphasizes environmental issues as well as the organization's economic viability in terms of choosing and buying pollution-reducing services and items (Saha, 2014). Investigating the environmental track record of IT service and software suppliers may be another green procurement activity (Molla, n.d.).

- Green operations: When the data center is deployed or configured to serve IT-based business end users, employees, practitioners, and managers, the strategy aims to save fire, and electricity and thus reduce carbon dioxide emissions to the world. As a result, data managers must understand how to use green projects to reduce information center energy usage (Saha, 2014). Sustainability initiatives also aim to improve the functionality of IT-based company resources by lowering energy consumption by cooling and increasing the energy efficiency of the company's data center assets, optimizing data center energy efficiency, and lowering carbon dioxide emissions explained by Huang (2009) caused by data centers and implementing low-energy technologies. evaluation of carbon-emitting businesses, energy applications, and, finally, the organization's overall environmental footprint (Molla, n.d.).

- Green waste management: Green waste offers IT-based organizations and information centers a flexible and easily searchable remedy for end-of-life IT equipment collection, waste disposal, and reuse (Brown et al., 2007). Thus, while the IT infrastructure industry prepares to process and reuse waste materials that are not recycled in green waste, other unwanted electronic products can be made ready for recycling (Nuttapon & Gabriel, 2012). Additionally, OGCIO Adservio team. (2022) and Molla (n.d.) advocate the destruction or repair of automated data and IT equipment. Several challenges are addressed by green computing (Nuttapon & Gabriel, 2012; Saha, 2014).

More specifically, the objectives are as follows:

1. To achieve energy efficiency, the energy consumption of the items should be limited to a minimum.
2. To mitigate the environmental damage caused by the use of hazardous equipment.
3. Increasing the product's living time.
4. Using the intellectual property for the life of the product.
5. Promotes recycling of old products and waste.

Therefore, the use of green applications in IT-based business information centers can reduce costs, improve energy efficiency, environmental management, and energy savings, and ultimately reduce CO_2 emissions through green production, green purchasing, green operations, and green environmental protection. to strive.

These computational methods can help with a variety of green procurement initiatives, including:

- To help businesses choose the most environmentally friendly option, LCA can be used to evaluate the environmental effects of various products.
- For businesses to choose suppliers who adhere to their environmental standards, SERS can be used to rate suppliers on how well they perform in terms of the environment.
- Performance monitoring: EMS can be used to monitor and manage environmental performance, allowing businesses to pinpoint problem areas and take appropriate action.

Organizations can gain a competitive advantage, lessen their impact on the environment, and increase their bottom line by using computational techniques for green procurement. Numerous other computational techniques can be used to support green purchasing and production in addition to the ones already mentioned. These methods consist of:

- Data analytics: By analyzing vast amounts of data, data analytics can be used to spot trends and patterns that can help with green procurement decisions.
- Simulation modeling: Different procurement and production scenarios can be used to simulate the effects on the environment.
- Algorithms for optimization: Algorithms for optimization can be used to find the best solution to a green procurement issue, such as the issue of reducing a product portfolio's environmental impact.

Thus computational techniques are becoming more and more crucial in green production and procurement. Organizations will need to employ these strategies to improve their environmental performance and make wise procurement decisions as the demand for sustainable goods and services rises.

10.7 Findings and conclusion

This study aims to focus on ecologically sound IT companies with significant energy usage and some concerns regarding the environment. The information technology industry is being transformed by green computing. The IT industry was previously only concerned with the processing of electronic products. Power, air conditioning, and data center requirements are unaffected. It also outlines the steps that businesses and organizations must take to create a healthy environment.

With increasing pollution from various sources, this strategy will help reduce the carbon footprint in space. Therefore, with little understanding of the value and necessity of green computing, we must act now, if not immediately. Green computing is a way of thinking about global warming. Business leaders can contribute to environmental care and protection by implementing green practices. So far, lowering power consumption has had a major effect in terms of performance evaluations. Optimization determines the degree of energy effectiveness for a device. Low voltage can have an impact on life due to frequent system shutdowns and restarts.

An in-depth investigation of the most recent techniques in manufacturing and procurement, including sustainable computational techniques, reveals that green algorithms and processes are nearing completion and have the potential to produce significant results. By contrast, the survey found that management in the IT industry is declining, making green practices most important.

Green computing should be used at organizational levels in particular to add and develop learning efficiency methods and strategies. The adaptability of green computing solutions, with a focus on flexibility and adaptability, as well as cost-effective processors, memory, and network resources for optimization, is their strength. This, however, requires the use of powerful and intelligent algorithms capable of evaluating overall performance and energy economy in a multifaceted optimization. Evaluating the emphasizes the significance of additional research into the effectiveness and optimization of the way work is done, such as delivery and uptime. Evaluating achievement and utilization of energy will result in greater effective green solutions that will be easier to implement for IT managers and produce better results.

Despite this, recent advancements in the IT industry show a strong desire and determination to address environmental concerns. This document summarizes current thinking and proposes components for a long-term IT strategy. Thus, the computer technology industry will play a critical role in the emerging green economy. The information technology sector is increasing the effectiveness of equipment and cooling through the creation of instruments that can track machinery and data centers, adjust IT energy demand, and monitor temperature equipment to aid in the identification and troubleshooting of calming data or load control cooling data.

Energy Star certification is already in place, and the European Union is adhering to aggressive green computing goals for energy-intensive data centers.

10.7.1 Managerial implications

In general, computational methods can be a useful tool for assisting green production and procurement. But to successfully use these strategies, top management must be on board, employees must be informed and trained, and a clear strategy must be in place.

The following are some management implications of computational techniques for green production and procurement:

- The need for top management's backing. The backing of top management is necessary for the effective implementation of computational techniques for environmentally friendly production and procurement. This is because these techniques may need a sizable investment and may need to be incorporated into already-in-place procurement and production processes.

- The requirement for training and awareness. Employees must be trained in using computational techniques to support green production and procurement, and they must be made aware of the advantages of these practices.

- A clear strategy is required. Clear green procurement and production strategies must be in place for businesses. The environmental objectives of the company should be specified in this strategy, along with the precise ways in which computational techniques can be applied to meet these objectives.

The use of computational techniques for green procurement and production can be impacted by a variety of other factors in addition to the managerial implications already mentioned. One of these elements is:

- Information accessibility: The use of computational techniques effectively depends on the availability of data on vendors, goods, and environmental impacts.

- Implementation costs: Some businesses may find it difficult to implement computational techniques due to the cost.

- The required technical knowledge: Utilizing computational methods might call for technical knowledge that not all businesses have.

Despite these difficulties, computational techniques have significant potential for use in green purchasing and production. Businesses can improve productivity, lower costs, and lessen their impact on the environment by implementing these strategies.

10.8 Outlook and future research

The field of computational techniques for environmentally friendly production and procurement is rapidly expanding, and new studies are being published regularly. Some of the most promising areas for this field's future research are listed below:

- New life cycle assessment (LCA) tools are being developed. LCA tools are used to evaluate the environmental impact of goods and services throughout their entire life cycles. However, using the LCA tools available today can be difficult and time-consuming. Future research in this field might aim to create LCA tools that are easier to use and more effective.

- New decision support tool development: Decision support tools can assist organizations in making knowledgeable choices regarding green production and procurement. These instruments could be used to evaluate the environmental effects of various procurement options, find affordable green procurement strategies, and monitor the development of an organization's green procurement initiatives.

- Creation of novel approaches to fusing economic and environmental concerns: Trade-offs between environmental and economic factors frequently occur in decisions about green production and procurement. New approaches to incorporating these two sets of factors into decision-making processes may be the subject of future research in this field.

- Development of new techniques for assessing and tracking the effectiveness of green production and procurement initiatives: It's critical to be able to assess and track the effectiveness of green production and procurement initiatives. The development of new techniques for measuring and monitoring the environmental impact of procurement decisions, the efficacy of green procurement strategies, and the overall performance of green production and procurement programs may be the main topics of future research in this field.

To further develop the field of computational techniques for green procurement and production, several more general challenges as well as these particular research areas must be addressed. These difficulties include:

- The requirement for better information on how products and services affect the environment.

- The requirement for more uniform techniques when carrying out LCAs and other environmental assessments.

- The requirement for easier-to-use and more effective decision support tools.

- The requirement for more effective decision-making processes that consider economic and environmental factors.

- The requirement for improved metrics for gauging and monitoring the efficiency of green production and procurement.

Regardless of these difficulties, computational techniques for environmentally friendly purchasing and production hold great promise. These methods could support businesses in making more informed and eco-friendly production and procurement decisions with further research and development.

There are numerous additional potential areas of research in this field beyond the ones already mentioned. For instance, future studies might concentrate on creating fresh techniques for:

- Recognizing and controlling environmental risks throughout the supply chain.
- Communicating to stakeholders the environmental advantages of green procurement.
- Including green purchasing in larger sustainability initiatives.

Thus, future research opportunities are abundant and exciting in the rapidly developing field of computational techniques for environmentally friendly purchasing and production. These methods mentioned above have the potential to significantly contribute to helping organizations lessen their environmental impact with further research and development.

List of abbreviations

ICT	information and communication technology
IT	information technology
EPDs	environmental product declarations
EMS	environmental management systems
SERS	surface-enhanced Raman spectroscopy
ISO	International Organization for Standardization
G7	Group of Seven
G8	Group of Eight
GHG	greenhouse gas
GPS	Global Positioning System
GHGs	greenhouse gases
GIT	Green Information Technology
CO₂	carbon dioxide
LED	light-emitting diodes
LCD	liquid crystal display
CRT	cathode ray tube
LCA	life cycle assessment
PC	personal computer
CPUs	Central Processing Unit
GPUs	Graphic Processing Unit
ACPI	Advanced Configuration and Power Interface
APM	application performance monitoring
OGCIO	Office of the Government Chief Information Officer

References

Adservio tam. (2022). *Green computing - approaches: challenges.* Adservio. Available at: https://www.adservio.fr/post/green-computing-approaches-challenges.

Agarwal, S., & Nath, A. (2011, June). Green computing-a new horizon of energy efficiency and electronic waste minimization: A global perspective. In 2011 *International Conference on Communication Systems and Network Technologies* (pp. 688–693). IEEE.

Ahuja, S. P., & Muthiah, K. (2021). *Advances in green cloud computing. Research anthology on architectures, frameworks, and integration strategies for distributed and cloud computing* (pp. 2651–2662). IGI global. [32] Bisoyi, B., & Das, B. (2016). Necessitate Green Environment for Sustainable Computing. Advances in Intelligent Systems and Computing. Available from https://doi.org/10.1007/978-81-322-2523-2_50.

Akhtar, R., Sultana, S., Masud, M. M., Jafrin, N., & Al-Mamun, A. (2021). Consumers' environmental ethics, willingness, and green consumerism between lower and higher income groups. *Resources, Conservation and Recycling, 168*105274.

Ali, A., Mahmood, A., Ikram, A., & Ahmad, A. (2020). Configuring the drivers and carriers of process innovation in manufacturing organizations. *Journal of Open Innovation: Technology, Market, and Complexity, 6*(4), 154. Available from https://doi.org/10.3390/joitmc6040154.

Anthony, B., Majid, M.A., & Romli, A. (2018). A descriptive study towards green computing practice application for data centers in IT Based Industries. *MATEC Web of Conferences, 150,* 05048.https://doi.org/10.1051/matecconf/201815005048.

Apple. (n.d.). *Environment.* Apple (India). Available at: https://www.apple.com/in/environment/.

Azadnia, A. H., Saman, M. Z. M., & Wong, K. Y. (2015). Sustainable supplier selection and order lot-sizing: An integrated multi-objective decision-making process. *International Journal of Production Research, 53*(2), 383–408.

Barquet, K., Järnberg, L., Rosemarin, A., & Macura, B. (2020). Identifying barriers and opportunities for a circular phosphorus economy in the Baltic Sea region. *Water Research, 171*115433.

Belhadi, A., Kamble, S. S., Zkik, K., Cherrafi, A., & Touriki, F. E. (2020). The integrated effect of Big Data Analytics, Lean Six Sigma and Green Manufacturing on the environmental performance of manufacturing companies: The case of North Africa. *Journal of CleanerProduction, 252*119903. Available from https://doi.org/10.1016/j.jclepro.2019.119903.

Bolla, R., Bruschi, R., Davoli, F., & Cucchietti, F. (2010). Energy efficiency in the future internet: A survey of existing approaches and trends in energy-aware fixed network infrastructures. *IEEE Communications Surveys & Tutorials, 13*(2), 223–244.

Bravo, A., Vieira, D., & Rebello, T. A. (2022). The origins, evolution, current state, and future of green products and consumer research: A bibliometric analysis. *Sustainability, 14*(17), 11022. MDPI AG. Retrieved from http://doi.org/10.3390/su141711022.

Brown, R.E., Incorporated, I., & Incorporated, E. (2007). *Report to congress on server and data center energy efficiency: public law* 109–431. Available from https://doi.org/10.2172/929723.

Büchi, G., Cugno, M., & Castagnoli, R. (2020). Smart factory performance and Industry 4.0. *Technological Forecasting and Social Change, 150*119790. Available from https://doi.org/10.1016/j.techfore.2019.119790.

Chang, Y., Iakovou, E., & Shi, W. (2019). Blockchain in global supply chains and cross border trade: A critical synthesis of the state-of-the-art, challenges and opportunities. *International Journal of Production Research, 58*(7), 2082–2099. Available from https://doi.org/10.1080/00207543.2019.1651946.

Chen, T., Kim, H., Pan, S., Tseng, P., Lin, Y., & Chiang, P. (2020). Implementation of green chemistry principles in circular economy system towards sustainable development goals: Challenges and perspectives. *Science of the Total Environment*, 716136998. Available from https://doi.org/10.1016/j.scitotenv.2020.136998.

Cherrafi, A., Elfezazi, S., Govindan, K., Garza-Reyes, J. A., Benhida, K., & Mokhlis, A. (2017). A framework for the integration of Green and Lean Six Sigma for superior sustainability performance. *International Journal of Production Research*, *55*(15), 4481−4515. Available from https://doi.org/10.1080/09537287.2018.1501808, 399.

Cherrafi, A., Elfezazi, S., Hurley, B., Kumar, V., Kumar, V., Anosike, A., & Batista, L. (2019). Green and lean: A Gemba−Kaizen model for sustainability enhancement. *Production Planning & Control*, *30*(5−6), 385.

Choudhary, S., Nayak, R., Dora, M., Mishra, N., & Ghadge, A. (2019). An integrated lean and green approach for improving sustainability performance: A case study of a packaging manufacturing SME in the U.K. *Production Planning & Control*, *30*(5−6), 353−368. Available from https://doi.org/10.1080/09537287.2018.1501811.

Cocacola company. (n.d.) *Sustainable packaging*. Available at: https://www.cocacolacompany.com/sustainability/packaging-sustainability.

Da Costa, R. L., Da Fonseca Resende, T. N. E., Dias, Á., Pereira, L., & António, N. (2020). Public sector shared services and the lean methodology: Implications on military organizations. *Journal of Open Innovation: Technology, Market, and Complexity*, *6*(3) 78. Available from https://doi.org/10.3390/joitmc6030078.

Dawson., & Probert. (2007). A sustainable product needing a sustainable procurement commitment: The case of green waste in Wales. *Sustainable Development*, *15*(2), 69−82.

Deng, W., Liu, F., Jin, H., Li, B., & Li, D. (2014). Harnessing renewable energy in cloud datacenters: Opportunities and challenges. *IEEE Network*, *28*(1), 48−55.

Duarte, S., & Machado, V. C. (2017). Green and lean implementation: An assessment in the automotive industry. *International Journal of Lean Six Sigma*, *8*(1), 65−88. Available from https://doi.org/10.1108/ijlss-11-2015-0041.

Emetere, M. (2019). Introduction to computational techniques. *Studies in Big Data*. Available from https://doi.org/10.1007/978-3-030-13405-1_2.

Global Gender Gap Report 2023. (2023, October 20). *World Economic Forum*. Available at: https://www.weforum.org/publications/global-gender-gap-report-2023.

Griffins. (2005). Emerging trends in green cleaning. *Journal of Distribution Sales and Management*, *46*(6), 26−30.

Harmon, R.R., & Auseklis, N. (2009, August). Sustainable IT services: Assessing the impact of green computing practices. In *PICMET'09−2009 Portland International Conference on Management of Engineering & Technology* (pp. 1707−1717). IEEE.

Hawking, P. (2018). Big Data Analytics and IoT in logistics: A case study. *The International Journal of Logistics Management*, *29*(2), 575−591. Available from https://doi.org/10.1108/ijlm-05-2017-0109IKEA and sustainability efforts. (n.d.). IKEA. https://www.ikea.com/gb/en/this-is- ikea/about-us/were-all-in-this-together-pubc8331c51.

Huang. (2009). *A model for environmentally sustainable information systems development*. ResearchGate. Available at: https://www.researchgate.net/publication/266202538_A_model_for_environmentally_sustainable_information_systems_development.

Inman, R. A., & Green, K. W. (2018). Lean and green combine to impact environmental and operational performance. *International Journal of Production Research*, *56*(14), 4802−4818. Available from https://doi.org/10.1080/00207543.2018.1447705.

Karkalos, N. E., Markopoulos, A. P., & Davim, J. P. (2019). General aspects of the application of computational methods in Industry 4.0. In Springer Briefs in applied sciences and technology. *Springer Berlin Heidelberg*. Available from https://doi.org/10.1007/978-3-319-92393-2-1.

Kazandjieva, M., Heller, B., Gnawali, O., Levis, P., & Kozyrakis, C. (2012, June). Green enterprise computing data: Assumptions and realities. In *2012 International Green Computing Conference (IGCC)* (pp. 1−10). IEEE.

Kirkire, M. S., Rane, S. B., & Singh, S. P. (2018). Integrated SEM-FTOPSIS framework for modeling and prioritization of risk sources in medical device development process. *Benchmarking: An International Journal, 25*(1), 178−200.

Klassen., & Vachon. (2003). Collaboration and evaluation in the supply chain: The impact on plant-level environmental investment. *Production and Operations Management, 12*(3), 336−352.

Kouhizadeh, M., Sarkis, J., & Zhu, Q. (2019). At the nexus of blockchain technology, the circular economy, and product deletion. *Applied Sciences, 9*(8), 1712. Available from https://doi.org/10.3390/app9081712.

Kumar, V. B., Reddy, N. V., & Reddy, C. Y. (2016). A review on globalization and green technologies to mitigate pollution. *International Journal of Advancce Researach in Science and Engineering*, 225−236.

Kurdve, M., Zackrisson, M., Wiktorsson, M., & Harlin, U. (2014). Lean and green integration into production system models − experiences from Swedish industry. *Journal of Cleaner Production, 85*, 180−190. Available from https://doi.org/10.1016/j.jclepro.2014.04.013.

Leong, W. D., Teng, S. Y., How, B. S., Ngan, S. L., Abd Rahman, A., Tan, C. P., & Lam, H. L. (2020). Enhancing the adaptability: Lean and green strategy towards the Industry Revolution 4.0. *Journal of Cleaner Production, 273*122870.

Li, L. (2019). Lean Smart Manufacturing in Taiwan—Focusing on the bicycle industry. *Journal of Open Innovation: Technology, Market, and Complexity, 5*(4), 79. Available from https://doi.org/10.3390/joitmc5040079.

Lozano, R. (2012). Towards better embedding sustainability into companies' systems: An analysis of voluntary corporate initiatives. *Journal of Cleaner Production, 25*, 14−26. Available from https://doi.org/10.1016/j.jclepro.2011.11.060.

Mani, V., Jabbour, C. J. C., & Mani, K. T. (2020). Supply chain social sustainability in small and medium manufacturing enterprises and firms' performance: Empirical evidence from an emerging Asian economy. *International Journal of Production Economics, 227*107656. Available from https://doi.org/10.1016/j.ijpe.2020.107656.

Marrucci, L., Daddi, T., & Iraldo, F. (2019). The integration of circular economy with sustainable consumption and production tools: Systematic review and future research agenda. *Journal of Cleaner Production, 240*118268.

Melville. (2012). *Environmental sustainability 2.0: Empirical analysis of environmental ERP implementation.*

Molla, A. (n.d.). The Reach and richness of green it: a principal component analysis. *AIS Electronic Library (AISeL)*. https://aisel.aisnet.org/acis2009/31.

Molla, A., Cooper, V., & Pittayachawan, S. (2011). The green IT readiness (G-readiness) of organisations: An exploratory analysis of a construct and instrument. *Communications of the Association for Information Systems, 29*(1), 67−96.

Murugesan, S. (2008). Harnessing green IT: Principles and practices. *IT Professional, 10*(1), 24−33. Available from https://doi.org/10.1109/mitp.2008.10.

Nuttapon, P., & Gabriel, C. (2012). *Analysis of Green Information Technology in DellAnd Toshiba Companies.* IDT: Malardalen University.

Oliveira, G. A., Tan, K. H., & Guedes, B. F. (2018). Lean and green approach: An evaluation tool for new product development focused on small and medium enterprises. *International Journal of Production Economics, 205*, 62−73. Available from https://doi.org/10.1016/j.ijpe.2018.08.026.

Ong, D., Moors, T., & Sivaraman, V. (2012, September). Complete life-cycle assessment of the energy/CO_2 costs of videoconferencing vs face-to-face meetings. In *2012 IEEE Online Conference on Green Communications (GreenCom)* (pp. 50−55). IEEE.

Parikka-Alhola, K. (2008). Promoting environmentally sound furniture by green public procurement. *Ecological Economics, 68*(1−2), 472−485.

Pawlish, M.J., & Varde, A.S. (2010, October). A decision support system for green data centers. In *Proceedings of the 3rd workshop on Ph. D. students in information and knowledge management* (pp. 47−56).

Perez-Castillo, D., & Vera-Martinez, J. (2021). Green behaviour and switching intention towards remanufactured products in sustainable consumers as potential earlier adopters. *Asia Pacific Journal of Marketing and Logistics, 33*(8), 1776−1797.

Prasad, S., Khanduja, D., & Sharma, S. (2016). An empirical study on applicability of lean and green practices in the foundry industry. *Journal of Manufacturing Technology Management, 27*(3), 408−426. Available from https://doi.org/10.1108/jmtm-08-2015-005.

Raza, K., Patle, V. K., & Arya, S. (2012). A review on Green Computing for Eco-Friendly and Sustainable IT. *Journal of Computational Intelligence and Electronic Systems, 1*(1), 3−16. Available from https://doi.org/10.1166/jcies.2012.1023.

Sagnak, M., & Kazancoglu, Y. (2016). Integration of green lean approach with six sigma: An application for flue gas emissions. *Journal of Cleaner Production, 127*, 112−118. Available from https://doi.org/10.1016/j.jclepro.2016.04.016.

Saha. (2014). Green computing. *International Journal of Computer Trends and Technology (IJCTT), 14*(2), 46−50.

Sarkis, J., Zhu, Q., & Lai, K. H. (2011). An organizational theoretic review of green supply chain management literature. *International Journal of Production Economics, 130*(1), 1−15.

Shokri, A., & Li, G. (2020). Green implementation of Lean Six Sigma projects in the manufacturing sector. *International Journal of Lean Six Sigma, 11*(4), 711−729, ISSN 2040-4166.

Siegel, R., Antony, J., Garza-Reyes, J. A., Cherrafi, A., & Lameijer, B. A. (2019). Integrated green lean approach and sustainability for SMEs: From literature review to a conceptual framework. *Journal of Cleaner Production, 240*118205. Available from https://doi.org/10.1016/j.jclepro.2019.118205.

Sinha, N., Garg, A. K., & Dhall, N. (2016). Effect of TQM principles on performance of Indian SMEs: The case of automotive supply chain. *The TQM Journal, 28*(3).

Song, H., & Xue-Xian, G. (2018). Green supply chain game model and analysis under revenue-sharing contract. *Journal of Cleaner Production, 170*, 183−192. Available from https://doi.org/10.1016/j.jclepro.2017.09.138.

Sun, Y., Li, T., & Wang, S. (2022). I buy green products for my benefits or yours": Understanding consumers' intention to purchase green products. *Asia Pacific Journal of Marketing and Logistics, 34*(8), 1721−1739.

Tjahjadi, B., Soewarno, N., Hariyati, H., Nafidah, L. N., Kustiningsih, N., & Nadyaningrum, V. (2020). The role of green innovation between green market orientation and business performance: Its implication for open innovation. *Journal of Open Innovation: Technology, Market, and Complexity, 6*(4), 173. Available from https://doi.org/10.3390/joitmc6040173.

Tripathi, V., Chattopadhyaya, S., Mukhopadhyay, A., Sharma, S., Singh, J., Pimenov, D. Y., & Giasin, K. (2021). An innovative agile model of smart Lean−Green approach for sustainability enhancement in industry 4.0. *Journal of Open Innovation: Technology, Market, and Complexity, 7*(4), 215. Available from https://doi.org/10.3390/joitmc7040215.

Tripathi, V., Chattopadhyaya, S., Mukhopadhyay, A.K., Sharma, S., Singh, J., Li, C., & Bona, G.D. (2022). *An agile model of a hybrid approach to enhance the sustainability of production management system in industry 4.0.* Available from https://doi.org/10.21203/rs.3.rs-.1258347/v1.

Uddin, M., Talha, M., Rahman, A. A., Shah, A., Khader, J. A., & Memon, J. (2012). Green Information Technology (IT) framework for energy efficient data centers using virtualization. *International Journal of Physical Sciences, 7*(13), 2052−2065.

Vachon (2015). *Linking supply chain strength to sustainable development: A country-level analysis.* Westernu. Available at:https://www.academia.edu/14839821/Linking_supply_chain.

Vachon., & Mao. (2008). Linking supply chain strength to sustainable development: A country-level analysis. *Journal of Cleaner Production, 16*(15), 1552−1560.

Wang, Y., Yang, Y., Qin, Z., Yang, Y., & Li, J. (2023). A literature review on the application of digital technology in achieving green supply chain management. *Sustainability, 15*(11), 8564. MDPI AG. Retrieved from Available from http://doi.org/10.3390/su15118564.

Wiprächtiger, M., Haupt, M., Heeren, N., Waser, E., & Hellweg, S. (2020). A framework for sustainable and circular system design: Development and application on thermal insulation materials. *Resources, Conservation and Recycling, 154*104631.

Zhu, X. (2020). *Application of green-modified value stream mapping to integrate and implement lean and green practices: A case study*. Available at: https://www.semanticscholar.org/paper/Application-of-green-modified-value-stream-mapping-Zhu-Zhang/09e937989d4c588488f20128c9572edfdcae3dd7.

CHAPTER ELEVEN

Predictive big data analytics for supply chain demand forecasting

Supriyo Ahmed[1,2], Ripon K. Chakrabortty[1] and Daryl L. Essam[1]
[1]School of Systems & Computing, UNSW Canberra at ADFA, Campbell, ACT, Australia
[2]Department of Electrical and Electronic Engineering, BRAC University, Dhaka, Bangladesh

11.1 Introduction

Inventory managers face critical questions regarding the quantity and timing of orders in inventory management. To address these questions, they utilize the Economic Order Quantity (EOQ) as a determining factor. However, accurate estimation of the EOQ relies on a precise understanding of customer demands (Mashud et al., 2022). The effectiveness of inventory models heavily depends on the accuracy of demand predictions. A reliable inventory model and precise demand data estimation enable businesses to manage and adapt to unexpected shortages effectively (Paul et al., 2023). Forecasting future demands is challenging and typically relies on historical or past data. The unpredictability in customers' demand patterns, the potential for mass customization among manufacturers, and the vast amount of customer data, such as big data Sarfaraz et al. (2023b), render traditional forecasting methods, including moving average, exponential smoothing, and naive methods, inefficient (Zhu et al., 2021).

The advent of Artificial Intelligence (AI) has brought about a significant transformation across various industries over the past decade. Concurrently, the concept of industrial evolution, exemplified by Industry 4.0 and Industry 5.0, strongly emphasizes digitizing business operations (Bassiouni et al., 2023). Recent developments have introduced Deep Learning (DL) approaches to accurately forecast customer demands (Pereira and Frazzon, 2021, Tirkolaee et al., 2021, Zhu et al., 2021). As the world becomes increasingly digitized, historical data related to supply chains have become more accessible, facilitating the use of DL models for future predictions (Chakrabortty et al., 2023). Leveraging the forecasting

Computational Intelligence Techniques for Sustainable Supply Chain Management.
DOI: https://doi.org/10.1016/B978-0-443-18464-2.00011-X

capabilities of DL models using historical data enables businesses to establish improved reorder points and lot sizes (Sarfaraz et al., 2023a). Businesses can operate at lower costs by adopting DL-based inventory management instead of traditional methods. Furthermore, precise predictions contribute to enhanced customer response times, thereby minimizing the impact of lost sales, holding costs, and order costs, ultimately reducing overall supply chain expenses (Seyedan and Mafakheri, 2020).

Despite the numerous time-series forecasting models developed by researchers using statistical methods, these models face limitations when dealing with complex market scenarios. Statistical forecasting models primarily rely on univariate analysis techniques (Weller and Crone, 2012) and struggle to perform effectively in situations characterized by nonlinear relationships among data, such as strong sales seasonality, trends, volatile demand, or limited historical data availability (Deng et al., 2021, Thomassey, 2010). In recent times, scholars have turned to neural network (NN) models to address various challenges within supply chain management (Ahmed et al., 2022). These approaches have been utilized for tasks such as supply chain risk forecasting (Kamble et al., 2021; Yan et al., 2019), supplier selection (Guo et al., 2009), demand estimation (Ni et al., 2020), and inventory management through lead time forecasting (Gumus et al., 2010).

Researchers are actively engaged in various aspects of sales estimation in the retail industry. To achieve better accuracy in sales estimation, Li et al. (2021) combined and explored approaches from NNs to enhance the accuracy of time series sales forecasting, specifically in the context of clothing sales. Additionally, Kantasa-Ard et al. (2021) employed simulation techniques in supply chain models, utilizing simple forecasting approaches from the NN group to minimize costs. While some scholars focus on enhancing prediction accuracy, others aim to reduce supply chain costs through modeling techniques. However, there is limited research evidence demonstrating the extent to which more accurate AI prediction reduces supply chain costs (Abdel-Basset et al., 2023). Furthermore, no universally accepted approach has emerged as superior in estimating short to long-term horizons with all-time series datasets. Distinguishing between different NN approaches proves challenging since all approaches are widely used and demonstrate higher accuracy and stability when forecasting nonlinear and non-stationary time series data with multiple features (Gu et al., 2020, Masini et al., 2021).

Hence, there remains a gap in the exploration of combining efficient forecasting models from the category of NNs. Furthermore, there is limited investigation into the predictive performance of such a model over an extended time horizon. Additionally, only a few studies have been conducted to demonstrate the effectiveness of an enhanced forecasting approach within the context of supply chains. This chapter proposes a hybrid forecasting model that can dynamically switch between different models based on a validation score to address these gaps. These models are trained to forecast different future periods. During the training phase, the parameter settings of NN models remain constant to save time. The practicality of improved predictions is subsequently demonstrated using a supply chain model to reduce overall costs effectively.

To address the aforementioned research gaps, this study conducted in this chapter employs the following methodology. DL is utilized to analyze historical sales data and forecast future sales. The projected sales figures are then input into a two-echelon supply chain model with dynamic lead time (LT). Instead of relying on traditional average sales values, the model calculates the reorder point (ROP) based on the predicted sales values during the lead time, using a fixed lot size (Q). The model follows a (r, Q) policy, where the decision variables (r) are computed from the predicted sales data while maintaining a fixed lot size (Q). The supply chain model operates daily, replenishing inventory when the inventory level and pending orders fall below the reorder point. To validate the model's effectiveness, conventional forecasting techniques such as moving average, exponential smoothing, and normal distribution are employed to generate predictions, from which the supply chain model's reorder point and associated costs can be calculated. Furthermore, the developed supply chain model is further tested using predictions from a switching-based forecasting network (SBFA-NN) to demonstrate that improved accuracy reduces supply chain costs.

Thus, in summary, this chapter makes several contributions:

- Introducing an SBFA-NN that combines various advanced forecasting methods from different neural networks. The selection of the forecasting approach is based on the lowest validation error observed across different datasets.
- Developing a two-echelon supply chain model with diverse attributes to assess and validate the proposed SBFA-NN against several traditional forecasting approaches. The reorder points of the supply chain model

are determined using predictions generated by DL forecasting methods.

- Analyzing the predictions derived from SBFA-NN and other approaches by calculating the overall cost of the supply chain.

The remaining sections of this chapter are structured as follows: Section 11.2 presents a concise overview of advanced DL-based forecasting methods. Section 11.3 describes the methodology employed by SBFA-NN, followed by the modeling and validation of our supply chain network in Section 11.4. Section 11.5 delves into the specifics of the experimental setup, while Section 11.6 outlines the findings obtained. Lastly, Section 11.7 concludes the chapter by presenting the conclusions drawn and discussing potential future directions.

11.2 Deep learning-based forecasting approaches

In this section, a brief overview is provided of the functioning principles of cutting-edge DL models. These explanations are derived from recently published research papers. Furthermore, this section discusses the slight modifications made to each approach, customizing them to generate sales predictions within a unified forecasting system. This integration of approaches represents the primary contribution of this chapter in the book.

11.2.1 Recurrent neural network

A recurrent neural network (RNN) serves as a simplified representation of an NN, particularly effective for handling time series or sequential data (Muhaimin et al., 2021). Its ability to predict future outputs based on past inputs is made possible by its looped structure, with these loops acting as hidden units or neurons within a specific hidden layer of the network. These loops enable the neurons to retain previous input information for a certain period, enabling the network, comprising multiple hidden layers, to generate output predictions.

As the iterations progress, the output of a recursive neuron is passed on to the next layer, allowing the network to retain and utilize past information for a longer duration, resulting in more comprehensive output predictions (Apaydin et al., 2020). Moreover, mean square errors are calculated and propagated backward to update the weights of the network. This study employs a two-layer stacked RNN architecture, with its specific structure illustrated in Fig. 11.1. The cell structure of the RNN is also depicted in Fig. 11.2.

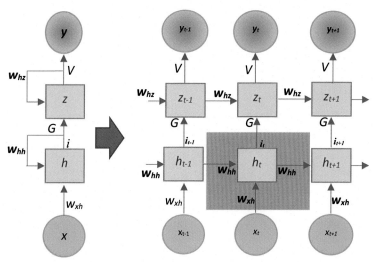

Figure 11.1 Folded and unfolded architecture of a 2-layer RNN. *RNN*, recurrent neural network.

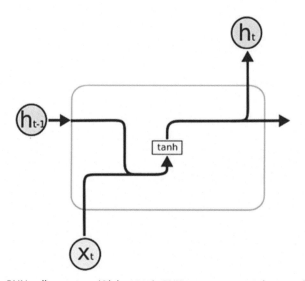

Figure 11.2 RNN cell structure (Olah, 2015). *RNN*, recurrent neural network.

The equation of a simple RNN cell (highlighted memory cell) can be represented by Eqs. (11.1 and 11.2)

$$h_t = \tanh(W_{hh}h_t - 1 + W_{xh}x_t) \tag{11.1}$$

$$i_t = Gi_t \tag{11.2}$$

where, Wxh: weights of the input layer to the hidden layer; Whh: own weights of the cell that change according to the change in cell state; G, V: weights of the connections between hidden layers and between the hidden layer and the output layer, respectively; ht: new state; $ht - 1$: previous state, while xt and yt are the current input and output respectively. tanh: activation function.

11.2.2 Long short term memory

RNN excels at capturing short-term data but faces limitations in retaining long-term memory. The LSTM (Long Short-Term Memory) approach is introduced to address this drawback, offering improved handling of long sequential data. This study employs a similar architecture, featuring two stacked layers of LSTM networks. This section provides a concise overview of how a single memory cell within an LSTM operates, and the architecture with interconnected components is depicted in Fig. 11.3. An LSTM cell consists of three gates: input, forget, and output. The gates operate in such a way that each line carries some particular information, as explained by Eqs. (11.3, 11.4, 11.5, 11.6, 11.7 and 11.8).

In RNN, the learning process relies on gradient descent, but as the network trains over a large sequence of cells, the gradient can diminish to the point where it has minimal impact on network learning. This phenomenon is known as the Vanishing Gradient Descent problem. LSTM

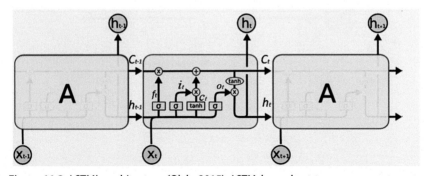

Figure 11.3 LSTM's architecture (Olah, 2015). *LSTM*, long short-term memory.

overcomes this issue through its architecture, which includes a memory line at the top and bottom to represent the output (ht), where $xt-1$ represents the previous block and $xt+1$ represents the next future block. This design allows for the input (xt) to be passed to future blocks. By using back-propagation, the learning process approximates the weights, enabling data to be stored or deleted within the cells. The transition equations of LSTM, as shown in Fig. 11.3, are provided below.

$$i_t = \sigma(W_i x_t + U_i h_{t-1} + V_i c_{t-1}) \quad (11.3)$$

$$f_t = \sigma\left(W_f x_t + U_f h_{t-1} + V_f c_{t-1}\right) \quad (11.4)$$

$$o_t = \sigma(W_o x_t + U_o h_{t-1} + V_o c_t) \quad (11.5)$$

$$\tilde{c}_t = \tanh(W_c x_t + U_c h_{t-1}) \quad (11.6)$$

$$c_t = f_t^i \odot c_{t-1} + i_t \odot \tilde{c}_t \quad (11.7)$$

$$h_t = o_t \odot \tanh(c_t) \quad (11.8)$$

where it denotes the input gate and ot denotes the output gate. $\odot$ characterizes a point-wise product. The forget gate, memory cell, and hidden state are denoted by ft, ct, and ht, respectively (Graves and Schmidhuber, 2005).

It is worth noting that the output of each block in LSTM is influenced by the output of its previous block. Each block consists of three input vectors and two output vectors. These variables contain arrays of values rather than single values. The sigmoid function controls the point-wise operations within the block, which determines whether information is allowed to pass through or not. Depending on the decision of the operation, the input value is either added to the memory or not, and if it is, it is passed to the next layer for further processing. The point-wise operations produce values between zero and one, and the output gate, in the final point-wise operation, determines how much of the memory is transmitted to the output. While there are different LSTM architectures available, the core functionalities remain consistent.

11.2.3 Gated recurrent unit

Gated recurrent unit (GRU) is a type of neural network that is similar to LSTM but is simpler in structure. The key difference between LSTM and GRU is that in GRU, the input and forget gates are merged to form a single update gate. Hence, there are fewer parameters in GRU than in LSTM, which makes training easier (Apaydin et al., 2020). Fig. 11.4 displays the structure of a single GRU cell.

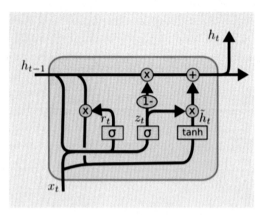

Figure 11.4 Gated recurrent unit (GRU) (Olah, 2015).

The output of GRU is *ht*. It can be computed in the following ways:

$$z_t = \sigma(W_z x_t + U_z h_{t-1} + b_z) \tag{11.9}$$

$$r_t = \sigma(W_r x_t + U_r h_{t-1} + b_r) \tag{11.10}$$

$$\tilde{h}_t = \tanh(W_h x_t + (r_t * h_{t-1})U_h) \tag{11.11}$$

$$h_t = (1 - z_t) * h_{t-1} + z_t * \tilde{h}_t \tag{11.12}$$

where *r* represents the reset gate and *z* is an update gate. An element-wise product among two vectors is implied by the operand *. The reset gate in GRU allows the current input to be combined with the memory. The update gate controls the amount of previous memory that should be retained. A value of 1 in the update gate indicates full preservation of previous memory, while a value of 0 indicates a complete disregard for previous memory. Unlike LSTM, where the forget gate determines the extent of information retention, GRU either retains all past information or discards it entirely. The experimental analysis demonstrates that GRU performs similarly to LSTM when given an equal number of parameters, with the advantage of shorter training time (Li et al., 2021).

11.2.4 Convolutional neural network

Convolutional Neural Network (CNN) is a DL model commonly employed for image analysis. However, it can also be applied to time series data, where the convolutional layers can capture important information and learn internal patterns within the data (Livieris et al., 2020). The convolution operation

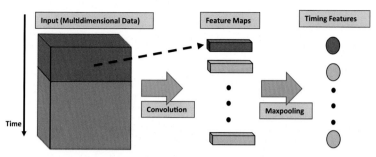

Figure 11.5 One dimensional CNN (Jin et al., 2020). *CNN*, convolutional neural network.

involves computing the inner product (multiplication and summation) between a window of data and a fixed set of weights known as a filter matrix. The max-pooling layer serves the purpose of reducing the dimensionality of the feature vectors, simplifying the computational complexity of the network, and extracting the most significant features. The architecture of a one-dimensional CNN model is depicted in Fig. 11.5.

In this chapter, a two-layer convolutional network with a max-pooling layer in between is utilized for experimentation purposes. The output from the last convolutional layer is then passed through a flattened layer, which serves the purpose of converting the output into a one-dimensional array, creating a single long feature vector. Finally, similar to other neural network approaches, a dense layer is employed to combine this single vector and generate the predicted outcome.

11.3 The proposed switching-based forecasting approach with neural networks

Advanced DL approaches are combined to form a switching-based approach, termed switching-based forecasting approach with neural networks (SBFA-NN). A layout diagram of the approach is illustrated in Fig. 11.6.

The diagram illustrates the input of three data subsets used to train four distinct advanced forecasting approaches, each making predictions for future time frames. It is observed that the performance of the algorithms varies depending on the forecasting duration. Consequently, the approach with the lowest validation error is selected for each prediction, ensuring optimal accuracy. To mitigate the impact of random weights and biases in

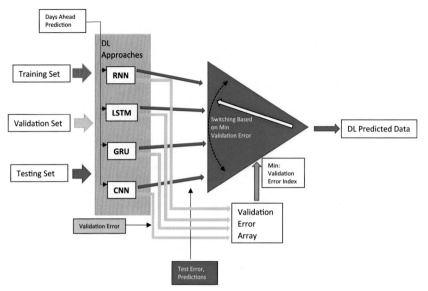

Figure 11.6 Switching-based forecasting approach using NN: SBFA-NN. *SBFA-NN,* switching-based forecasting approach with neural networks.

the networks, each approach is trained, validated, and tested ten times for a one-day-ahead prediction. Consequently, there are ten different test RMSEs (Root Mean Squared Errors) for each specific prediction duration, and their average is computed using Eq. (11.13).

$$\text{RMSE} = \sqrt{\left(\frac{1}{n}\right) \sum_{i=1}^{n} (\hat{\gamma} - \gamma_i)^2} \qquad (11.13)$$

where $\hat{y}$ is the predicted value, yi is the true value and n is the number of observations in both the above-mentioned equations.

11.4 The considered supply chain model

A two-stage supply chain model is considered in this study, consisting of a retailer and a distributor (supplier). The model is designed based on several assumptions: the retailer deals with a single type of product, there is no cost associated with back orders, but costs are incurred for lost sales when the retailer fails to meet customer demand. The supplier stores and provides finished products to the retailer upon receiving an order. The decision to place an order is determined by the reorder point (ROP), which is calculated based on the predicted demand during the lead time (LTPD) generated by machine learning (ML) approaches. Once an order is placed, the retailer's inventory is replenished after a specific lead time. The timing of order placement depends on the sum of the amount in delivery and the current inventory level. The model allows for both fixed and dynamic lead times. In this context, a fixed lead time refers to the predetermined duration between when the supplier receives an order and when the units are fully delivered to the retailer's inventory. The fixed lead time becomes dynamic when the delivery time varies within a specified range, although it is known at the time of order placement.

The supply chain model developed in this study incorporates the estimation of LTPD, which is obtained from ML forecasting algorithms. By considering the variability in previous lead times and the predicted demand, the model operates with a designated level of safety stock (ss). Safety stock represents an additional quantity of goods stored in inventory to prevent stock-out situations. The reorder point (ROP) is determined by the sum of LTPD and ss. The model also tracks the total quantity of orders currently in transit. Whenever the ROP exceeds the sum of the current inventory level and orders in transit, an order is placed with the supplier to replenish the stock. The ROP can be adjusted by modifying the level of safety stock, providing flexibility to the model. Furthermore, inspired by the work of Gonçalves et al. (2020), Eqs. 11.14 and 11.16 are employed to determine the safety stock and the reorder point, respectively, enhancing the realism of the model.

$$ss = Z\sqrt{(\sigma_d^2)\overline{L} + (\sigma_L^2)\overline{d}}$$ (11.14)

where, σd = standard deviation of predicted demand; σL = standard deviation of Lead time; $\overline{L}$ = average lead time; $\overline{d}$ = average predicted demand; Z = number of standard deviations corresponding to service level probability.

Safety stocks encompass two key aspects: the variability in demand and the variability in supplier lead time. Demand variations can be derived from the predictions obtained through ML/DL techniques. However, when it comes to lead time, there are two distinct scenarios to consider: fixed lead time and dynamic lead time. When the lead time is fixed,

$\sigma L = 0$, and hence Eq. (11.14) is simplified to Eq. (11.15). The symbols in Eq. (11.15) have their conventional meanings.

$$ss = Z\sigma_d\sqrt{L} \tag{11.15}$$

$$ROP = LTPD + ss \tag{11.16}$$

The model encompasses four distinct costs: purchasing costs, ordering costs, holding costs, and lost sales. The cumulative sum of these costs represents the total cost (TC) of the supply chain, as indicated by Eq. (11.17).

The assumptions regarding the associated costs are as follows:

Purchasing cost (PC): The cost per unit of product is known and remains constant. Ordering Cost (OC): It is known, fixed, and does not depend on the quantity of the order. Holding Cost (HC): It varies linearly based on the amount of inventory held in stock.

Lost Sales (LS): It is assumed to be fixed and increases linearly with the number of items that are out of stock. This cost is relatively the highest per unit among all other costs, as the loss incurred from a single lost sale can be significantly higher than the gross profit.

$$TC = \sum (PC, OC, HC, LS) \tag{11.17}$$

The parameters of the supply chain model are adjusted slightly to ensure accurate triggering based on varying sales levels in different datasets. Table 11.1 provides an overview of the parameters used for the experimentation.

By examining Table 11.1, it is evident that certain parameters have been modified, including the retailer's initial inventory, and lot size Q, while the ordering cost, purchasing cost, holding cost, and lost sales cost remain unchanged. These parameters have been selected to conduct a comprehensive analysis using advanced DL approaches, aiming to

Table 11.1 Input data for the supply chain model.

	Dataset 1	Dataset 2
Ordering Cost (per order)	$20	$20
Purchasing Cost (per unit)	$2	$2
Holding Cost (per unit per year)	$5	$5
Lost Sales Cost (per unit)	$35	$35
Retailer's Beginning Inventory (units)	$200	$1000
Retailer Lot Size Q (units)	$150	$500

determine the impact on forecasting accuracy within the supply chain model. Consequently, parameters related to both Dataset 1 and Dataset 2 are taken into consideration.

11.5 Experimental design for the proposed switching-based forecasting approach with neural networks

This section explains the methodology of applying the proposed SBFA-NN for the supply chain model.

11.5.1 Data collection

Two data sets were obtained from an open-source platform to conduct experiments with the proposed SBFA-NN method. The primary data set consisted of daily sales data for various products, which was sourced from the Kaggle ML Dataset Repository. From this primary data set, a specific product's data was extracted and deemed useful for the experimentation. Additionally, a secondary data set was obtained from the same source, which contained a greater number of time steps compared to the primary data set.

The first dataset primarily serves as an educational resource for data scientists, whereas the secondary dataset was originally used in a competition focused on predicting sales levels for different products across various stores. Both datasets consist of time series data, with Dataset 1 covering the period from January 2, 2017, to May 17, 2018, and Dataset 2 spanning from January 1, 2013, to August 31, 2017. It should be noted that these datasets and their respective date ranges do not reflect recent times, as recent sales data was not available in the current literature. To conduct our experimental analysis, publicly accessible data was collected from an open-source platform and filtered to include only the sales data of a specific item from a particular store, excluding other information.

Once the data features were extracted, the dataset was divided into three equal parts, namely the train, validation, and test sets. This division was carried out for cross-validation, as detailed in Section 5.2. To ensure consistency throughout the experiments, all the DL approaches mentioned in this chapter were provided with the same set of features, where the target variable was set as the sales.

11.5.1.1 Feature extraction

Properly extracting and organizing features is a crucial step in preparing data for DL models. To train the DL models effectively, significant features including year, month, week, day, trend, and seasonality were extracted. Python's built-in function to datetime was utilized to obtain the year, month, week, and day from the data. Additionally, the seasonal decompose function from the statsmodels library was employed to extract the trend and seasonality components from the sales data. As a result, the data frame now comprises seven features, as illustrated in Fig. 11.7, with the sales variable serving as the target.

11.5.1.2 Data structuring

The features of the data have been scaled between 0 and 1 using a min-max scalar normalization function (Geron, 2019). This scaling is performed to facilitate the handling of the data, particularly when sigmoid activation functions are employed in the output layer of neural networks. To achieve this normalization, the function is calculated according to the formula provided in Eq. (11.18).

$$x_{norm} = \frac{x - min(x)}{max(x) - min(x)} \qquad (11.18)$$

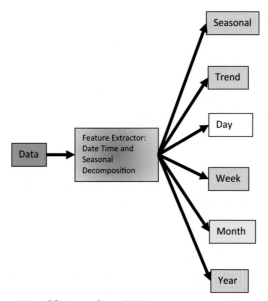

Figure 11.7 Extraction of features from data.

In the subsequent step, the scaled data frame was organized sequentially to facilitate multivariate predictions. The sales, which serve as the dependent variable, rely on various independent variables such as year, month, week, day, trend, and seasonality. The independent variables are denoted as X, while the dependent variable is represented by y. The data from the data frame is then structured into a two-dimensional array, incorporating both the independent and dependent variables, as outlined in Table 11.2.

In this way, three different segments were selected from the data frame for training, validation, and testing. For training, X train was formed from the independent variables while y train was the target column. Similarly, X val, and y val were taken for validation, and finally, X test, and y test were for testing. If, at any time, there is no data at the beginning, then only the available time steps are considered. For neural networks (e.g., LSTM, RNN, GRU, and CNN), these X train, X val, and X test are converted to suitable 3D arrays using a sliding window approach (Selvin et al., 2017). In the experiments with NN, a window of size 10 (working days of two consecutive weeks) is used to predict one day into the future (short term) and up to 30 days ahead (long term) sales.

11.5.1.3 Hyperparameters
Hyperparameters are an integral part of any DL/ML-based forecasting approach. In this chapter, four different advanced neural network-based forecasting approaches are applied to making sales predictions. The NNs are trained based on some fixed sets of parameters and those are selected based on the best possible default values acquired from Keras documentation.

Table 11.3 shows the list of the possible parameters used for forecasting sales data. It is presumed that the default parameter values are the most appropriate. The default values are mostly collected from the documentation of the individual approaches (Buitinck et al., 2013).

11.5.2 Cross-validation

In essence, both datasets are divided into three equal sections as indicated in Table 11.4. A total of 12 experiments were conducted, utilizing various combinations of the two datasets. One portion of the data was allocated for training the networks, another portion for validation, and the remaining portion for testing. This approach allowed for cross-validation using different combinations of the segmented data during the experiments.

Table 11.2 Data structuring.

Independent variables		Target variable
Variable X1	Variable X2	y
$X1_{t - n - days\_ahead\_prediction}$	$X2_{t - n - days\_ahead\_prediction}$	$y_{t - n + daysaheadprediction}$
$\cdots$	$\cdots$	$\cdots$
$\cdots$	$\cdots$	$\cdots$
$X1_{t - 2 - days\_ahead\_prediction}$	$X2_{t - 2 - days\_ahead\_prediction}$	$y_{t - 2 + days\_ahead\_prediction}$
$X1_{t - 1 - days\_ahead\_prediction}$	$X2_{t - 1 - days\_ahead\_prediction}$	$y_{t - 1 + days\_ahead\_prediction}$
$X1_{t - days\_ahead\_prediction}$	$X2_{t - days\_ahead\_prediction}$	$y_{t + days\_ahead\_prediction}$

where t, current time; n, number of past time steps; *days ahead prediction*, forecasting horizon.

Table 11.3 Hyperparameter values.

Algorithm	Hyperparameters
LSTM	Number of Layers: 2
RNN	Hidden Units Size: 64
GRU	Learning Rate: 0.001 (Default)
	Dropout: 0 (Default)
	Epochs: 150
	Activation Function: hard sigmoid (Layer 1), tanh
	(Layer 2), Optimiser: ADAM
	Loss: MSE
	Dense Layer: Hidden Units:1, Activation Function: Linear
CNN	Filters: 64, kernel size: 1
	Layers: 2 Conv1D layers
	MaxPooling1D layer in between Conv1D layers, Last Layers:
	Flatten Layer
	Dense Layer (Hidden Units: 1, Activation Function: Linear)
	Other parameters remain the same as before

LSTM, long short-term memory; *RNN*, recurrent neural network; *GRU*, gated recurrent unit; *CNN*, convolutional neural network.

Table 11.4 Subsets of dataset.

	Subset A	Time-steps	Subset B	Time-steps	Subset C	Time-steps
Dataset 1	2 Jan 2017−23 June 2017	118	24 June 2017−1 Dec 2017	118	2 Dec 2017−17 May 2018	118
Dataset 2	1 Jan 2013−22 July 2014	406	23 July 2014−10 Feb 2016	406	11 Feb 2016−31 Aug 2017	406

The combinations used for cross-validating the networks with the two datasets are presented in Table 11.5.

Here, in Table 11.5, "Subset A" of Dataset 1 represents a range of 118 time steps starting from 2 Jan 2017 until 23 June 2017. On the other hand, "Subset C" of Dataset 2 refers to a span of 406 time steps from 11 Feb 2016 until 31 Aug 2017. For "Expt 1" using Dataset 1, Subset A was utilized to train the network, followed by Subset B for validation, and finally, Subset C was used for testing. The order of these subsets was randomized to facilitate different experiments for cross-validation with each of the DL models.

Table 11.5 Experiments performed using cross-validation of dataset.

Dataset 1	Training dataset	Validation dataset	Testing dataset
Expt 1	Subset A	Subset B	Subset C
Expt 2	Subset A	Subset C	Subset B
Expt 3	Subset B	Subset A	Subset C
Expt 4	Subset B	Subset C	Subset A
Expt 5	Subset C	Subset B	Subset A
Expt 6	Subset C	Subset A	Subset B
Dataset 2			
Expt 1	Subset A	Subset B	Subset C
Expt 2	Subset A	Subset C	Subset B
Expt 3	Subset B	Subset A	Subset C
Expt 4	Subset B	Subset C	Subset A
Expt 5	Subset C	Subset B	Subset A
Expt 6	Subset C	Subset A	Subset B

11.6 Result analyses for the proposed switching-based forecasting approach with neural networks

This section presents an evaluation of the performance of the SBFA-NN approach in comparison to other DL approaches, using different datasets and assessing the RMSE. The predictions generated by various notable approaches, including the proposed SBFA-NN, were then assessed using the developed supply chain model, as described in Section 11.4. Furthermore, it demonstrates that the proposed SBFA-NN yields lower supply chain costs compared to the standalone approaches.

11.6.1 A comparative analysis of switching-based forecasting approach with neural networks with existing deep learning approaches

Every DL approach discussed in this chapter, including SBFA-NN, underwent training, validation, and testing using various subsets of data according to Table 11.5, considering different days for future predictions (up to 30 days). For each forecasting horizon, all models were trained 10 times using different random seeds. Subsequently, the average predicted values and their corresponding RMSEs were calculated for each specific forecasting horizon. The experimental results for both datasets are presented in Tables 11.6 and 11.7. Fixed seed values were utilized to ensure reproducibility of the results.

Table 11.6 Average test RMSE (Root Mean Squared Errors) of 10 Runs for 30 days ahead of prediction using Dataset 1.

Dataset 1	Proposed: SBFA-NN	RNN	LSTM	GRU	CNN
Expt 1	0.2229	0.2246	0.2221	0.2227	0.2267
Expt 2	0.1962	0.1981	0.1964	0.1970	0.1994
Expt 3	0.2208	0.2218	0.2206	0.2214	0.2236
Expt 4	0.2048	0.2065	0.2047	0.2057	0.2068
Expt 5	0.2046	0.2066	0.2059	0.2051	0.2050
Expt 6	0.1970	0.1978	0.1978	0.2008	0.1964
Mean RMSE	**0.2077**	0.2092	0.2079	0.2088	0.2096

The bold value corresponds to the lowest RMSE value compared to the other competitive approaches.
LSTM, long short-term memory; RNN, recurrent neural network; GRU, gated recurrent unit; CNN, convolutional neural network.

Table 11.7 Average test RMSE (Root Mean Squared Errors) of 10 Runs for 30 days ahead of prediction using Dataset 2.

Dataset 2	Proposed: SBFA-NN	RNN	LSTM	GRU	CNN
Expt 1	0.4326	0.4411	0.5061	0.4421	0.4562
Expt 2	0.3886	0.3982	0.4418	0.4073	0.4052
Expt 3	0.0639	0.0673	0.0731	0.0628	0.0595
Expt 4	0.1974	0.1796	0.1711	0.2221	0.1793
Expt 5	0.2956	0.3386	0.3220	0.2995	0.2986
Expt 6	0.1224	0.1607	0.1242	0.1323	0.1217
Mean RMSE	**0.2501**	0.2642	0.2731	0.2610	0.2534

The bold value represents the best RMSE (i.e., the lowest RMSE) obtained among all other competing algorithms.
LSTM, long short-term memory; RNN, recurrent neural network; GRU, gated recurrent unit; CNN, convolutional neural network.

Table 11.6 reveals that, in the case of Dataset 1, the average Test RMSE of the proposed SBFA-NN is 0.2077. Despite LSTM being the top-performing standalone algorithm with an RMSE of 0.2079, it falls short of the performance achieved by the proposed switching-based approach. Therefore, the proposed SBFA-NN stands out as the superior choice among all approaches for Dataset 1.

The findings from the experiments conducted on Dataset 2 are presented in Table 11.7. The proposed switching-based technique (SBFA-NN) continues to yield the lowest test RMSE of 0.2501, surpassing all other standalone approaches. Among the standalone approaches, CNN performs the best with an RMSE of 0.2534. To enhance the visual appeal

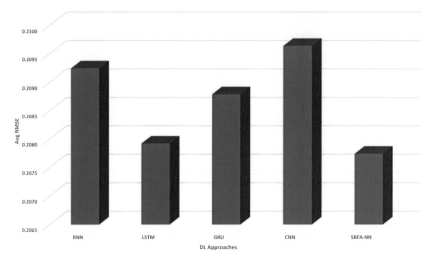

Figure 11.8 Average RMSE of different approaches including SBFA-NN for Dataset 1. *RMSE*, root mean squared errors; *SBFA-NN*, switching-based forecasting approach using multiple neural networks.

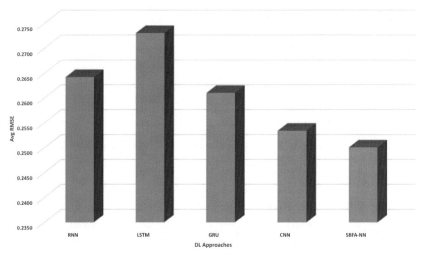

Figure 11.9 Average RMSE of different approaches including SBFA-NN for Dataset 2. *RMSE*, root mean squared errors; *SBFA-NN*, switching-based forecasting approach using multiple neural networks.

for readers, bar charts illustrating the average RMSEs of different approaches are included in Figs. 11.8 and 11.9.

When comparing Figs. 11.8 and 11.9, practitioners face a challenge in selecting a single standout standalone approach. Certain approaches demonstrate better prediction accuracy on one dataset but exhibit poorer performance on the other dataset. For instance, LSTM performs exceptionally well on Dataset 1 but experiences a decline in performance when applied to Dataset 2. The same observation can be made for CNN. Therefore, it can be concluded that there is no single approach that consistently delivers the best predictions across all datasets. This highlights the significance of the SBFA-NN, which consistently outperforms other approaches regardless of the dataset's characteristics.

11.6.2 Application of switching-based forecasting approach using multiple neural networks to the designed supply chain model

The predictions generated by the Proposed SBFA-NN, as well as predictions from other DL approaches, were integrated into the supply chain system for additional assessment, considering both static and dynamic lead time scenarios. While the supply chain (SC) parameters for Dataset 1 remained unchanged, certain adjustments were made for Dataset 2. Since Dataset 2 has a different range of values compared to Dataset 1, modifications were necessary to ensure accurate timing for order placement. The specific parameter changes are detailed in Table 11.1.

11.6.2.1 Fixed lead time with variable safety stock

In the preliminary analysis, the supply chain model is configured with a fixed lead time of 30 days, and various levels of safety stock are incorporated for both Dataset 1 and Dataset 2. The outcomes of utilizing predictions from SBFA-NN and other DL approaches in the supply chain model are presented in Tables 11.8 and 11.9.

After conducting six experiments on two datasets, it is observed that the proposed SBFA-NN achieves the second lowest average supply chain cost for Dataset 1, with CNN producing the lowest cost in this case. However, for Dataset 2, the SBFA-NN achieves the lowest supply chain cost. This demonstrates that improved forecasting accuracy can directly contribute to reducing overall supply chain costs. The percentage differences in cost between the proposed SBFA-NN, CNN, GRU, and RNN (the top two standalone approaches from each dataset) are presented in Table 11.10, calculated using Eq. (11.19). Additionally, Table 11.10 also provides the overall average reduction in supply chain cost across the two datasets.

$$\%_{diff} = \frac{SC_{SBFA-NN} - SC_{Estbd}}{SC_{Estbd}} \times 100 \qquad (11.19)$$

Table 11.8 SC Cost (Fixed Lead time with variable safety stock) associated with different advanced forecasting approaches using Dataset 1.

		SC cost using Dataset 1 ($)				
		RNN	LSTM	GRU	CNN	SBFA-NN
Expt 1	No SS	16,587	15,246	15,316	14,608	14,604
	95% SS	15,252	14,472	14,481	14,497	14,475
	99% SS	15,262	14,481	14,494	14,501	14,410
	99.9% SS	14,481	14,488	14,501	14,429	14,354
Expt 2	No SS	10,706	10,929	10,917	10,955	10,699
	95% SS	10,299	10,961	10,718	10,292	10,292
	99% SS	10,324	10,712	10,292	10,299	10,299
	99.9% SS	10,337	10,731	10,299	10,305	10,318
Expt 3	No SS	15,402	15,402	15,418	14,586	15,402
	95% SS	14,309	14,296	14,302	14,628	14,315
	99% SS	14,227	14,296	14,309	14,309	14,321
	99.9% SS	14,166	14,208	14,154	14,208	14,221
Expt 4	No SS	14,199	14,181	14,201	14,212	14,194
	95% SS	14,226	14,226	14,239	14,239	14,239
	99% SS	14,251	14,245	14,251	14,277	14,264
	99.9% SS	14,270	14,264	14,270	11,317	14,276
Expt 5	No SS	14,279	14,266	14,273	14,125	14,279
	95% SS	14,193	14,162	14,175	14,186	14,180
	99% SS	14,201	14,175	14,183	14,224	14,193
	99.9% SS	14,245	14,206	14,226	14,237	14,232
Expt 6	No SS	10,292	10,718	10,299	10,961	10,712
	95% SS	10,337	10,731	10,311	10,292	10,299
	99% SS	10,343	10,311	10,117	10,305	10,311
	99.9% SS	10,469	10,311	10,443	10,324	10,324
	Avg	13,194	13,168	13,091	**12,930**	13,050
	Rank (Lowest Cost)	5	4	3	1	2

The bold value corresponds to the lowest supply chain costs compared to the other competitive approaches. *LSTM*, long short-term memory; *RNN*, recurrent neural network; *GRU*, gated recurrent unit; *CNN*, convolutional neural network.

where, %*diff*: Percentage difference between SC costs produced by different approaches; *SCSBF A − NN*: Average SC Cost from Proposed: SBFA-NN; *SCEstbd*: Average SC Cost from Established Approach (CNN/GRU/RNN).

From Table 11.10, it is evident that the predictions made by the proposed SBFA-NN result in a slight increase in supply chain cost for Dataset 1 compared to CNN. However, for Dataset 2, there is a significant cost reduction, accounting for a 1.64% decrease. When compared to GRU, the cost reduction for Dataset 1 is 0.31%, and for Dataset 2, it is 0.40% in

Table 11.9 SC Cost (Fixed lead time with variable safety stock) associated with different advanced forecasting approaches using Dataset 2.

		SC Cost using Dataset 2 ($)				
		RNN	LSTM	GRU	CNN	SBFA-NN
Expt 1	No SS	46,295	48,090	46,325	47,518	46,184
	95% SS	47,462	48,121	47,505	47,727	47,388
	99% SS	47,561	48,134	47,555	47,819	47,493
	99.9% SS	47,629	48,134	47,635	47,912	47,561
Expt 2	No SS	120,890	119,985	121,043	118,608	120,846
	95% SS	121,007	120,016	121,117	118,762	118,454
	99% SS	121,025	120,022	121,154	118,830	118,491
	99.9% SS	121,105	120,047	121,197	118,867	118,559
Expt 3	No SS	76,996	89,987	76,974	67,160	73,638
	95% SS	67,091	85,089	75,341	60,650	65,506
	99% SS	65,482	83,481	73,688	60,690	63,887
	99.9% SS	62,221	81,855	68,749	57,362	60,609
Expt 4	No SS	272,322	272,344	338,166	305,239	305,243
	95% SS	239,581	256,026	288,786	288,793	272,340
	99% SS	223,089	239,583	288,823	272,507	255,988
	99.9% SS	206,679	223,132	272,474	272,501	239,591
Expt 5	No SS	453,297	436,827	420,357	420,364	420,364
	95% SS	425,860	420,389	403,906	403,915	387,465
	99% SS	420,403	403,908	387,595	387,555	383,896
	99.9% SS	404,046	387,563	371,114	371,130	371,136
Expt 6	No SS	144,675	149,831	142,912	146,930	146,942
	95% SS	142,909	148,962	140,969	144,578	144,601
	99% SS	140,296	144,541	141,060	144,620	144,644
	99.9% SS	140,358	144,577	138,902	140,953	140,972
	Avg	173,262	176,694	179,306	175,458	**172,575**
	Rank (Lowest Cost)	2	4	5	3	1

The bold value corresponds to the lowest supply chain costs compared to the other competitive approaches.
LSTM, long short-term memory; RNN, recurrent neural network; GRU, gated recurrent unit; CNN, convolutional neural network.

Table 11.10 Differences of SC (supply chain) cost in terms of percentage.

Approaches	Dataset 1	Dataset 2	Average	SC Cost Status
Proposed: SBFA-NN vs CNN	0.93%	−1.64%	−**0.35%**	decrease
Proposed: SBFA-NN vs GRU	−0.31%	-		
Proposed: SBFA-NN vs RNN	-	−0.40%		

The bold value represents the average percentage of cost decreased should a practitioner use the SBFA for predicting data.
LSTM, long short-term memory; RNN, recurrent neural network; GRU, gated recurrent unit; CNN, convolutional neural network.

contrast to RNN. On average, using the SBFA-NN approach leads to a 0.35% decrease in cost. Therefore, employing a more accurate prediction system like SBFA-NN can effectively reduce business expenses, especially in cases with a static lead time policy. Furthermore, as the safety stock level increases, the supply chain cost associated with static lead time and different forecasting systems is further reduced. This observation is demonstrated in Fig. 11.10, which focuses on the cost associated with Dataset 1 and Expt 1. The figure clearly shows that the supply chain cost related to SBFA-NN consistently outperforms RNN, LSTM, GRU, and CNN.

From Fig. 11.10, it can be seen that the supply chain costs with varying levels of safety stock associated with SBFA-NN are significantly lower than those of RNN, LSTM, GRU, and CNN. When the lead time is fixed, the associated standard deviation (σL) becomes zero. This means that predictions from RNN, LSTM, GRU, and CNN have limited variation (hence, lower σd), which leads to a negligible impact on safety stock even with higher z-values. As a result, these approaches have lower safety stock (ss), which in turn increases the risk of lost sales leading to a higher cost. On the other hand, the proposed SBFA-NN, presented in this chapter, can capture sufficient variation (relatively higher σd) in the test data, leading to a higher reorder point (ROP) and subsequently lower supply chain costs compared to other approaches.

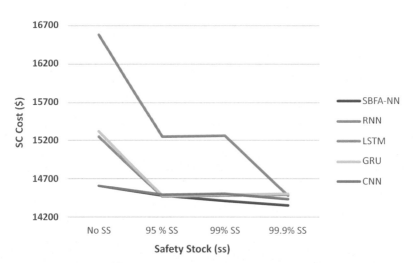

Figure 11.10 Impact of forecasting approaches on supply chain cost with static lead time and different levels of safety stock based on Expt 1 of Dataset 1.

11.6.2.2 Dynamic lead time with variable safety stock

The impact of SBFA-NN is further examined in business models that follow a dynamic lead time policy. In this analysis, it is assumed that the replenishment of the retailer's inventory takes between 27 and 30 days, resulting in a dynamic lead time that includes an additional 3-day delay for replenishment. Using the same set of six experiments with cross-validation, Dataset 1 is used to evaluate the costs associated with different forecasting approaches in a supply chain model operating under the dynamic lead time policy. The results of these experiments are presented in Table 11.11.

Table 11.11 SC (supply chain) cost (Dynamic Lead time with variable safety stock) associated with different advanced forecasting approaches using Dataset 1.

		SC cost using Dataset 1 ($)				
		RNN	LSTM	GRU	CNN	SBFA-NN
Expt 1	No SS	13,397	13,390	13,052	13,414	13,395
	95% SS	13,100	13,170	13,170	13,113	12,608
	99% SS	13,176	12,608	12,608	12,576	12,627
	99.9% SS	12,627	12,652	12,646	12,678	12,652
Expt 2	No SS	10,257	11,244	11,244	11,352	10,162
	95% SS	9610	10,095	9585	9585	9585
	99% SS	9693	9629	9629	9648	9642
	99.9% SS	9705	9680	9680	9705	9705
Expt 3	No SS	13,535	13,516	13,145	13,503	13,516
	95% SS	13,196	12,614	12,614	12,627	12,633
	99% SS	12,684	12,652	12,652	12,652	12,652
	99.9% SS	12,748	12,709	12,748	12,703	12,748
Expt 4	No SS	14,029	14,036	10,102	9440	9370
	95% SS	9125	9087	9273	9292	9106
	99% SS	9305	9318	9324	9324	9324
	99.9% SS	9375	9362	9362	9375	9375
Expt 5	No SS	14,044	14,038	14,226	14,045	14,044
	95% SS	9453	14,049	9465	9453	9440
	99% SS	9061	9453	9465	9087	9465
	99.9% SS	9138	9100	9119	9125	9125
Expt 6	No SS	10,070	10,257	10,070	10,257	10,251
	95% SS	9667	9610	9570	9597	9610
	99% SS	9602	9576	9583	9616	9557
	99.9% SS	9634	9608	9614	9602	9608
	Avg	11,093	11,311	10,914	10,907	**10,842**
	Rank (Lowest Cost)	4	5	3	2	1

The bold value represents the best value among all other competing algorithms.
LSTM, long short-term memory; RNN, recurrent neural network; GRU, gated recurrent unit; CNN, convolutional neural network.

From Table 11.11, it can be observed that SBFA-NN achieves the lowest average cost of $10,842 for Dataset 1, in comparison to CNN which yields a cost of $10,907. Other approaches, such as LSTM and GRU, result in higher overall supply chain costs of $11,311 and $10,914, respectively. To further investigate the influence of SBFA-NN on a supply chain model operating with dynamic lead time, the impact of SBFA-NN on Dataset 2 is examined. The model assumptions remain the same, with slight parameter adjustments for Dataset 2 as indicated in Table 11.1. The results of the impact of SBFA-NN and other forecasting approaches on the supply chain model, considering dynamic lead time and utilizing Dataset 2, are presented in Table 11.12.

Table 11.12 SC (supply chain) cost (Dynamic Lead time with variable safety stock) associated with different advanced forecasting approaches using Dataset 2.

		SC cost using Dataset 2 ($)				
		RNN	LSTM	GRU	CNN	SBFA-NN
Expt 1	No SS	44,877	46,667	44,883	45,081	44,797
	95% SS	46,630	47,443	46,710	46,839	46,568
	99% SS	47,006	47,800	47,043	47,246	46,901
	99.9% SS	47,375	48,139	47,338	47,554	47,258
Expt 2	No SS	114,907	116,401	115,005	115,091	114,827
	95% SS	116,592	110,454	116,733	109,961	116,487
	99% SS	110,041	110,768	110,257	110,386	109,949
	99.9% SS	110,497	112,219	110,614	110,829	110,423
Expt 3	No SS	82,766	96,645	83,482	75,342	82,806
	95% SS	58,955	82,853	63,103	55,070	55,001
	99% SS	55,033	63,885	55,020	46,923	55,076
	99.9% SS	46,849	55,007	50,225	47,064	46,935
Expt 4	No SS	288,831	272,385	339,995	334,630	321,832
	95% SS	223,164	223,284	272,393	255,938	255,991
	99% SS	190,334	195,813	255,980	239,673	223,159
	99.9% SS	173,965	173,995	223,183	206,820	192,213
Expt 5	No SS	453,249	438,678	420,327	420,362	420,330
	95% SS	403,956	404,010	354,495	370,971	354,484
	99% SS	387,470	371,016	339,973	338,108	338,138
	99.9% SS	354,663	338,154	321,768	321,791	321,791
Expt 6	No SS	144,565	154,922	144,895	155,019	152,988
	95% SS	123,921	134,594	123,938	134,674	134,674
	99% SS	124,095	128,279	122,750	118,315	118,315
	99.9% SS	122,051	119,985	112,147	112,366	112,005
	Avg	161,325	162,225	163,427	161,086	159,290
	Rank (Lowest Cost)	3	4	5	2	1

LSTM, long short-term memory; *RNN*, recurrent neural network; *GRU*, gated recurrent unit; *CNN*, convolutional neural network.

Table 11.13 Differences of SC cost in terms of percentage.

Approaches	Dataset 1	Dataset 2	Average	SC cost status
Proposed: SBFA-NN vs. CNN	− 0.60%	− 1.11%	−0.91%	Decrease
Proposed: SBFA-NN vs. RNN	− 0.67	− 1.26%		

It can be observed from Table 11.12 that SBFA-NN once again achieves the lowest average cost of $159,290, while CNN follows closely with a slightly higher cost of $161,086, securing the second position in terms of cost reduction. On the other hand, RNN, LSTM, and GRU yield even higher costs, as indicated in the table. To assess the overall effectiveness of SBFA-NN in comparison to other approaches, the average costs from the top three forecasting methods (SBFA-NN, CNN, and RNN from Tables 11.11 and 11.12) are combined for the dynamic lead time scenario. The percentage difference in supply chain costs is calculated (using Eq. 11.19) to highlight the comparative advantages of SBFA-NN. The results are presented in Table 11.13.

From Table 11.13, it is evident that the proposed SBFA-NN consistently achieves the lowest overall average supply chain cost when dynamic lead time is implemented, surpassing both CNN and RNN. In comparison to CNN, SBFA-NN demonstrates a reduction of 0.60% in average cost for Dataset 1 and 1.11% for Dataset 2. Moreover, SBFA-NN outperforms RNN with cost reductions of 0.67% and 1.26% for both Dataset 1 and Dataset 2, respectively. Overall, SBFA-NN exhibits an average cost reduction of 0.91% when compared to a few top-performing approaches.

11.7 Conclusion

This chapter introduced a forecasting approach called SBFA-NN (Switching-Based Forecasting Approach with Neural Networks), which is capable of generating forecasts for both short and long-term horizons. SBFA-NN incorporates state-of-the-art DL/ML techniques and employs validation error as a discriminator to select the most suitable approach. Through cross-validation analysis, it has been demonstrated that SBFA-NN consistently achieves lower RMSE values compared to other approaches. The managerial implications of the predictions were further explored using a comprehensive supply chain (SC) model, where the reorder point is determined based on the predicted values. The findings from various scenarios, including static and dynamic lead times, as well as different safety stock levels, reveal that improved prediction accuracy directly leads to reduced overall SC costs. SBFA-NN, as proposed in this study, emerges as an effective solution for achieving these cost savings.

In the future, there are several potential extensions to enhance the SBFA-NN research. One possibility is to incorporate additional approaches within the SBFA-NN framework. It would be valuable to explore the impact of grid search on neural networks (NN) as well. Another interesting avenue for investigation is the implementation of transfer learning, which could enable practitioners to predict sales for newly launched products.

References

Abdel-Basset, M., Chakrabortty, R. K., & Gamal, A. (2023). *Soft computing for smart environments: Techniques and applications.* CRC Press.

Ahmed, S., Chakrabortty, R. K., Essam, D. L., & Ding, W. (2022). Poly-linear regression with augmented long short term memory neural network: Predicting time series data. *Information Sciences, 606,* 573−600.

Apaydin, H., Feizi, H., Sattari, M. T., Colak, M. S., Shamshirband, S., & Chau, K.-W. (2020). Comparative analysis of recurrent neural network architectures for reservoir inflow forecasting. *Water, 12*(5), 1500.

Bassiouni, M. M., Chakrabortty, R. K., Hussain, O. K., & Rahman, H. F. (2023). Advanced deep learning approaches to predict supply chain risks under covid-19 restrictions. *Expert Systems with Applications, 211,* 118604.

Buitinck, L., Louppe, G., Blondel, M., Pedregosa, F., Mueller, A., Grisel, O., Niculae, V., Prettenhofer, P., Gramfort, A., Grobler, J., Layton, R., VanderPlas, J., Joly, A., Holt, B., & Varoquaux, G. (2013). API design for machine learning software: Experiences from the scikit-learn project. In *ECML PKDD Workshop: Languages for Data Mining and Machine Learning,* pp. 108−122.

Chakrabortty, R. K., Rahman, M. H. F., & Ding, W. (2023). Guest editorial: Special section on developing resilient supply chains in a post-covid pandemic era: Application of artificial intelligent technologies for emerging industry 5.0. *IEEE Transactions on Industrial Informatics, 19*(3), 3296−3299.

Deng, Z., Liu, C., & Zhu, Z. (2021). Inter-hours rolling scheduling of behind-the-meter storage operating systems using electricity price forecasting based on deep convolutional neural network. *International Journal of Electrical Power & Energy Systems, 125,* 106499.

Geron, A. (2019). *Hands-on machine learning with Scikit-Learn, Keras, and TensorFlow: Concepts, tools, and techniques to build intelligent systems.* O'Reilly Media, Inc.

Gonçalves, J. N. C., Sameiro Carvalho, M., & Cortez, P. (2020). Operations research models and methods for safety stock determination: A review. *Operations Research Perspectives,* 100164.

Graves, A., & Schmidhuber, J. (2005). Framewise phoneme classification with bidirectional lstm and other neural network architectures. *Neural Networks, 18*(5−6), 602−610.

Gu, S., Kelly, B., & Xiu, D. (2020). Empirical asset pricing via machine learning. *The Review of Financial Studies, 33*(5), 2223−2273.

Gumus, A. T., Guneri, A. F., & Ulengin, F. (2010). A new methodology for multi-echelon inventory management in stochastic and neuro-fuzzy environments. *International Journal of Production Economics, 128*(1), 248−260.

Guo, X., Yuan, Z., & Tian, B. (2009). Supplier selection based on hierarchical potential support vector machine. *Expert Systems with Applications, 36*(3), 6978−6985.

Jin, X., Yu, X., Wang, X., Bai, Y., Su, T., & Kong, J. (2020). *Prediction for time series with cnn and lstm. Proceedings of the 11th International Conference on Modelling, Identification and Control (ICMIC2019)* (pp. 631−641). Springer.

Kamble, S. S., Gunasekaran, A., Kumar, V., Belhadi, A., & Foropon, C. (2021). A machine learning based approach for predicting blockchain adoption in supply chain. *Technological Forecasting and Social Change, 163*, 120465.

Kantasa-Ard, A., Nouiri, M., Bekrar, A., Cadi, A. A., & Sallez, Y. (2021). Machine learning for demand forecasting in the physical internet: A case study of agricultural products in Thailand. *International Journal of Production Research, 59*(24), 7491−7515.

Li, Y., Yang, Y., Zhu, K., & Zhang, J. (2021). Clothing sale forecasting by a composite gru−prophet model with an attention mechanism. *IEEE Transactions on Industrial Informatics, 17*(12), 8335−8344.

Livieris, I. E., Pintelas, E., & Pintelas, P. (2020). A cnn−lstm model for gold price time-series forecasting. *Neural computing and applications, 32*(23), 17351−17360.

Mashud, A. H. M., Miah, S., Daryanto, Y., Chakrabortty, R. K., Mahmudul Hasan, S. M., & Tseng, M. −L. (2022). Inventory decisions on the transportation system and carbon emissions under covid-19 effects: A sensitivity analysis. *Computers & Industrial Engineering, 171*, 108393.

Masini, R. P., Medeiros, M. C., & Mendes, E. F. (2021). Machine learning advances for time series forecasting. *Journal of Economic Surveys.*

Muhaimin, A., Prastyo, D. D., & Lu, H. H. S. (2021). Forecasting with recurrent neural network in intermittent demand data. In *2021 11th International Conference on Cloud Computing, Data Science & Engineering (Confluence)*, pp. 802−809. IEEE.

Ni, D., Xiao, Z., & Lim, M. K. (2020). A systematic review of the research trends of machine learning in supply chain management. *International Journal of Machine Learning and Cybernetics, 11*(7), 1463−1482.

Olah, C. (2015). Understanding lstm networks. Available at: http://colah.github.io/posts/2015-08-Understanding-LSTMs/.

Paul, S. K., Moktadir, M. A., Sallam, K., Choi, T.-M., & Chakrabortty, R. K. (2023). A recovery planning model for online business operations under the covid-19 outbreak. *International Journal of Production Research, 61*(8), 2613−2635.

Pereira, M. M., & Frazzon, E. M. (2021). A data-driven approach to adaptive synchronization of demand and supply in omni-channel retail supply chains. *International Journal of Information Management, 57*, 102165.

Sarfaraz, A., Chakrabortty, R. K., & Essam, D. L. (2023a). The implications of blockchain-coordinated information sharing within a supply chain: A simulation study. *Blockchain: Research and Applications, 4*(1), 100110.

Sarfaraz, A., Chakrabortty, R. K., & Essam, D. L. (2023b). Reputation based proof of cooperation: An efficient and scalable consensus algorithm for supply chain applications. *Journal of Ambient Intelligence and Humanized Computing*, 1−17.

Selvin, S., Vinayakumar, R., Gopalakrishnan, E. A., Menon, V. K., & Soman, K. P. (2017). *Stock price prediction using lstm, rnn and cnn-sliding window model. 2017 international conference on advances in computing, communications and informatics (ICACCI)* (pp. 1643−1647). IEEE.

Seyedan, M., & Mafakheri, F. (2020). Predictive big data analytics for supply chain demand forecasting: Methods, applications, and research opportunities. *Journal of Big Data*, 7(1), 1−22.

Thomassey, S. (2010). Sales forecasts in clothing industry: The key success factor of the supply chain management. *International Journal of Production Economics*, *128*(2), 470−483.

Tirkolaee, E. B., Sadeghi, S., Mooseloo, F. M., Vandchali, H. R., & Aeini, S. (2021). Application of machine learning in supply chain management: A comprehensive overview of the main areas. *Mathematical Problems in Engineering, 2021.*

Weller, M., & Crone, S. F. (2012). Supply chain forecasting: Best practices & benchmarking study. *Lancaster University.*

Yan, W., He, J., & Trappey, A. J. C. (2019). Risk-aware supply chain intelligence: AI-enabled supply chain and logistics management considering risk mitigation. *Advanced Engineering Informatics*, *42*(4), 2. Available from https://doi.org/10.1016/j. aei.2019.100976.

Zhu, X., Ninh, A., Zhao, H., & Liu, Z. (2021). Demand forecasting with supply-chain information and machine learning: Evidence in the pharmaceutical industry. *Production and Operations Management*, *30*(9), 3231−3252.

Bayesian network based on cross bow-tie to analyze differential effects of internal and external risks on sustainable supply chain

Gholamreza Khojasteh[1], Mustafa Jahangoshai Rezaee[1], Ripon K. Chakrabortty[2] and Morteza Saberi[3]

[1]Faculty of Industrial Engineering, Urmia University of Technology, Urmia, Iran
[2]School of Systems & Computing, UNSW Canberra at ADFA, Campbell, ACT, Australia
[3]School of Computer Science, University of Technology Sydney, Ultimo, NSW, Australia

12.1 Introduction and motivation

The current global market's fierce competition, coupled with short-ened product life cycles and elevated customer demands, has prompted organizations to give precedence to sustainable supply chain management. Yet, factors like political challenges, fluctuations in demand, technological advancements, financial instability, and natural calamities have heightened uncertainty and risk in supply chains. As a result, the implementation of effective supply chain risk management strategies has become imperative. This research puts forth a holistic risk assessment framework that considers both internal and external risk factors, aiming to establish a sustainable supply chain. The conventional bow-tie (BT) method is enhanced into a cross BT approach, allowing for simultaneous consideration of risks of internal and external origins and their interactions. Bayesian networks, Dempster—Shafer theory, and fuzzy sets are integrated to handle the dynamic and uncertain nature of supply chain risks, particularly in the context of Iran's manufacturing industries facing challenges posed by sanctions and the COVID-19 pandemic. The research aims to provide valuable insights into managing supply chain risks more effectively and sustainably.

In today's competitive environment, organizations must embrace sustainable supply chain management to address the complexities and risks

Computational Intelligence Techniques for Sustainable Supply Chain Management.
DOI: https://doi.org/10.1016/B978-0-443-18464-2.00005-4

posed by the global market. However, external factors and uncertainties have made risk management in supply chains more difficult. This study introduces a novel risk assessment framework to improve the understanding and management of risks in Iran's manufacturing industries, taking into account both internal and external risk factors.

The study emphasizes the importance of supply chain risk management in the face of dynamic challenges. This paper examines the effects of different risk factors on the profitability, reliability, and sustainability of supply chains. It emphasizes the importance of accurate risk assessment to create effective risk management strategies.

In this research, three categories of risks will be analyzed: normal risks, risks caused by sanctions, and risks caused by the COVID-19 pandemic. Some risks overlap and others intensify each other's effect. Examples of these risks and their effects on each other are presented below.

1. Logistics Disruptions: Sanctions and COVID-19 impact transportation and customs, causing delays.
2. Financial Constraints: Sanctions and the pandemic-induced economic downturn affect cash flow.
3. Supplier Reliability: Sanctions and COVID-19 disrupt sourcing options and suppliers.
4. Inventory Management: Sanctions can affect product availability, and COVID-19 can alter demand patterns.
5. Demand Fluctuations: Sanctions and the pandemic can change consumer behavior and preferences.
6. Workforce Disruptions: Sanctions and COVID-19 restrictions can impact labor productivity.
7. Technology and Communication: Sanctions can impede access to technology, and remote work can affect coordination.

These examples illustrate how regular, sanctions, and COVID-19 risks can intertwine and create complex challenges for the supply chain in Iran. Effectively managing these overlapping risks requires a comprehensive approach that takes into account the unique characteristics of each risk type and their potential synergistic effects on the supply chain's resilience and sustainability.

To overcome the limitations of conventional methods, this study proposes a cross BT model. This novel approach enables the simultaneous consideration of two categories of risks: those of internal and external origins. By examining the interaction between these risk types, organizations can develop more comprehensive risk mitigation strategies.

Given the uncertain nature of supply chain risks, this study integrates Bayesian networks and Dempster—Shafer theory. The use of Bayesian networks allows for dynamic risk analysis, accommodating changes in risk occurrence over time, while Dempster—Shafer's theory enhances the handling of uncertainty in expert evaluations.

The proposed risk assessment framework is applied to Iran's manufacturing industry chain, which currently faces challenges arising from economic sanctions and the COVID-19 pandemic. The study provides valuable insights into managing supply chain risks in this specific context.

The research's contributions lie in its simultaneous consideration of internal and external risks, the introduction of the cross BT approach, and the use of Bayesian networks for dynamic risk analysis. These innovations offer promising implications for achieving sustainable supply chain management.

This scientific study presents a robust risk assessment framework to enhance sustainable supply chain management in Iran's manufacturing industries. By integrating the cross BT approach, Bayesian networks, Dempster—Shafer theory, and fuzzy sets, organizations can gain a deeper understanding of supply chain risks and develop effective risk management strategies to navigate uncertainties and achieve sustainable success in the ever-changing global landscape.

The innovations of this research can be summarized as follows:
- Simultaneously considering risks of internal and external origin, as well as the interaction between them.
- Developing the BT method as a cross BT to simultaneously consider two categories of risks.
- Designing a Bayesian network in accordance with the proposed cross BT method and examining different scenarios dynamically.
- Analyzing the manufacturing industry chain in Iran under sanctions and pandemic.

The remainder of this research is structured as follows. The initial section, Section 12.2, offers a literature review encompassing supply chain risk management, risk assessment utilizing BT analysis, and risk assessment employing the Bayesian network. Subsequently, Section 12.3 expounds on the methods utilized in this study. Section 12.4 delves into the proposed methodology for risk analysis in the manufacturing industry's supply chain in Iran, considering both internal and external risk factors. Section 12.5 focuses on the application of the proposed approach and the analysis of the results. Lastly, Section 12.6 provides a discussion on the conclusion.

12.2 Literature review and state of the art

The current section summarizes the background related to supply chain risk management, risk assessment using the BT analysis, risk assessment using the Bayesian network, and finally, risk assessment using integrated methods.

In today's business world, supply chains may expand worldwide to provide the product with the lowest cost and highest quality. Therefore, they are threatened by a whole new set of factors that can cause chaos and disruption. Wagner and Bode (2008) described supply chain risk as a combination of (1) an unexpected and abnormally irritating event that occurs in a part of the supply chain or its environment; and (2) a subsequent situation that significantly threatens the normal business activities of supply chain companies. Supply chain risk management has never been more challenging than it is today. As more companies outsource their production to overseas units, the number of nodes in the supply chain has increased, and network complexity has grown exponentially. Risk assessment is one of the important elements of risk management. The objective is to evaluate risks using different indicators such as severity and probability of occurrence. The accuracy of the outcomes in this stage is crucial in determining the efficacy of the risk management process. One of the major weaknesses in previous studies is the lack of simultaneous consideration of risks of internal and external origin separately, while also considering the interactions between them. To address these two categories of risks, models will need to be developed, which will be addressed in the following sections.

12.2.1 Supply chain risk management

Schoenherr et al. (2008) outlined a procedure employed by an American manufacturing company to evaluate supply chain risks in the context of cross-border sourcing decision-making. The results of this study can enhance cross-border research efforts and improve risk management in procurement and supply, while also assisting in decision-making amid uncertain conditions. Another study was conducted by Thun and Hoenig (2011) in automotive manufacturing plants, based on a survey in collaboration with 67 plants in Germany. The study aimed to identify supply chain risks through the probabilistic assessment of their likelihood and potential impact on the supply chain.

Berenji and Anantharaman (2011) introduced a framework aimed at identifying and prioritizing supply chain risks through the application of the Fuzzy Network Process and Fuzzy TOPSIS techniques. The primary objective of this framework was to assess and rank the identified risks within the supply chain using the Fuzzy Network Process, followed by the implementation of Fuzzy TOPSIS to prioritize supply chain members. This research led to the development of novel approaches and recommendations for the evaluation and management of supply chain risks. Diabat et al. (2012) developed a model utilizing Structural Interpretive Modeling to analyze and interpret different risks present within a food supply chain. The model enabled the decomposition and analysis of various risks that occur in the food supply chain, providing valuable insights for risk management and mitigation strategies.

Another research in SCM risk analysis was provided by Curkovic et al. (2013). This research primarily focused on exploratory investigations, demonstrating how the Failure Mode and Effects Analysis (FMEA) method could play a crucial role in the process of supply chain risk management through supplier evaluation and selection. By assessing and selecting suppliers using FMEA, the study showed the significant impact it had on effectively managing supply chain risks.

A quantitative modeling and risk analysis of the supply chain through the application of Bayesian theory was presented by Badurdeen et al. (2014). The main objective of this study was to present a developed method and tool for modeling and analyzing multi-layered supply chain risks. The proposed approach aimed to provide an effective framework for comprehensively assessing and managing supply chain risks utilizing Bayesian techniques. In a separate study, Aqlan and Lam (2015) put forward an integrated fuzzy-based framework for supply chain risk assessment. The framework consisted of three key elements: data collection through surveys, butterfly analysis, and a fuzzy inference system. The proposed framework aimed to provide a comprehensive and robust approach for evaluating supply chain risks, utilizing the advantages of fuzzy logic to handle uncertainties and complexities within the assessment process. Mangla et al. (2015) employed a fuzzy analytical hierarchy process approach to analyze risks associated with the adoption and effective implementation of green supply chain practices. This study focused on analyzing risks related to the adoption and execution of efficient green supply chain methods from an industrial perspective. The research aimed to provide insights and recommendations for enhancing sustainability and risk management strategies in the context of green supply chains. Kumar et al. (2015) extended a Bayesian network model for supply chain risk assessment. The objective of this study was to develop a risk assessment tool for evaluating and identifying risks faced by a supply chain. The model aimed to provide an effective way of assessing and determining the various risks that could potentially impact a supply chain, facilitating better risk management and decision-making processes in supply chain operations.

Giannakis and Papadopoulos (2016) investigated the characteristics of supply chain risks associated with sustainability and differentiated them from conventional supply chain risks. They devised an analytical approach for handling these specific risks. The study employed the FMEA technique to evaluate the relative significance of the designated risks, identify their causes and potential consequences, and investigate potential associations among the identified risks. The objective was to offer insights and tactics for the efficient management of sustainability-related risks within supply chains.

Silva et al. (2021) researched the assessment of risks in the supply chain of electric vehicles using a combined fuzzy evaluation approach. The primary objective of this study was to identify and evaluate potential risk factors in the electric vehicle supply chain in China, under conditions of uncertainty. The research aimed to provide insights into the risks faced by the electric vehicle supply chain and offer valuable recommendations for effective risk management strategies in this domain. Khan et al. (2021) utilized the Fuzzy Best-Worst method to prioritize risk dimensions and related risk elements in the context of supply chain management of solvents. The significant contribution of this research was the compilation of a comprehensive list of risks associated with the supply chain management of solvents and their prioritization to facilitate effective risk management strategies. The study aimed to enhance risk management practices in the solvent supply chain and promote better decision-making in this critical domain. The study by Ghadir et al. (2022) focused on the risk assessment of the supply chain, taking into account the impact of COVID-19, through the application of FMEA and the Best-Worst Method (BWM). By conducting a comprehensive literature review on supply chain risk management, the researchers identified and listed risks into seven categories: demand, supply, procurement, political, production, financial, and informational. The combined use of FMEA and BWM provides a systematic and effective approach for evaluating and prioritizing the identified risks, enabling supply chain managers to make informed decisions and implement proactive risk mitigation strategies in the face of uncertainties induced by the COVID-19 pandemic. The research conducted by Karmakar et al. (2023) sought to identify, prioritize, and examine the interconnections among supply chain risk factors using an integrated methodology that incorporates the Fuzzy Technique for Order of Preference by Similarity to Ideal Solution (TOPSIS), Interpretive Structural Modeling (ISM), and Matriced' Impacts Cruoses Multiplication Applique a un Classement (MICMAC) within a fuzzy environment. This study focused on supply chain management within small and medium enterprises in emerging economies. By applying these integrated methodologies, the study provides a comprehensive assessment of supply chain risks, considering the uncertainties and complexities inherent in the domain. The findings offer valuable insights into understanding the dynamics of supply chain risks and facilitate effective decision-making for managing and mitigating risks in these economic contexts.

The theoretical framework suggested by Pellegrino et al. (2022) is based on an extensive literature review that examines the connections between the effects of COVID-19 and strategies for mitigating supply chain risks. Subsequently, they empirically tested the proposed framework using Interpretive Structural Modeling and Bayesian Belief Network to support decision-making approaches in supply chain management. The study aimed to shed light on the complex interrelationships between the pandemic's effects and various risk reduction strategies, considering the uncertainties and disruptions caused by COVID-19. By applying integrated methodologies, this research provides valuable insights into understanding the dynamics of supply chain risks and offers practical guidance for effective risk management strategies in response to the unprecedented challenges posed by the COVID-19 pandemic. Babu and Yadav (2023) used the theory of fuzzy sets to develop a comprehensive conceptual framework for evaluating supply chain risk in small and medium-sized enterprises (SMEs). The main objectives were to create a coherent framework for assessing the overall risk index of the organization's supply chain using the fuzzy logic approach. Additionally, the study aimed to assess all identified supply chain risk variables/features and identify the key obstacles in Supply Chain Risk Management (SCRM) based on their fuzzy performance importance index in the post-COVID-19 era. By applying fuzzy logic, the proposed framework enables SMEs to systematically evaluate and prioritize the identified risks and effectively identify and address critical SCRM challenges in the aftermath of the COVID-19 pandemic.

12.2.2 Bow-tie method

The BT method is an efficient and simple tool used to analyze risk and control paths in selected scenarios. The name "Bow-tie" is derived from the shape of the created diagram. BT analysis provides a clear and comprehensible view of the barriers to preventing the causes involved in the occurrence of the critical event, as well as the response measures to reduce the severity of the consequences of the occurrence of the critical event. BT Analysis can be used to demonstrate control, countermeasures, and mitigation of risks, and can be applied to many industries, services, and commercial sectors (Popov et al., 2016).

In a study aimed at evaluating semi-quantitative occupational risks, the BT analysis technique was employed (Jacinto, C. and C. Silva). The primary focus of the initial qualitative analysis was centered around the BT diagram technique, while concurrently incorporating concepts and classification schemes defined by the European Statistical Agency for workplace accidents within the European Statistical Project. The study aimed to enhance the understanding and management of workplace risks through the integration of qualitative and statistical approaches, enabling better risk assessment and prevention strategies in the context of occupational safety and health.

In another study, a simplified BT approach was introduced for risk assessment in the workplace (Targoutzidis, 2010). The main objective was to provide a straightforward methodological tool that incorporates human factors into the risk assessment process. By simplifying the BT technique, the research aimed to enhance the practicality and usability of risk assessments, ensuring a more comprehensive consideration of human–related factors in managing workplace risks. Another study used BT diagrams for identifying hazards and assessing risks in low-tier industries (Saud et al., 2014). The research discussed the evolution of the risk-based approach in the United States and demonstrated how the BT model is well-suited for risk management processes in projects and facilities within low-tier industries. The study highlighted the effectiveness of BT diagrams in enhancing risk management practices and promoting safety measures in low-tier industrial settings.

Aqlan and Ali (2014) introduced a framework that combined Lean principles and fuzzy BT analysis for risk assessment in chemical industries. The research utilized Lean tools, such as FMEA, for risk analysis and mitigation. Additionally, the fuzzy BT analysis was employed to assess the risks associated with process failures in chemical industries. The study aimed to provide an effective approach for evaluating and reducing the impact of risks in chemical processes, ensuring improved safety and reliability in the chemical manufacturing sector. Another study conducted by Babaei et al. (2018), focused on evaluating the risk of human injuries resulting from medium-voltage electrical shocks. The main objective was to calculate the risk using the BT model in a fuzzy environment. By applying the BT model in a fuzzy setting, the research aimed to provide a comprehensive and probabilistic risk assessment for medium-voltage electrical shock incidents, which could contribute to improving safety measures and mitigating potential hazards in electrical systems.

Analouei et al. (2020) developed a framework for the risk assessment of industrial wastewater treatment plants using the BT analysis technique. The BT model was employed to determine the risk of an industrial wastewater treatment system exceeding the standard limits for the quality of its effluent parameters. The primary objective of this research was to provide a comprehensive risk assessment approach for industrial wastewater treatment plants, allowing for better understanding and management of potential hazards related to the quality of treated effluent, thereby contributing to enhanced environmental protection and regulatory compliance.

12.2.3 Bayesian network

A Bayesian network (BN) is a graphical model that represents a network of causes and effects, enabling precise quantification of risks and their interrelationships. BNs are widely recognized as an exciting and powerful technology for addressing risk assessment, uncertainty, and decision-making (Fenton & Neil, 2018; Rezaee et al., 2020). In the field of risk assessment, the Bayesian network has recently received more attention due to its high ability in diagnostic analysis and prediction. So far, various studies have contributed to the field of risk assessment by using Bayesian networks in different fields such as medicine, oil, and gas, nuclear industries, and transportation.

Lee and Lee (2006) introduced a framework for quantitative risk assessment, combining Bayesian network inference with traditional probabilistic risk analysis. The main goal was to predict the consequences of environmental changes caused by nuclear waste disposal facilities. By combining Bayesian network modeling with conventional probabilistic risk analysis, the research aimed to provide a robust and comprehensive risk assessment approach for evaluating the potential environmental impacts associated with nuclear waste disposal sites, aiding in informed decision-making and risk management in the nuclear waste management field. Trucco et al. (2008) created a Bayesian belief network to model the maritime transportation system, considering multiple stakeholders such as ship-owners, ship operators, ports, and regulatory bodies, and analyzing their interrelationships. The main objective of this research was to integrate human and organizational factors into risk analysis, providing a comprehensive understanding of the complex dynamics and interdependencies within the maritime transportation system. The Bayesian belief network allowed for a holistic assessment of risk by considering the influences and relationships among different stakeholders, leading to more effective risk management strategies in the maritime domain.

In another research, Yun et al. (2009) conducted a risk assessment of liquefied natural gas (LNG) import terminals by integrating Bayesian network modeling with layers of protection analysis. The main objective was to combine the Bayesian network and layers of protection analysis to compensate for insufficient data and uncertainty in the context of system failure events related to LNG import terminals. By using this integrated approach, the research aimed to enhance the risk assessment process and provide more accurate and comprehensive evaluations of the risks associated with LNG import terminal operations, leading to improved safety measures and risk management strategies in the LNG industry. Bayesian network was employed for the analysis of occupational fall accidents (Martin et al., 2009). It was used as a tool for analyzing workplace accidents to identify the most significant factors contributing to these incidents and determine the existing relationships between these factors. The study aimed to provide insights into the root causes of fall accidents and highlight the interdependencies among the identified factors, facilitating a better understanding of the complex dynamics associated with occupational fall accidents.

Kalantarnia et al. (2009) presented a framework demonstrating the utilization of Bayesian networks in quantitative risk assessment, highlighting its efficacy as a valuable instrument in dynamic risk evaluation. The framework involved the identification and introduction of potential incident scenarios using an event tree, enabling a systematic analysis of potential events and their consequences. By employing Bayesian networks in this context, the research aimed to provide a robust and flexible approach for assessing and managing risks dynamically, accommodating changing conditions and uncertainties in risk scenarios. Another study conducted by Chen and Leu (2014); focused on the risk assessment of fall incidents in cable-stayed bridge construction projects using a BN based on fault tree transformation. The proposed model was initiated by constructing a fault tree based on the problem domain and then transforming this fault tree into a Bayesian network to obtain a foundational Bayesian network. The research aimed to provide an effective and comprehensive risk assessment tool by combining the strengths of fault tree analysis and Bayesian networks, enabling a systematic analysis of fall-related risks and their associated probabilities in cable-stayed bridge construction projects.

The risk assessment of high-pressure jet engine turbine assembly was conducted by combining Bayesian belief networks and the Analytic Hierarchy Process (AHP) method (Pereira et al., 2016). Bayesian belief networks and AHP were used to integrate and evaluate the potential risks involved in the turbine assembly process. The research aimed to provide a comprehensive and systematic approach for analyzing and prioritizing the various risks associated with the assembly process, allowing for informed decision-making and risk management strategies to ensure the safe and efficient assembly of high-pressure jet engine turbines. Baksh et al. (2018) focused on the risk assessment of maritime transportation using Bayesian networks, proposing a new risk model specifically designed for the Northern Sea Route. The model was developed to examine the likelihood of maritime accidents, including collisions and sinkings, along this particular shipping route. By utilizing Bayesian networks, the research aimed to provide a comprehensive and probabilistic approach to evaluating the risks associated with maritime transportation in the Northern Sea Route, contributing to enhanced safety measures and risk management strategies for navigating this challenging region.

A novel method for assessing the risk of production lines, considering operational risks that significantly impact the production line, was proposed by Punyamurthula and Badurdeen (2018). The approach involved the creation of a hybrid model combining Bayesian belief networks and system dynamics. This hybrid model allowed for capturing dynamic causality mechanisms within a complex system, handling uncertainties between risk events, and evaluating the long-term effects of operational risks on the production line. By utilizing this hybrid approach, the study aimed to provide a comprehensive and sophisticated risk assessment tool, assisting in identifying potential vulnerabilities, optimizing production line performance, and implementing effective risk management strategies to ensure a resilient and efficient production process. A hybrid model combining Bayesian networks and the Dempster—Shafer evidence theory was developed for assessing school bus accidents, considering influential factors such as human error, vehicle malfunction, environmental effects, and management shortcomings (Wu et al., 2019; Wu, Jia, et al., 2019). By integrating both probabilistic and evidence-based reasoning approaches, the model aimed to provide a more comprehensive and robust risk assessment framework for school bus accidents. This approach enables a more accurate and reliable evaluation of the contributing factors and their interactions, facilitating better decision-making and risk management strategies to enhance the safety and security of school bus transportation.

Li, Wang, et al. (2019) and Li, Liu, et al. (2019) presented a framework for risk assessment of combustion resources in mining operations, utilizing a fuzzy Bayesian network. The framework consisted of three main stages. In the first stage, risk factors were assessed, and a Bayesian network model was developed. The second stage involved fuzzification and defuzzification processes. Lastly, the third stage involves the computation of the likelihood of potential risk events and the probability distribution of risk factors. By employing this framework, the researchers aimed to provide an effective tool for assessing and managing potential risks associated with combustion resources in mining operations, aiding in decision-making and risk mitigation strategies. In another study, an integrated method for dynamic qualitative risk assessment was introduced, utilizing the Decision Making Trial and Evaluation Laboratory (DEMATEL) and Bayesian networks (Meng et al., 2019). The approach aimed to assess system vulnerabilities and predict the likelihood of leak-related incidents in offshore oil and gas platforms. The framework combined the strengths of both methodologies to provide a more robust and accurate risk assessment.

Li, Wang, et al. (2019) and Li, Liu, et al. (2019) developed a model for dynamic risk assessment in the field of healthcare, utilizing Bayesian network approaches. The model combined a fault tree to illustrate potential risk scenarios with Bayesian dynamic networks and Bayesian inference, to analyze and predict the performance of medical devices, taking into account their failures, maintenance, and human errors over time. This model enables the dynamic analysis and assessment of risks associated with medical systems, providing healthcare authorities and decision-makers with valuable insights to improve performance and mitigate risks. Abbaspour Onari et al. (2021) presented a self-assessment decision support system (DSS) to differentiate COVID-19 severity among confirmed cases and optimize patient care. The DSS combined Data-Driven BN and Fuzzy Cognitive Map (FCM) techniques, extracting evidence-based paired relationships between symptoms and impact probabilities from patient data. Results showed that common symptoms may not always indicate severity. The proposed DSS enhances the precision of COVID-19 severity assessment, thereby aiding healthcare providers in making informed decisions regarding patient care and resource distribution amid the pandemic.

12.2.4 Integrated methods

How Bayesian networks can effectively overcome the limitations of fuzzy analytical techniques was demonstrated by Khakzad et al. (2013). Additionally, the flexibility of Bayesian networks in conducting dynamic safety analyses across a diverse range of incident scenarios has been highlighted. Their adaptable structure has made the integration of Bayesian networks into risk assessment procedures a valuable asset, aiding in the management of intricate uncertainties and the refinement of safety-related dynamics.

Zarei et al. (2017) proposed a unified and comprehensive methodology for dynamically assessing safety risks in the dimethylamine storage system within a gasoline refinery, employing Bayesian networks. The modeling of incident scenarios was accomplished through the use of a fault tree, establishing a clear cause-and-effect relationship for accidents. To facilitate a thorough risk assessment process and address uncertainties, the researchers combined fuzzy analysis with Bayesian networks, facilitating effective probabilistic reasoning.

A systematic approach for assessing the risk of fire and explosion in pipeline systems, employing Bayesian networks, fault tree analysis, and fuzzy logic was done by Bilal et al. (2017). The integration of these methodologies allows for a comprehensive evaluation of the potential hazards and their associated consequences. By combining probabilistic reasoning with cause-effect modeling and fuzzy inference, the proposed approach offers a robust risk assessment framework for enhancing the safety of pipeline operations.

Rezaee et al. (2018) presented a comprehensive risk analysis framework for sequential processes in the food industry by combining Multi-Stage FCM (MFCM) and process FMEA. The aim is to identify, prioritize, and examine the interrelationships among risk factors within the sequential stages of food production. The MFCM is employed to model the dynamic causal relationships between different risk elements, considering the inherent uncertainties and complexities of the food supply chain. Subsequently, process FMEA is applied to assess the potential failure modes and their corresponding effects at each stage of the process. The integration of these two methodologies enables a more holistic approach to risk assessment, providing valuable insights for improving the safety and quality of food production processes while also enhancing the overall risk management in the food industry. Chang et al. (2019) introduced a novel approach for the dynamic risk assessment of leaks in a hydrogen production unit by combining fault tree analysis and a Bayesian network. The dynamic BN is employed to address potential uncertainties and the dynamic nature of the risk environment associated with hydrogen production unit leaks. By integrating these two powerful techniques, the research presents a comprehensive and effective methodology for continuously evaluating and mitigating the risks associated with hydrogen production unit leaks.

Yousefi et al. (2020) investigated the causal effects of logistics process risks in manufacturing industries using a Multi-Stage FCM approach. The study presents a case study where FCM is applied to analyze the interrelationships and impact of various risk factors on the logistics processes. The sequential nature of FCM allows for a comprehensive understanding of how these risks propagate and influence the overall performance of the manufacturing supply chain. The findings provide valuable insights for decision-makers in identifying critical risk areas and developing effective risk management strategies in the manufacturing sector.

Aliabadi et al. (2020) focused on the risk assessment of hydrogen gases by employing a combination of fuzzy fault tree analysis, Bayesian network, and fuzzy logic. The study presents a comprehensive approach to assess and analyze the potential risks associated with hydrogen gases. By integrating these methodologies, the research offers a robust and versatile framework for evaluating the risks and uncertainties linked to hydrogen gases, enabling more effective risk management strategies in various industrial applications.

Keshtiban et al. (2022) proposed an enhanced risk assessment approach for manufacturing production processes by integrating FMEA with Sequential FCM (SFCM). The objective of the research is to improve the accuracy and efficiency of risk analysis in the manufacturing sector. The FMEA method is utilized to identify potential failure modes and their associated effects within the production process. Concurrently, SFCM is employed to capture the dynamic interdependencies and uncertainties among various risk factors throughout the sequential stages of production. The combined framework allows for a more comprehensive evaluation of risks, considering both individual failure modes and the collective influence of interconnected factors. The proposed approach provides manufacturers with a powerful tool to identify critical areas, prioritize mitigation strategies, and optimize their risk management practices to ensure the smooth and safe functioning of their production processes.

As mentioned above, various methods have been used for risk analysis and assessment. Each of these methods has its weaknesses. Additionally, in the risk assessment section, the source of the risks has not been investigated, which may be due to the lack of an appropriate approach for this type of vision. Therefore, in Section 12.4, a cross BT method is presented to address this issue. On the other hand, since this method is static and cannot update risks in different situations, the Bayesian network is used to consider different scenarios over time. The details of the developed methods and results are presented in the next sections. A summary of the literature review has been presented in Table 12.1.

12.3 Application definition

In the following, the methods used in this article are briefly reviewed. First, the conventional structure of the BT method is investigated. Then, the Bayesian Belief Network will be provided. Finally, the Dempster–Shafer Theory will be presented. It should be noted that the development of the BT method and how to integrate these methods are presented in the next section.

12.3.1 Bow-tie method

The BT method, as outlined by Bilal et al. (2017), is an integrated probabilistic approach used for analyzing incident scenarios, focusing on the assessment of probabilities and potential paths of occurrence. It aids in the prevention, control, and mitigation of unintended events by establishing a

Table 12.1 The summary of the literature review in risk management.

	Study	Classification of risk	Approach/method
Supply chain risk management	Schoenherr et al. (2008)	Product and Environment Partner	AHP
	Thun and Hoenig (2011)	Internal and External	Field study
	Berenji Aliabad Anantharaman (2011)	Supply, operational, demand, economic/competitive, control and plan, and social/political	Fuzzy AHP/Fuzzy, TOPSIS
	Diabat et al. (2012)	Macro-level, demand management, supply management, product/ service management and information management	Modeling
	Curkovic et al. (2013)	Parts or product, occupation and communication	FMEA
	Badurdeen et al. (2014)	Organizational, industrial and environmental	Bayesian Theory
	Aqlan Aliabad Lam (2015)	Vendor, buyer, process & control, technology, product, occupation, culture, shipping	Fuzzy Theory
	Mangla et al. (2015)	Operational, supply, product recovery, finance, demand	AHP Fuzzy
	Kumar et al. (2015)	Firm, industry, supply chain, and public environment	BN
	Giannakis Aliabad Papadopoulos (2016)	Environmental, social, financial, and economic	FMEA
	Wu, Fang, et al. (2019), Wu, Jia, et al. (2019)	Market, technical, and environmental	Fuzzy Theory
	Silva et al. (2021)	Demand, process, strategic, supply, operational, capacity, financial, economic, and information	Simulation and –conditional probability
	Khan et al. (2021)	Program, sourcing, production, logistics and outsourcing, market, information and technology, and Sustainability	Fuzzy BWM
	Ghadir et al. (2022)	Demand, supply, logistics, political, manufacturing, financial, and information	FMEA/BWM
	Karmakar et al. (2023)	Price fluctuation, Demand uncertainty, Lack of knowledge, Difficulty in maintaining sustainable relationships with the suppliers, etc.	Pareto Analysis/ TISM/ MICMAC/ Fuzzy Theory

	Pellegrino et al. (2022)	Demand-related disruption, supply-related disruption, financial-related disruption, production-related disruption, and logistics-related disruption	BN/ISM
	Babu and Yadav (2023)	Environmental, information/technology, supply, process, transportation, delay, and demand	Fuzzy set theory
Bow-tie method	Jacinto and Silva (2010)	Semi-quantitative assessment of occupational risks	BT analysis
	Targoutzidis (2010)	Workplace risks assessment	BT analysis
	Saud et al. (2014)	Downstream industry risk assessment	BT analysis
	Aqlan and Ali (2014)	Chemical process risk assessment	BT analysis, FMEA, Fuzzy Theory
	Babaei et al. (2018)	Assessing the risk of human injuries due to medium voltage surges	BT analysis, Fuzzy theory
	Analouei et al. (2020)	Risk assessment of wastewater treatment plant	BT analysis
Bayesian network	Lee and Lee (2006)	Assess the potential risk of nuclear waste disposal	BN, Monte Carlo simulation
	Trucco et al. (2008)	Integration of human and organizational factors in risk analysis	BN
	Yun et al. (2009)	Risk assessment of LNG importation terminals	LOPA, BN
	Martin et al. (2009)	Analysis of occupational accidents falling from a height with Bayesian networks	BN
	Kalantarnia et al. (2009)	Dynamic risk assessment to prevent accidents and enhance overall system performance	BN, ETA
	Chen and Leu (2014)	Fall risk assessment of cantilever bridge projects	BN, FTA
	Pereira et al. (2016)	Risk assessment of jet engine high-pressure turbine assembly	BN, AHP
	Zarei et al. (2018)	Risk assessment of process industry accident	BN
	Baksh, Abbassi et al. (2018)	Maritime transport risk assessment	BN

(Continued)

Table 12.1 (Continued)

Study	Classification of risk	Approach/method
Punyamurthula and Badurdeen (2018)	Production line risk assessment	BN, System dynamics
Wu, Fang, et al. (2019), Wu, Jia, et al. (2019)	Assessment of school bus accidents	BN, Dempster–Shafer theory
Li, Wang, et al. (2019), Li, Liu, et al. (2019)	Risk assessment of mine ignition sources	BN, Fuzzy theory
Meng et al. (2019)	Assess system vulnerabilities and predict the probability of leakage accidents at offshore oil and gas rigs	BN, DEMATEL
Li, Wang, et al. (2019), Li, Liu, et al. (2019)	Dynamic risk assessment in healthcare	BN
Abbaspour Onari et al. (2021)	Predicting the severity level of COVID-19	BN, FCM
Khakzad et al. (2013)	Dynamic analysis of the safety of process systems	BN, BT analysis
Zarei et al. (2017)	Dynamic evaluation of process systems	BN, BT analysis
Bilal et al. (2017)	Risk assessment of fire and explosion of pipelines	BN, BT analysis, Fuzzy theory
Rezaee et al. (2018)	Risk analysis of sequential processes in the food industry	PFMEA, Sequential FCM
Chang et al. (2019)	Dynamic risk assessment of leakage in the hydrogen production unit	BN, BT analysis
Aliabadi et al. (2020)	Risk assessment related to hydrogen gases	BN, BT analysis, Fuzzy theory
Yousefi et al. (2020)	Logistics processes risks in manufacturing industries	PFMEA, Multi-stage FCM
Keshtiban et al. (2022)	Risk assessment of the auto parts manufacturing process	PFMEA, Multi-stage FCM

Integrated Methods

coherent linkage between the causes and consequences. The BT diagram comprises five fundamental elements: Basic Events, Fault Tree, Main Event, Event Tree, and Output Events. The Fault Tree, located on the left side of the BT diagram, initiates from the critical event (such as the main event) and branches out as the primary or intermediate causes are defined, utilizing logical gates based on the root events. On the right side of the BT diagram, the Event Tree is established, beginning with the critical event as the initiator event and then progressing through a series of events (consequences) to determine potential output events. By using both the Event Tree and the Fault Tree, the BT diagram identifies all the causes and consequences associated with a critical event.

After constructing the BT diagram, it becomes possible to carry out quantitative analysis for Event Tree Analysis (ETA) and Fault Tree Analysis (FTA) (Shahriar et al., 2012):

$$P_{OE} = \prod_{i=1}^{n} P_i \text{ for intersection in event tree analysis} \qquad (12.1)$$

$$P_{AND} = \prod_{i=1}^{n} P_i \text{ for intersection in fault tree analysis} \qquad (12.2)$$

$$P_{OR} = 1 - \prod_{i=1}^{n}(1 - P_i) \text{ for conjunction in fault tree analysis} \qquad (12.3)$$

12.3.2 Bayesian network

In a Bayesian network (BN) model, there are typically two stages: a qualitative stage and a quantitative stage. The qualitative stage involves a directed acyclic graph, while the quantitative stage encompasses the conditional and prior probabilities associated with the Bayesian network nodes. In an acyclic graph, nodes are visual representations of random variables typically depicted as circles. In contrast, directed arcs illustrate the cause-and-effect relationships between the nodes. In a BN, when there is an arc from node X to node Y, it signifies that node X is the parent of node Y in the network. The parent nodes have a direct effect on the child nodes, and each child node has a conditional probability distribution defined as $\Pr\{X_i|Parents(X_i)\}$ that quantizes the parental influence on the child node. If a node does not have a parent, it is known as a root node or leaf node, respectively. Using the assumptions of conditional independence of Bayesian networks, the joint probability distribution of a set of random variables $\{X_1, X_2, X_3, ..., X_n\}$ using a chain rule can be determined as follows:

$$\Pr\{X_1, X_2, X_3, ..., X_n\} = \prod_{i=1}^{n} \Pr(X_i|\text{Parent}(X_i)) \tag{12.4}$$

BNs offer a powerful probabilistic framework for reasoning under conditions of uncertainty. They can integrate diverse information sources, enabling a comprehensive and precise evaluation. Bayesian network models can perform both forward (prediction) and backward (diagnostic) analysis. The analysis in the forward direction involves tracing the network arcs from the root nodes to the leaf nodes. In contrast, the backward analysis of the Bayesian network entails following the arcs in the opposite direction, thereby deducing effects from their causes. Diagnostic analysis of the Bayesian network model employs Bayes' theorem. If A and B are two random events and B has occurred, the posterior probability of event A, given that B has occurred, is as follows:

$$\Pr(A|B) = \frac{\Pr(B|A)\Pr(A)}{\Pr(B)} \tag{12.5}$$

Where $\Pr(A)$ and $\Pr(B)$ are the prior probabilities of Events A and B, respectively.

12.3.3 Dempster–Shafer theory

Dempster (1967) introduced the Dempster–Shafer theory. Later it developed and was recognized by Shafer (1976). The Dempster–Shafer theory is related to the Bayesian theory, meaning that both deal with subjective beliefs. Dempster–Shafer theory is widely used in various areas (Awasthi & Chauhan, 2011). The structure of Dempster–Shafer's theory has been founded based on the following four definitions.

Definition 1. It is assumed that Ω is a set of mutually unique and collectively complete of E_i, shown as:

$$\Omega = \{E_1, E_2, ..., E_i, ..., E_N\} \tag{12.6}$$

Ω is called frame of discernment. The power set Ω is displayed by 2^Ω.

$$2^\Omega = \{\varnothing, \{E_1\}, ..., \{E_N\}, \{E_1, E_2\}, ..., \{E_1, E_2, .., E_i\}, ..., \Omega\} \tag{12.7}$$

If $A \in 2^\Omega$, A is a proposition.

Definition 2. A mass function is a mapping of m from 2^{Ω} to $[0, 1]$ which is defined as follows:

$$m: 2^{\Omega} \to [0, 1] \tag{12.8}$$

Which meets the following condition:

$$\sum_{A \in P(\theta)} m(A) = 1, m(\phi) = 0 \tag{12.9}$$

Definition 3. If $m(A) > 0$, A is called a focal element and the set of all focal elements is called a body of evidence. When several bodies of evidence are available independently, the Dempster combination rule can be used to obtain the combined evidence as follows:

$$m(A) = \frac{\sum_{B,C \subseteq \Omega, B \cap C = A} m_1(B)m_2(C)}{1 - K} \tag{12.10}$$

where $K = \sum_{B \cap C = \varnothing} m_1(B)m_2(C)$ is a Normalization Constant which is called inconsistency. Relation 10 only makes sense when $m_{\oplus}(\varnothing) \neq 1$, otherwise the rule is meaningless.

Definition 4. Discounted Combination.

The classic Dempster's combination rule does not work when evidence is in strong conflict with each other. Many methods have been proposed to address the problem of combining conflicting evidence (Han & Deng, 2018). It can be considered a reliability coefficient, which reflects the reliability of the information source. Once the reliability coefficients have been determined, the next step is to combine them in the fusion process. To control the conflict between information sources, a discounting rule is introduced as follows:

$$m_i^{\gamma_i}(A) = \gamma_i \times m_i(A), \forall A \subset \Theta, A \neq \varnothing \tag{12.11}$$

$$m_i^{\gamma_i}(\Theta) = \gamma_i \times m_i(\Theta)' + 1 - \gamma_i \tag{12.12}$$

where $m_i(.)$ is a presentation of basic probability assignment and γ_i is the discounting coefficient, respectively (Li & Chen, 2019).

12.4 Proposed solution

This section introduces a risk assessment framework based on the integration of BT analysis, Bayesian Network, Dempster—Shafer Theory, and Fuzzy Set Theory. The combination of Fuzzy Set and Evidence Theories is used to better manage the uncertainty caused by the lack of knowledge of experts and to integrate the personal evaluations of the expert team into a group evaluation. Furthermore, to overcome the limitations of the BT analysis, such as its static structure and highly subjective and uncertain nature, the combination of this method with the Bayesian Network is used. The proposed framework can be summarized in the following steps.

12.4.1 Stage 1. Bow-tie analysis

Step 1. By utilizing historical data, past experiences, and the opinions of a team of experts, both internal and external risk factors and sub-factors of each supply chain management process are identified, as well as any potential barriers and consequences.

Step 2. To perform a BT analysis, the probability of occurrence and non-occurrence of each risk factor is calculated based on fuzzy linguistic variables based on the opinions of experts. This fuzzy rating is used to reduce the uncertainty in estimating the probability of risk factors. In this rating, seven linguistic variables including "very high," "high," "nearly high," "medium," "low," "very low" and "impossible" with scores ranging from 7 to 1 are used, as per Li, Wang, et al. (2019) and Li, Liu, et al. (2019).

Step 3. Conversion of estimated probabilities by a team of experts to basic probability assignments.

To transform the judgments of experts about the probability of occurrence and non-occurrence of risk factors into a possible variable, which is expressed as a triangular fuzzy number, the process of defuzzification is used. Several defuzzification methods have been proposed, such as the mean of maxima, the center of gravity, the height method, and the center of maxima. Zhang et al. (2014) summarized the whole defuzzification process and finally obtained Eq. (12.13).

$$Val(F) = \frac{a + 2r + b}{4} \tag{12.13}$$

where $F(a, r, b)$ is an arbitrary triangular fuzzy number and $Val(F)$ presents the defuzzified value.

Step 4. Using the Dempster combination rule.

In this study, the identification framework is considered as $\Theta = \{o, n\}$, where o indicates the occurrence and n indicates the non-occurrence of the risk factor. After obtaining basic probability assignments (BPAs) using expert evaluations and considering the inefficiency of the Dempster combination rule in cases where the evidence is inconsistent with each other, the discounting rule introduced in the "Dempster Composition Law" is used. After obtaining the discounted BPAs (using Eqs. (12.11) and (12.12)), the Dempster combination rule is then used to combine the evidence.

Step 5. Use FTA to quantify the left side of the BT diagram (fault tree).

After determining the likelihood of the occurrence of each of the risk factors according to the logical relationship between the risk factors and sub-factors, the likelihood of occurrence of each risk factor, and finally, the likelihood of a critical event through the rules of gate combination using Eq. (12.3) (FTA) was obtained.

Step 6. Using ETA to quantify the right side of the BT diagram (Event tree).

The likelihood of the occurrence and non-occurrence of any of the inhibitory barriers in the event tree is estimated based on the linguistic expressions (Hosseini et al., 2020). The purpose of using fuzzy set theory is to reduce uncertainty and increase the accuracy of probability estimation. After collecting expert evaluations, Eq. (12.13) is used for defuzzification of these evaluations. Finally, using Eq. (12.1) and for each outcome event, the probability of occurrence is calculated.

12.4.2 Stage 2. Mapping algorithm from the bow-tie diagram to Bayesian network

Step 1. Mapping from the fault tree to the Bayesian network.

First, for each root event in the fault tree, a root node is considered in the BN. If an event is repeated multiple times in the fault tree, only one node in the BN is required. Additionally, the intermediate nodes in the fault tree form the corresponding intermediate nodes in the Bayesian network, and the final failure in the fault tree becomes a leaf node in the BN. Then, each root node in the Bayesian network is assigned a value of the corresponding probability from that node in the fault tree.

Step 2. Mapping from the event tree to the BN.

In this step, each barrier in the event tree is presented by a node in the BN that has two modes, one for failure and one for success. An outcome node that has states equal to the number of consequences in the event tree is also added to the network. When mapping from the event tree to BN, a barrier node is connected to the previous barrier node only if the probability of failure of this node depends on the previous barrier node. Additionally, a barrier node is connected to the outcome node if the probability of the outcome node states is affected by the failure or success of the intended barrier node.

Step 3. Mapping from a BT diagram to the BN.

After the development of Bayesian networks that are equivalent to fault trees and event trees, they are connected through a critical event as a central node. The central node is also connected to the outcome node. Fig. 12.1 shows the steps for creating a Bayesian network equivalent to a BT diagram.

12.4.3 Stage 3. Bayesian network analysis

In this stage, the designed Bayesian network is analyzed, and based on different scenarios, the effects of nodes on each other, as well as the outcome nodes, are examined.

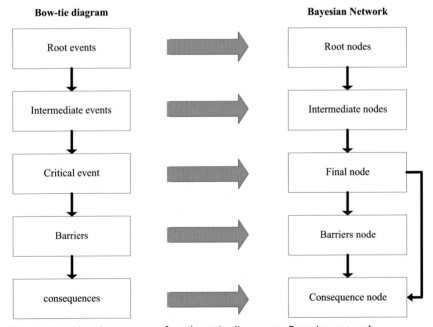

Figure 12.1 Mapping structure from bow tie diagram to Bayesian network.

12.5 Analysis

This section first deals with the problem statement, and then the proposed cross BT method is implemented on the problem. Following this, the use of the Dempster—Shafer theory and Bayesian network and the relevant results are presented, and sensitivity analysis based on the BN is examined. Manufacturing industries in the Iranian economy play an important role in creating added value and job creation, however, they are also associated with high diversity, which increases the possibility of unexpected events occurring in the supply chain and resulting in significant losses for the company. Since 2020, Iran, like all other countries, has been affected by the COVID-19 outbreak, which has disrupted logistics systems and supply chains around the world, and the diverse policies in different countries to deal with this issue have caused a lot of inconsistency in communication between suppliers and buyers. Additionally, the inefficient management of the pandemic in Iran and the restrictions that have arisen have caused internal risks for manufacturing industries. On the other hand, due to the economic and trade sanctions against Iran, the supply of materials and goods has become so difficult that it has increased costs, extended supply time, or reduced quality. Therefore, the sustainable supply chain of manufacturing industries in Iran since 2020 has been affected by a combined consequence of sanctions and the COVID-19 pandemic. Despite the interaction between them, not distinguishing between the two categories can lead to an inadequate and unreliable analysis. A cross BT model is introduced that can simultaneously examine their impact on each other while analyzing the risks in its group. Fig. 12.2 presents the structure of the proposed cross BT model.

To describe the factors involved in the occurrence of manufacturer loss (critical event) and the barriers and consequent events, a cross BT model (see Fig. 12.2) has been created for the internal and external phases simultaneously. Through the documents and reports provided by the responsible organizations, as well as reviewing the work processes and risk management documents of manufacturing companies, internal and external risks have been identified and separated. This has led to the identification of 34 internal and 26 external risks (levels risks) for the sustainable supply chain processes of manufacturing industries in

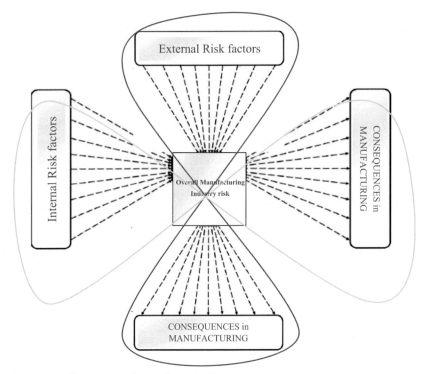

Figure 12.2 The proposed cross bow-tie diagram.

Iran, respectively. Additionally, 24 internal and 14 external outcomes (high-level risks) for related risks have been determined.

The category of risks is mentioned in the first and last columns of Tables 12.2 and 12.3 and indicates from which category each risk received the most impact. It included three categories: Regular, Sanctions, and Pandemic. Additionally, some risks are common in some processes due to the possibility of their occurrence in different processes. To control and prevent the increase in manufacturer loss, five inhibitory barriers are used in the event tree, which is selected and prioritized based on the opinions of experts. For this purpose, the five barriers are defined as Increasing Price (Sb01), Decreased Production (Sb02), Decreased Quality (Sb03), Receiving Facilities from the Bank (Sb04), and Workforce Adjustment (Sb05). On the other hand, six consequences are defined for the occurrence of manufacturer loss in both internal and external phases, depending on the occurrence or non-occurrence of successive inhibitory barriers. These include Decreased Demand and Maneuverability in the Market (C1),

Table 12.2 Classification of levels of risk factors of supply chain management processes.

Risk category	External		Process	Internal		Risk category
		Symbol		Symbol	Description	
	Description					
S	The emergence of new competitors	Out01	Planning	In01	Incorrect estimation of the required space in the warehouse	R
				In02	Incorrect estimate of the number of staff required	R
S	Lack of manpower due to problems between the manufacturer and the contractors	Out02		In03	Lack of proper estimation of required production machinery and equipment	R
				In04	Inaccurate estimates of required raw materials	R
R	Customs staff strikes	Out03	Procurement	In05	Wrong choice of suppliers	S
P	Widespread closures	Out04		In06	Lack of funds to buy	S
S	Inflation	Out05		In07	Failure to place an order on time and failure to properly assess the economic value of the order	P,S
S	Procurement of equipment and raw materials with low quality and high cost	Out06		In08	Non-payment of customs duties and non-clearance for more than four months	S
S	Cyber attacks	Out07	Manufacturing	In09	Lack of accurate estimation of customer needs	R
P	Disruption of the production process due to nationwide closure due to the disease pandemic	Out08		In10	Incorrect estimation of production capacity	R
S	Threat of war	Out09		In11	Lack of response to customer needs and lack of emergency plans for specific situations	S
S	Water, electricity, or gas outages	Out10		In12	Lack of production planning	R

(Continued)

Table 12.2 (Continued)

Risk category	External Description	Symbol	Process	Symbol	Internal Description	Risk category
S,P	Failure to provide the necessary equipment and raw materials	Out11	Packing	In13	Failure to follow the packaging schedule due to carelessness or replacement of the packaging staff	R
S	Rising exchange rates	Out12		In14	Lack of packing due to lack of pallets	S
				In15	Failure to implement pallet management effectively	R
S,P	Rising prices for raw materials	Out13		In16	Lack of integration of existing pallets and existence of different pallets	S
S	Deliberate sabotage	Out14	Warehousing	In17	Inadequate storage conditions	R
R	Natural disasters (floods, earthquakes, ...)	Out15		In18	Failure to follow the FIFO method and timely production of raw materials	R
				In19	Lack of proper layout in storage space	R
R	Theft	Out16		In20	No ID card on the parts	R
S,P	Internal riots and disturbances	Out17	Sending	In21	Defective loading equipment and failure to perform periodic testing services	S
				In22	Incomplete documents required for sending such as invoice sheet	R
S	War broke out	Out18		In23	Lack of timely loading (lack of access to loading equipment, lack of loading staff and lack of complete documentation)	S
				In24	Lack or sufficiency of emergency plans for the sending process	S

Code	Risk	Type
Out19	Problems between insurance companies and manufacturers	S,P
Out20	Natural disasters (floods, earthquakes, ...)	R
Out21	Accidents	R
Out22	Seizure of cargo	S
Out23	Cargo theft	R
Out24	Prevent the unloading of cargo to pass new laws	P
Out25	Internal riots and disturbances	S,P
Out26	Seizure or confiscation by governments	S

Shipping

Delivery

Code	Risk	Type
In25	No loading due to lack of staff	R
In26	Damage to cargo during loading	R
In27	Vehicle breakdown on the road	S
In28	Failure to use the proper path	S
In29	Inappropriate bin packing	R
In30	Failure to complete the sending checklist	R
In31	Lack of equipment such as forklifts and lack of replacement equipment	S
In32	No receipt from the customer	S
In33	Carelessness and lack of necessary staff skills	R
In34	Lack of relevant staff to delivery	R

S, Sanctions; P, Pandemic; R, Regular.

Table 12.3 Classification of high levels risk factors of supply chain management processes.

Risk category	Description	Symbol	Process	Symbol	Description	Risk category
					External … **Internal**	
S	Defects in the planning	Se07	Planning	Ie01	Delay in timely delivery to the customer	R
				Ie02	Stop the production line	S
				Ie17	Defects in planning	R
S	Lack of timely supply of required raw materials	Se01	Procurement	Ie03	Lack of timely supply of parts	S
S	Deficit in raw material inventory	Se02		Ie04	Parts inventory deficit	R
S,P	Defects in the procurement process	Se08		Ie18	Defects in the procurement process	S,P
P	Failure to achieve the planned production rate	Se03	Manufacturing	Ie05	Delay in timely delivery to customer	S
S,P	Inability to deliver orders on time	Se04		Ie06	Failure to respond to orders and reduced on-time delivery	S
S	Defects in the manufacturing process	Se09		Ie19	Defects in the manufacturing process	S
R	Defects in the packaging process	Se10	Packing	Ie07	Reduce timely delivery to the customer	S
				Ie08	Failure to respond to orders and reduced on-time delivery	S
				Ie20	Defects in packaging process	R
R	Defects in the warehousing process	Se11	Warehousing	Ie09	Increased waste	R
				Ie10	Wrong use of parts	R
				Ie21	Defects in the warehousing process	R

S,P	Defects in the sending process	Se12	Sending	Ie11	Delayed loading	S
				Ie12	Failure to send the shipment and timely delivery to the customer	R
R	Damage to cargo	Se05	Shipping	Ie22	Defects in the sending process	R
				Ie13	Lack of timely delivery to the customer and lack of stock	S
S,P	Failure to deliver the cargo on time	Se06		Ie14	Damage to products and their return	R
S	Defects in the Shipping process	Se13		Ie23	Defects in the shipping	S,P
S,P	Defects in the delivery process	Se14	Delivery	Ie15	unloading of goods	S,P
				Ie16	Waiting to evacuate and increase shipping costs	S
				Ie24	Defects in the delivery process	S,P

S, Sanctions; P, Pandemic; R, Regular.

Loss of Market Share (C2), Decreased Desire to Buy and Loss of Regular Customers (C3), Increase in Debt and Exposure to Bankruptcy (C4), Lack of Skilled Manpower and Closure of Production Lines (C5), and Decreased Profitability (C6).

To describe the factors involved in the occurrence of manufacturer loss (critical event) and the barriers and consequences of events, two BT models are created for the internal and external phases separately. Furthermore, the proposed cross BT for the hybrid phase is provided. The hybrid BT models created for the internal and the external phases simultaneously are shown in Fig. 12.3.

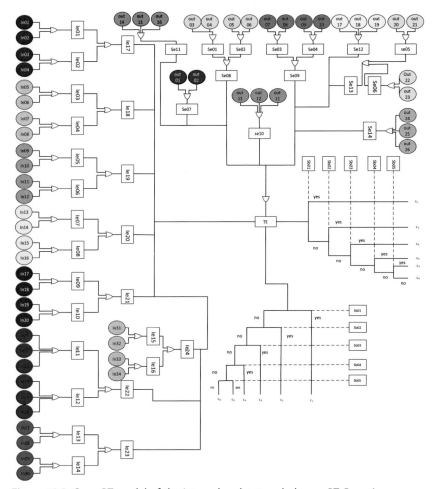

Figure 12.3 Cross BT model of the internal and external phases. *BT,* Bow-tie.

The possibility of the occurrence and non-occurrence of 34 internal and 26 external risk factors is assessed by three experts (TM1, TM2, and TM3) based on linguistic variables. Due to the large volume of data, an example of this assessment is provided for the planning process for internal risk factors. These are converted to crisp values using Eq. (12.13). The experts' assessments, as well as the equivalent converted values, are presented in Table 12.4.

After receiving the assessment of experts, evidence theory is used to integrate these opinions. For example, assume that $m_i(o)$ is the assignment of the probability presented by expert i in assessing the possibility of occurrence of In01 and $m_i(n)$ is the assignment of the probability provided by expert i in assessing the possibility of non-occurrence of In01 ($i = 1, 2, 3$). New BPAs are calculated using Eqs. (12.11) and (12.12) as follows:

$$m_1^{0.3}(o) = 0.3 \times 0.65 = 0.195 \quad m_1^{0.3}(n) = 0.3 \times 0.025 = 0.0075$$

$$m_1^{0.3}(\Theta) = 0.3 \times 0.325 + 1 - 0.3 = 0.7975 \quad m_2^{0.3}(o) = 0.3 \times 0.65 = 0.195$$

$$m_2^{0.3}(n) = 0.3 \times 0.15 = 0.045 \quad m_2^{0.3}(\Theta) = 0.3 \times 0.2 + 1 - 0.3 = 0.76$$

$$m_3^{0.3}(o) = 0.3 \times 0.65 = 0.195 \quad m_3^{0.3}(n) = 0.3 \times 0.35 = 0.105$$

$$m_3^{0.3}(\Theta) = 0.3 \times 0 + 1 - 0.3 = 0.7$$

Table 12.5 shows the results of the fusion of assessment derived from information sources (TM1, TM2, and TM3) using Dempster's combination rule (see Eq. 12.10). Additionally, the inconsistency coefficient is calculated as follows:

The final results of the fusion of assessments received from information sources are as follows:

$$m_1^{0.3} \oplus m_2^{0.3} \oplus m_3^{0.3}(o) = (0.068 + 0.245 + 0.12)/0.95 = 0.40$$

$$m_1^{0.3} \oplus m_2^{0.3} \oplus m_3^{0.3}(n) = (0.005 + 0.04 + 0.06)/0.95 = 0.20$$

$$m_1^{0.3} \oplus m_2^{0.3} \oplus m_3^{0.3}(\Theta) = 0.42/0.95 = 0.40$$

According to the process outlined above, the results of fusing expert opinions using evidence theory for the internal and external phases are calculated (see Appendix Table 12.A1). The probability of occurrence calculated for the risk factors is used to calculate the probability of the occurrence of intermediate events and the critical event in the bow tie diagram based on Eq. (12.3). Tables 12.6 and 12.7 show the calculated for the

Table 12.4 A sample of experts' assessment of the internal risk factors of the planning process.

Process	Symbol	TM1		TM2		TM3	
		Occurrence	Non-occurrence	Occurrence	Non-occurrence	Occurrence	Non-occurrence
Planning	In01	Extremely high (0.65)	Impossible (0.025)	Extremely high (0.65)	Extremely low (0.15)	Extremely high (0.65)	Low (0.35)
	In02	Moderate (0.5)	Impossible (0.025)	Extremely high (0.65)	Extremely low (0.15)	Moderate (0.5)	Extremely low (0.15)
	In03	High (0.8)	Extremely low (0.15)	Extremely high (0.93)	Impossible (0.025)	Extremely high (0.65)	Low (0.35)
	In04	Extremely low (0.15)	Extremely high (0.65)	Impossible (0.025)	Extremely high (0.93)	Moderate (0.5)	Moderate (0.5)

Table 12.5 Fusion of information provided by TM1, TM2, and TM3.

	$m_3^{0.3}(o) = 0.195$	$m_3^{0.3}(n) = 0.105$	$m_3^{0.3}(\Theta) = 0.7$
$m_1^{0.3} \oplus m_2^{0.3}(o) = 0.35$	$(o)0.195 \times 0.195 = 0.068$	$(\Theta)0.195 \times 0.0075 = 0.037$	$(o)0.195 \times 0.7975 = 0.245$
$m_1^{0.3} \oplus m_2^{0.3}(n) = 0.05$	$(\Theta)0.045 \times 0.195 = 0.0098$	$(n)0.045 \times 0.195 = 0.005$	$(n)0.045 \times 0.7975 = 0.04$
$m_1^{0.3} \oplus m_2^{0.3}(\Theta) = 0.6$	$(o)0.76 \times 0.195 = 0.12$	$(n)0.76 \times 0.0075 = 0.06$	$(\Theta)0.76 \times 0.7975 = 0.42$

Table 12.6 Calculated probabilities of occurrence of intermediate events and critical events (internal phase).

Process	Symbol	Description	Probability of occurrence
Planning	Ie01	Delay in timely delivery to the customer	0.670
	Ie02	Stop the production line	0.591
	Ie17	Defects in planning	0.865
Procurement	Ie03	Lack of timely supply of parts	0.644
	Ie04	Parts inventory deficit	0.283
	Ie18	Defects in the sourcing process	0.745
Manufacturing	Ie05	Delay in timely delivery to the customer	0.098
	Ie06	Failure to respond to orders and reduced on-time delivery	0.672
	Ie19	Defects in the manufacturing process	0.704
Packing	Ie07	Reduce timely delivery to the customer	0.657
	Ie08	Failure to respond to orders and reduced on-time delivery	0.585
	Ie20	Defects in the packaging process	0.857
Warehousing	Ie09	Increased waste	0.585
	Ie10	Wrong use of parts	0.527
	Ie21	Defects in the warehousing process	0.804
Sending	Ie11	Delayed loading	0.744
	Ie12	Failure to send the shipment and timely delivery to the customer	0.784
	Ie22	Defects in the submission process	0.945
Shipping	Ie13	Lack of timely delivery to the customer and lack of stock	0.654
	Ie14	Damage to products and their return	0.710
	Ie23	Defects in the Shipping	0.90
Delivery	Ie15	unloading of goods	0.693
	Ie16	Waiting to evacuate and increase Shipping costs	0.672
	Ie24	Defects in the delivery process	0.899
Critical event	Te	manufacturer loss	0.999

internal and external phases, respectively. For example, the probability of the occurrence of an intermediate event (Se01) is shown below.

$$P(Se01) = 1 - \prod_{i=1}^{2}(1 - P_i) = P_o(Out01) + P_o(Out02) - (P_o(Out01) \times P_o(Out02))$$
$$= 0.157 + 0.53 - (0.157 \times 0.53) = 0.604$$

To calculate the probability of consequential events occurring, first, the possibility of occurrence and non-occurrence of inhibitory barriers is

Table 12.7 Calculated probabilities of occurrence of intermediate events and critical events (external phase).

Process	Symbol	Description	Probability of occurrence
Planning	Se07	Defects in the planning	0.513
Procurement	Se01	Lack of timely supply of required raw materials	0.604
	Se02	The deficit in raw material inventory	0.746
	Se08	Defects in the sourcing process	0.899
Manufacturing	Se03	Failure to achieve the planned production rate	0.414
	Se04	Inability to deliver orders on time	0.534
	Se09	Defects in the manufacturing process	0.727
Packing	Se10	Defects in the packaging process	0.854
Warehousing	Se11	Defects in the warehousing process	0.711
Sending	Se12	Defects in the sending process	0.709
Shipping	Se05	Damage to cargo	0.789
	Se06	Failure to deliver the cargo on time	0.322
	Se13	Defects in the shipping process	0.857
Delivery	Se14	Defects in the delivery process	0.444
Critical event	Te	manufacturer loss	0.999

Table 12.8 The calculated probability of occurrence of consequence events for the internal phase.

		Internal phase	External phase
Symbol	Description	Probability of occurrence	Probability of occurrence
C_1	Decreased demand and maneuverability in the market	0.75	0.7
C_2	Loss of market share	0.15	0.19
C_3	Decreased desire to buy and loss of regular customers	0.08	0.09
C_4	Increase in debt and exposure to bankruptcy	0.015	0.01
C_5	Lack of skilled manpower and closure of some production lines	0.004	0.006
C_6	Decreased profitability	0.001	0.004

expressed based on expert assessment and using fuzzy linguistic expressions. Based on the estimated probability of occurrence for barriers and critical events, the probability of occurrence of consequential events is calculated by Eq. (12.1). Table 12.8 shows these values of outcome events.

Examination of the results of the BT model shows that in the internal phase, "defect in the sending process" (Ie22) is identified as the most probable event and "defect in the shipping process" (Ie23) and "defect in the delivery process" (Ie24) are jointly ranked next highest. In the external phase, "defects in the procurement process" (Se08), "defects in the shipping process" (Se13), and "defects in the manufacturing process" (Se09) are recognized as the most probable events, respectively. In terms of outcome events, both internal and external phases, reduced demand for products, and reduced maneuverability in the market are identified as the most likely consequences, respectively. Due to the prevailing conditions in the country's manufacturing industries increasing production costs and the inability of these industries to increase the price of products as the easiest option to compensate for losses, as well as inflation and recession that have overshadowed Iran (one of its consequences being the reduction of people's purchasing power), the reduction of demand for products as the most probable consequence can be justified.

12.5.1 Bayesian network model risk assessment

In the previous section, a cross BT model was introduced, as well as internal and external phases were analyzed using FTA and ETA. Conditional probabilities of the nodes of BN were completed using expert assessment. For this purpose, the probabilities provided in Appendix Table 12.A1 were considered the prior probabilities of the root nodes. After determining the conditional probabilities, the Bayesian network was developed to assess the risk of manufacturer loss for the internal and external phases. As mentioned, after drawing a cross-sectional BT model, FTA and ETA should be used to perform quantitative calculations of the BT analysis method. However, due to the high number of root events as well as the number of branches, the calculation of the likelihood of occurrence of intermediate events, critical events, and consequence events was associated with a lot of uncertainty and it was practically impossible to calculate these probabilities accurately. Because of the type of relationships, the values of all of them became 1 or very close to 1. Therefore, to overcome this problem as well as other weaknesses in the BT analysis method, the mapping of this cross BT model to the BN and performing risk analysis using the intrinsic characteristics of this model was used. Fig. 12.4 shows the cross BN model drawn in the same way as the proposed cross BT model to assess the risk of manufacturer loss for the hybrid internal and external phases.

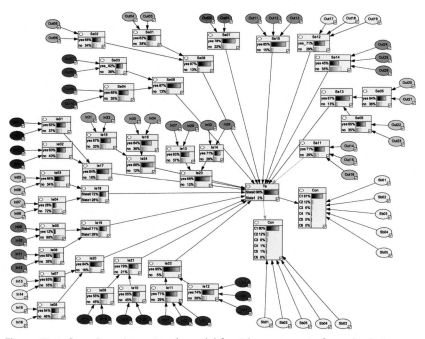

Figure 12.4 Cross Bayesian network model for risk assessment of supply chain management processes.

In the next step, the Bayesian network model is examined separately for both the internal and external phases. The comparison between the results provides valuable insights into the importance of considering conditional dependence between events with common causes in supply chain risk assessment.

In the internal phase, the BT analysis identifies the "defect in the sending process" (Ie22) as the most probable event, indicating its critical role in driving potential risks within this phase. Closely following, the "defect in the delivery process" (Ie24) and "defects in the shipping process" (Ie23) are ranked next, suggesting that these events also significantly contribute to the occurrence of manufacturer loss during the internal phase.

On the other hand, the Bayesian network analysis for the external phase reveals that the "defect in the transport process" (Se13) is identified as the most probable event, indicating its pivotal role in influencing risks during this stage of the supply chain. Additionally, "defects in the procurement process" (Se08) and "defects in the manufacturing process" (Se09) are ranked next, emphasizing their potential impact on the occurrence of manufacturer loss in the external phase.

The key reason for the difference between the results of the two analyses lies in the consideration of conditional dependence between events with common causes in the Bayesian network approach. Unlike the BT analysis method, which uses FTA and ETA, the Bayesian network allows for a more nuanced examination of how events may interact and influence each other.

By incorporating conditional dependence, the Bayesian network method can account for complex relationships between events, even when they share common causes. This is particularly important in supply chain risk management, where interconnected processes and events can lead to cascading effects. The BT analysis is useful for visualizing potential causes and consequences of risk events, but it may not capture the interdependencies and conditional probabilities between events. Therefore, the Bayesian network analysis complements the BT approach by providing a more comprehensive and probabilistic assessment of supply chain risk. Overall, the integration of both methods offers a robust and multi-faceted approach to supply chain risk analysis, enabling businesses to identify critical risk drivers, prioritize mitigation efforts, and strengthen their resilience against potential disruptions in the supply chain.

12.5.2 Probabilities update (posterior probabilities)

To perform a dynamic risk analysis, the deductive reasoning ability of Bayesian networks is utilized. The characteristics of the Bayesian network allow for the updating of the values of probabilities of root and intermediate nodes by getting new information and evidence, which makes the results of the proposed model more accurate, reduces model uncertainty, and provides a dynamic risk evaluation and assessment model (see Eq. 12.5). Updating the probabilities in BN model makes it possible to identify the events that have the greatest influence on the occurrence of the main event. Table 12.9 displays the initial occurrence probabilities and the recurrence probabilities of the intermediate and root nodes of the Bayesian network model for internal and external phases.

By comparing columns 2 and 3 of Table 12.9 in the internal phase and columns 5 and 6 of this table for the external phase, the following results are deduced.

The analysis presented the significant impact of various factors on the occurrence of manufacturer loss in the supply chain. Notably, "defect in the packing process" (Ie20) stands out as the primary driver behind

Table 12.9 Probability of occurrence of root and intermediate nodes.

Internal phase			External phase		
BN nodes	basic	posterior	BN nodes	Basic	posterior
In01	0.45	0.45	Out01	0.45	0.45
In02	0.4	0.4	Out02	0.115	0.115
In03	0.53	0.53	Out03	0.157	0.157
In04	0.13	0.13	Out04	0.53	0.53
In05	0.17	0.17	Out05	0.53	0.53
In06	0.57	0.57	Out06	0.46	0.46
In07	0.157	0.157	Out07	0.306	0.306
In08	0.15	0.15	Out08	0.157	0.157
In09	0.07	0.07	Out09	0.13	0.13
In10	0.03	0.03	Out10	0.464	0.464
In11	0.33	0.33	Out11	0.58	0.58
In12	0.51	0.51	Out12	0.457	0.458
In13	0.48	0.48	Out13	0.36	0.36
In14	0.34	0.34	Out14	0.26	0.26
In15	0.53	0.53	Out15	0.54	0.54
In16	0.116	0.116	Out16	0.15	0.15
In17	0.53	0.53	Out17	0.52	0.52
In18	0.116	0.116	Out18	0.054	0.054
In19	0.316	0.316	Out19	0.36	0.36
In20	0.31	0.31	Out20	0.55	0.55
In21	0.464	0.464	Out21	0.53	0.53
In22	0.457	0.457	Out22	0.23	0.23
In23	0.12	0.12	Out23	0.12	0.12
In24	0.457	0.457	Out24	0.27	0.27
In25	0.44	0.44	Out25	0.16	0.16
In26	0.29	0.29	Out26	0.094	0.094
In27	0.34	0.34	Se07	0.778	0.771
In28	0.476	0.476	Se01	0.623	0.623
In29	0.5	0.5	Se02	0.663	0.663
In30	0.42	0.42	Se08	0.867	0.868
In31	0.41	0.41	Se03	0.622	0.622
In32	0.48	0.48	Se04	0.65	0.65
In33	0.37	0.37	Se09	0.866	0.867
In34	0.48	0.48	Se10	0.573	0.575
Ie01	0.632	0.632	Se11	0.714	0.718
Ie02	0.596	0.596	Se12	0.71	0.711
Ie17	0.84	0.827	Se05	0.642	0.642
Ie03	0.665	0.665	Se06	0.654	0.654
Ie04	0.28	0.28	Se13	0.88	0.881
Ie18	0.722	0.705	Se14	0.449	0.438
Ie05	0.123	0.123			
Ie06	0.648	0.649			
Ie19	0.709	0.703			
Ie07	0.646	0.646			
Ie08	0.543	0.544			
Ie20	0.84	0.848			
Ie09	0.553	0.553			
Ie10	0.545	0.545			
Ie21	0.787	0.778			
Ie11	0.712	0.712			
Ie12	0.742	0.742			
Ie22	0.948	0.947			
Ie13	0.634	0.634			
Ie14	0.705	0.705			
Ie23	0.881	0.879			
Ie15	0.667	0.667			
Ie16	0.645	0.645			
Ie24	0.883	0.881			

increased occurrences of manufacturer loss in the internal phase. This finding underscores the critical importance of ensuring quality and efficiency in the packing process to mitigate potential risks.

Following the defect in the packing process, the study reveals that "Failure to respond to orders and reduced on-time delivery" in both Packing (Ie08) and manufacturing (Ie06) play key roles in amplifying the occurrence of manufacturer loss. This underscores the importance of prompt and reliable order fulfillment and delivery processes to avoid potential disruptions.

However, the analysis also sheds light on potential areas of improvement to reduce manufacturer loss occurrences in the internal phase. Specifically, "Defects in the sourcing process" (Ie18), "Defects in planning" (Ie17), and "Defects in the warehousing process" (Ie21) are identified as critical factors that can be addressed to lower the risks associated with the supply chain.

Moving on to the external phase, the study reveals that "defects in the warehousing process" (Se11) and "defects in the packaging process" (Se10) have the largest share in increasing the occurrence of manufacturer loss. This emphasizes the need for robust warehousing and packaging procedures to safeguard products and minimize potential damages during the external transportation phase.

On the other hand, the analysis highlights "Defects in the delivery process" (Se14) and "Defects in planning" (Se07) as crucial factors that can effectively reduce the occurrence of manufacturer loss in the external phase. Streamlining delivery operations and enhancing planning accuracy can contribute significantly to risk mitigation.

The proposed approach, which separately considers both internal and external phases, presents a comprehensive and insightful method for assessing supply chain risk. This detailed examination allows for a more targeted and effective risk management strategy to be implemented.

In particular, the Bayesian network derived from the proposed cross BT analysis identifies the "defect in sending process" (Ie22) as the most probable event in the supply chain risk assessment. This result suggests that addressing potential issues in the sending process could yield significant risk reduction benefits.

Overall, the findings from this scientific analysis provide valuable insights into the complex nature of supply chain risk and underscore the importance of adopting a holistic approach to risk management. By identifying key risk drivers and proposing targeted mitigation strategies, businesses can enhance their resilience and safeguard against potential losses in their supply chains.

12.6 Discussion

To perform the sensitivity analysis obtained for the internal phase, the likelihood of occurrence of the intermediate nodes Ie17, Ie18, Ie19, Ie20, Ie21, Ie22, Ie23, and Ie24, is considered 0%. It is assumed that these risks have been eliminated or substantially mitigated. Also, for the external phase, the occurrence of intermediate nodes Se07, Se08, Se09, Se10, Se11, Se12, Se13, and Se14 is considered 0%. In this case, too, it is assumed that the effects of these risks have been eliminated or substantially mitigated. Then, the effect of these conditions on the main (final) node (manufacturer loss) is investigated. The case that further reduces the likelihood of the occurrence of the main node is selected as the factor whose mitigation has the greatest effect. The first part of Table 12.10 shows the results of the sensitivity analysis performed for the internal and external phases separately.

The results of sensitivity analysis show that the mitigation of "defects in the shipping process" (Ie23) has the greatest effect on reducing the occurrence of manufacturer loss (see the second part of Table 12.10). Also, "defects in the delivery" (Ie24) and "warehousing" (Se11) processes are in the next ranks, respectively.

In this probabilistic analysis of manufacturer loss within the context of supply chain management, the focus lies on updating probabilities based on observed occurrences, ultimately highlighting the root causes responsible for the loss. The manufacturer loss is considered as a loss function,

Table 12.10 Sensitivity analysis of the Bayesian network based on the cross bow-tie.

Cross				Separately			
Internal phase		External phase		Internal phase		External phase	
Input nodes	Probability of final node occurrence (o)	Input nodes	Probability of final node occurrence (o)	Input nodes	Probability of final node occurrence (o)	Input nodes	Probability of final node occurrence (o)
Ie17	0.998	Se07	0.988	Ie17	0.985	Se07	0.956
Ie18	0.997	Se08	0.994	Ie18	0.963	Se08	0.929
Ie19	0.999	Se09	0.994	Ie19	0.926	Se09	0.936
Ie20	0.998	Se10	0.99	Ie20	0.861	Se10	0.914
Ie21	0.99	Se11	0.981	Ie21	0.947	Se11	0.913
Ie22	0.992	Se12	0.987	Ie22	0.926	Se12	0.922
Ie23	0.971	Se13	0.984	Ie23	0.923	Se13	0.927
Ie24	0.973	Se14	0.983	Ie24	0.927	Se14	0.943

emphasizing the significance of minimizing its value for improved performance. By iteratively updating the probabilities of intermediate and root nodes, and considering their associations with final node events, we gain insights into the likelihood of various factors contributing to the loss. The results of this update reveal that "defects in the shipping process" (Ie23) exhibit the highest share in the occurrence of manufacturer loss, indicating that addressing issues in the shipping process could lead to substantial improvements. Following closely are "defects in the delivery process" (Ie24) and "defects in the warehousing process" (Se11) as the next leading contributors to manufacturer loss. These findings provide valuable guidance to supply chain managers, enabling them to strategically allocate resources to reduce losses and enhance overall supply chain performance effectively.

The overlapping of risks, particularly those associated with sanctions and the COVID-19 pandemic, intensifies their impact. Addressing one factor alone is not effective, as all contributing factors must be resolved. For instance, consider the risk labeled "Defects in the procurement process," which may arise due to both sanctions and COVID-19 restrictions. If the COVID-19 restrictions are lifted, it will not alleviate this risk in Iran compared to other countries, as it is also affected by sanctions. As evident from the results, such risks carry significant weight, but in a broader context, important risks in Iran are primarily related to sanctions, necessitating measures beyond conventional risk management approaches. Tackling these multifaceted challenges requires comprehensive and strategic actions to mitigate their adverse effects effectively.

The research discussed the concept of risk overlap and highlighted its critical implications, particularly in the context of sanctions and the COVID-19 pandemic. By emphasizing that addressing a single factor may not suffice, the text emphasizes the complexity of managing risks that have multiple underlying causes. Furthermore, the analysis pointed out the unique challenges faced by Iran due to the intersection of sanctions and the pandemic. Lifting COVID-19 restrictions in Iran may not fully resolve certain risks, such as "Defects in the procurement process," because they are compounded by the enduring effects of sanctions. This underscores the need for a nuanced and holistic approach to addressing the complexities of managing such intertwined risks. Additionally, the analysis highlighted the significance of prioritizing actions concerning risks associated with sanctions, as they appear to have a more substantial impact in the Iranian context. This finding emphasizes the importance of formulating specialized strategies to cope with the unique challenges posed by sanctions and mitigate their detrimental effects on various aspects of the economy and society.

Overall, the research and analysis underscored the importance of considering risk overlap and interconnectedness in risk management strategies. It stressed the need for a comprehensive approach that accounts for multiple factors contributing to risks, particularly in scenarios where the confluence of different risk sources may magnify their effects. In the context of Iran, the analysis calls for innovative measures and policy considerations to address the compounded risks arising from the combination of sanctions and the COVID-19 pandemic effectively.

12.7 Conclusion

In this research, a comprehensive and innovative approach has been developed to evaluate the dynamic supply chain risk in Iranian manufacturing industries, considering both internal and external risk factors related to sustainable supply chain management processes. The proposed approach combines various methodologies, including BT analysis, BN, Dempster–Shafer evidence, and fuzzy set theories, to offer a robust framework for risk assessment. The results demonstrate the effectiveness of utilizing both BN and BT analysis in assessing supply chain risk. By integrating these techniques, the model provides a holistic view of critical events' root causes and their preventive effects. The Bayesian network's ability to handle deductive and inductive reasoning helps reduce uncertainty in calculating probabilities, leading to more reliable risk evaluations. Additionally, the dynamic model of risk assessment enhances the accuracy of results, ensuring a closer alignment with real-world scenarios.

However, the research acknowledges certain limitations that open avenues for future research. One key limitation lies in the dynamic nature of manufacturing industries and their operating environment, leading to constant changes in internal and external risk factors. To address this issue, the researchers suggest developing a decision support system that automatically monitors identified risk factors and identifies new ones to assess them promptly and effectively. Moreover, the broad nature of supply chain risk poses challenges in providing a comprehensive risk assessment model. To enhance accuracy and reduce uncertainty, the integration of simulation methods, such as system dynamics, could be explored to extract effective parameters for a more accurate risk assessment. Finally, the study recommends separately assessing the share of risks associated with sanctions and the COVID-19 pandemic. Considering different scenarios for future conditions can further enhance the understanding of these risks' impacts and aid in developing targeted risk management strategies.

In conclusion, this research presents a novel and comprehensive approach to evaluating supply chain risk in Iranian manufacturing industries. By incorporating multiple methodologies, the study provides valuable insights into risk assessment and preventive measures. Addressing the mentioned limitations through future research can lead to further advancements in risk management of the supply chain, contributing to the sustainability and resilience of the manufacturing sector in Iran.

Appendix 12.A

See (Tables 12.A1, 12.A2)

Table A1. Fusion experts' opinions for the internal risk factors.

Process	Symbol	o	n	Θ
Planning	In01	0.40	0.20	0.40
	In02	0.40	0.07	0.53
	In03	0.53	0.09	0.48
	In04	0.124	0.463	0.413
Procurement	In05	0.17	0.35	0.48
	In06	0.57	0.05	0.38
	In07	0.157	0.457	0.386
	In08	0.15	0,46	0,39
Manufacturing	In09	0.07	0.53	0.40
	In10	0.03	0.59	0.38
	In11	0.33	0.26	0.41
	In12	0.51	0.07	0.42
Packing	In13	0.48	0.10	0.42
	In14	0.34	0.27	0.39
	In15	0.53	0.07	0.40
	In16	0.116	0.496	0.388
Warehousing	In17	0.53	0.07	0.40
	In18	0.116	0.496	0.388
	In19	0.316	0.265	0.419
	In20	0.31	0.27	0.42

(Continued)

Table A1. (Continued)

Process	Symbol	o	n	Θ
Sending	In21	0.464	0.135	0.401
	In22	0.457	0.157	0.386
	In23	0.12	0.47	0.41
	In24	0.457	0.157	0.386
	In25	0.44	0.14	0.42
	In26	0.29	0.31	0.4
Shipping	In27	0.34	0.27	0.39
	In28	0.476	0.098	0.426
	In29	0.50	0.04	0.46
	In30	0.42	0. 20	0.38
Delivery	In31	0.19	0. 41	0.40
	In32	0.48	0.10	0.42
	In33	0.37	0.17	0.46
	In34	0.48	0.42	0.10

Table A2. Fusion experts' opinions for the external risk factors.

Process	Symbol	o	n	Θ
Planning	Out01	0.45	0.01	0.54
	Out02	0.115	0.367	0.518
Procurement	Out03	0.157	0.457	0.386
	Out04	0.53	0.09	0. 38
	Out05	0.53	0.09	0.38
	Out06	0.46	0.08	0.46
Manufacturing	Out07	0.305	0.306	0.389
	Out08	0.157	0.457	0.386
	Out09	0.13	0.46	0.41
	Out10	0.464	0.135	0.401
Packing	Out11	0.58	0.01	0.41
	Out12	0.457	0.157	0.386
	Out13	0.36	0.21	0.43
Warehousing	Out14	0.26	0.34	0.40
	Out15	0.54	0.06	0.40
	Out16	0.15	0.43	0.42
Sending	Out17	0.52	0.04	0.44
	Out18	0.054	0.565	0.381
	Out19	0.36	0.21	0.43
Shipping	Out20	0.55	0.03	0.42
	Out21	0.53	0.09	0.38
	Out22	0.23	0.38	0.39
	Out23	0.12	0.47	0.41
Delivery	Out24	0.27	0.34	0.39
	Out25	0.16	0.40	0.44
	Out26	0.094	0.504	0.402

References

Abbaspour Onari, M., Yousefi, S., Rabieepour, M., Alizadeh, A., & Rezaee., M. J. (2021). A medical decision support system for predicting the severity level of COVID-19. *Complex & Intelligent Systems, 7*, 2037−2051.

Aliabadi, M. M., Pourhasan, A., & Mohammadfam, I. (2020). Risk modelling of a hydrogen gasholder using Fuzzy Bayesian Network (FBN). *International Journal of Hydrogen Energy, 45*(1), 1177−1186.

Analouei, R., Taheriyoun, M., & Safavi, H. R. (2020). Risk assessment of an industrial wastewater treatment and reclamation plant using the bow-tie method. *Environmental Monitoring and Assessment, 192*, 1−16.

Aqlan, F., & Ali, E. M. (2014). Integrating lean principles and fuzzy bow-tie analysis for risk assessment in chemical industry. *Journal of Loss Prevention in the process Industries, 29*, 39−48.

Aqlan, F., & Lam, S. S. (2015). A fuzzy-based integrated framework for supply chain risk assessment. *International Journal of Production Economics, 161*, 54−63.

Awasthi, A., & Chauhan, S. S. (2011). Using AHP and Dempster−Shafer theory for evaluating sustainable transport solutions. *Environmental Modelling & Software, 26*(6), 787−796.

Babaei, M. M., Jabbari, M., & Babaei, A. A. (2018). Human injuries risk assessment of medium voltage electrocution using bow tie model in fuzzy environment (case study: golestan province electricity distribution company). *Iranian Journal of Health, Safety and Environment, 5*(2), 997−1006.

Babu, H., & Yadav, S. (2023). A supply chain risk assessment index for small and medium enterprises in Post COVID-19 era. *Supply Chain Analytics*, 100023.

Badurdeen, F., Shuaib, M., Wijekoon, K., Brown, A., Faulkner, W., Amundson, J., Jawahir, I. S., Goldsby, T. J., Iyengar, D., & Boden, B. (2014). Quantitative modeling and analysis of supply chain risks using Bayesian theory. *Journal of Manufacturing Technology Management, 25*(5), 631−654.

Baksh, A.-A., Abbassi, R., Garaniya, V., & Khan, F. (2018). Marine transportation risk assessment using Bayesian Network: Application to Arctic waters. *Ocean Engineering, 159*, 422−436.

Berenji, H. R., & Anantharaman, R. N. (2011). Supply chain risk management: Risk assessment in engineering and manufacturing industries. *International Journal of Innovation, Management and Technology, 2*(6), 452.

Bilal, Z., Mohammed, K., & Brahim, H. (2017). Bayesian network and bow tie to analyze the risk of fire and explosion of pipelines. *Process Safety Progress, 36*(2), 202−212.

Chang, Y., Zhang, C., Shi, J., Li, J., Zhang, S., & Chen, G. (2019). Dynamic Bayesian network based approach for risk analysis of hydrogen generation unit leakage. *International Journal of Hydrogen Energy, 44*(48), 26665−26678.

Chen, T.-T., & Leu, S.-S. (2014). Fall risk assessment of cantilever bridge projects using Bayesian network. *Safety science, 70*, 161−171.

Curkovic, S., Scannell, T., & Wagner, B. (2013). Using FMEA for supply chain risk management. *Modern management science & Engineering, 1*(2), 251−265.

Dempster, A. P. (1967). Upper and lower probabilities induced by a multivalued mapping. *The Annals of Mathematical Statistics, 38*, 325−339.

Diabat, A., Govindan, K., & Panicker, V. V. (2012). Supply chain risk management and its mitigation in a food industry. *International Journal of Production Research, 50*(11), 3039−3050.

Fenton, N., & Neil, M. (2018). *Risk assessment and decision analysis with Bayesian networks.* CRC Press.

Ghadir, A. H., Vandchali, H. R., Fallah, M., & Tirkolaee, E. B. (2022). Evaluating the impacts of COVID-19 outbreak on supply chain risks by modified failure mode and effects analysis: a case study in an automotive company. *Annals of Operations Research*, 1−31.

Giannakis, M., & Papadopoulos, T. (2016). Supply chain sustainability: A risk management approach. *International Journal of Production Economics*, *171*, 455−470.

Han, Y., & Deng, Y. (2018). An enhanced fuzzy evidential DEMATEL method with its application to identify critical success factors. *Soft Computing*, *22*, 5073−5090.

Hosseini, N., Givehchi, S., & Maknoon, R. (2020). Cost-based fire risk assessment in natural gas industry by means of fuzzy FTA and ETA. *Journal of Loss Prevention in the Process Industries*, *63*, 104025.

Jacinto, C., & Silva, C. (2010). A semi-quantitative assessment of occupational risks using bow-tie representation. *Safety Science*, *48*(8), 973−979.

Kalantarnia, M., Khan, F., & Hawboldt, K. (2009). Dynamic risk assessment using failure assessment and Bayesian theory. *Journal of Loss Prevention in the Process Industries*, *22*(5), 600−606.

Karmakar, R., Mazumder, S. K., Hossain, M. B., Illes, C. B., & Garai, A. (2023). Sustainable green economy for a supply chain with remanufacturing by both the supplier and manufacturer in a varying market. *Logistics*, *7*(3), 37.

Keshtiban, P. M., Onari, M. A., Shokri, K., & Rezaee, M. J. (2022). Enhancing risk assessment of manufacturing production process integrating failure modes and sequential fuzzy cognitive map. *Quality Engineering*, *34*(2), 191−204.

Khakzad, N., Khan, F., & Amyotte, P. (2013). Dynamic safety analysis of process systems by mapping bow-tie into Bayesian network. *Process Safety and Environmental Protection*, *91*(1−2), 46−53.

Khan, S., Haleem, A., & Khan, M. I. (2021). Assessment of risk in the management of Halal supply chain using fuzzy BWM method. *Supply Chain Forum: An International Journal*, *22*(1), 57−73, Taylor & Francis.

Kumar, S., Satyendra., & Sharma, S. (2015). Developing a Bayesian network model for supply chain risk assessment. *Supply Chain Forum: An International Journal*, *16*(4), 50−72, Taylor & Francis.

Lee, C.-J., & Lee, K. J. (2006). Application of Bayesian network to the probabilistic risk assessment of nuclear waste disposal. *Reliability Engineering & System Safety*, *91*(5), 515−532.

Li, M., Wang, D., & Shan, H. (2019). Risk assessment of mine ignition sources using fuzzy Bayesian network. *Process Safety and Environmental Protection*, *125*, 297−306.

Li, Z., & Chen, L. (2019). A novel evidential FMEA method by integrating fuzzy belief structure and grey relational projection method. *Engineering Applications of Artificial Intelligence*, *77*, 136−147.

Li, M., Liu, Z., Li, X., & Liu, Y. (2019). Dynamic risk assessment in healthcare based on Bayesian approach. *Reliability Engineering & System Safety*, *189*, 327−334.

Mangla, S. K., Kumar, P., & Barua., M. K. (2015). Risk analysis in green supply chain using fuzzy AHP approach: A case study. *Resources, Conservation and Recycling*, *104*, 375−390.

Martin, J. E., Rivas, T., Matías, J. M., Taboada, J., & Argüelles, A. (2009). A Bayesian network analysis of workplace accidents caused by falls from a height. *Safety Science*, *47*(2), 206−214.

Meng, X., Chen, G., Zhu, G., & Zhu, Y. (2019). Dynamic quantitative risk assessment of accidents induced by leakage on offshore platforms using DEMATEL-BN. *International Journal of Naval Architecture and Ocean Engineering*, *11*(1), 22−32.

Pellegrino, R., Gaudenzi, B., & Qazi, A. (2022). COVID-19 pandemic: Supply chain risk management by integrating Interpretive Structural Modeling and Bayesian belief network. *Ifac-Papersonline, 55*(10), 667−672.

Pereira, J. C., Fragoso, M. D., & Todorov, M. G. (2016). Risk assessment using bayesian belief networks and analytic hierarchy process applicable to jet engine high pressure turbine assembly. *IFAC-PapersOnLine, 49*(12), 133−138.

Popov, G., Lyon, B. K., & Hollcroft, B. D. (2016). *Risk assessment: A practical guide to assessing operational risks.* John Wiley & Sons.

Punyamurthula, S., & Badurdeen, F. (2018). Assessing production line risk using bayesian Belief networks and system dynamics. *Procedia Manufacturing, 26*, 76−86.

Rezaee, M. J., Sadatpour, M., Ghanbari-Ghoushchi, N., Fathi, E., & Alizadeh., A. (2020). Analysis and decision based on specialist self-assessment for prognosis factors of acute leukemia integrating data-driven Bayesian network and fuzzy cognitive map. *Medical & Biological Engineering & Computing, 58*, 2845−2861.

Rezaee, M. J., Yousefi, S., Valipour, M., & Dehdar, M. M. (2018). Risk analysis of sequential processes in food industry integrating multi-stage fuzzy cognitive map and process failure mode and effects analysis. *Computers & Industrial Engineering, 123*, 325−337.

Saud, Y. E., Israni, K., & Goddard, J. (2014). Bow-tie diagrams in downstream hazard identification and risk assessment. *Process Safety Progress, 33*(1), 26−35.

Schoenherr, Ts, Rao Tummala, V. M., & Harrison, T. P. (2008). Assessing supply chain risks with the analytic hierarchy process: Providing decision support for the offshoring decision by a US manufacturing company. *Journal of Purchasing and Supply Management, 14*(2), 100−111.

Shafer, G. (1976). *A mathematical theory of evidence* (Vol. 42). Princeton University Press.

Shahriar, A., Sadiq, R., & Tesfamariam, S. (2012). Risk analysis for oil & gas pipelines: A sustainability assessment approach using fuzzy based bow-tie analysis. *Journal of Loss Prevention in the Process Industries, 25*(3), 505−523.

Silva, L. M. F., de Oliveira, A. C. R., Leite, M. S. A., & Marins, F. A. S. (2021). Risk assessment model using conditional probability and simulation: Case study in a piped gas supply chain in Brazil. *International Journal of Production Research, 59*(10), 2960−2976.

Targoutzidis, A. (2010). Incorporating human factors into a simplified "bow-tie" approach for workplace risk assessment. *Safety Science, 48*(2), 145−156.

Thun, J.-H., & Hoenig, D. (2011). An empirical analysis of supply chain risk management in the German automotive industry. *International Journal of Production Economics, 131*(1), 242−249.

Trucco, P., Cagno, E., Ruggeri, F., & Grande, O. (2008). A Bayesian Belief Network modelling of organisational factors in risk analysis: A case study in maritime transportation. *Reliability Engineering & System Safety, 93*(6), 845−856.

Wagner, S. M., & Bode, C. (2008). An empirical examination of supply chain performance along several dimensions of risk. *Journal of Business Logistics, 29*(1), 307−325.

Wu, J., Fang, W., Tong, X., Yuan, S., & Guo, W. (2019). Bayesian analysis of school bus accidents: A case study of China. *Natural Hazards, 95*, 463−483.

Wu, Y., Jia, W., Li, L., Song, Z., Xu, C., & Liu, F. (2019). Risk assessment of electric vehicle supply chain based on fuzzy synthetic evaluation. *Energy, 182*, 397−411.

Yousefi, Sl, Rezaee, M. J., & Moradi, A. (2020). Causal effect analysis of logistics processes risks in manufacturing industries using sequential multi-stage fuzzy cognitive map: A case study. *International Journal of Computer Integrated Manufacturing, 33*(10−11), 1055−1075.

Yun, G. W., Rogers, W. J., & Mannan, M. S. (2009). Risk assessment of LNG importation terminals using the Bayesian–LOPA methodology. *Journal of Loss Prevention in the Process Industries*, *22*(1), 91–96.
Zarei, E., Azadeh, A., Aliabadi, M. M., & Mohammadfam, I. (2017). Dynamic safety risk modeling of process systems using bayesian network. *Process Safety Progress*, *36*(4), 399–407.
Zarei, E., Mohammadfam, I., Azadeh, A., Khakzad, N., & Aliabadi, M. M. (2018). Dynamic risk assessment of chemical process systems using Bayesian Network. *Iran Occupational Health*, *15*(3), 103–117.
Zhang, L., Skibniewski, M. J., Wu, X., Chen, Y., & Deng, Q. (2014). A probabilistic approach for safety risk analysis in metro construction. *Safety Science*, *63*, 8–17.

CHAPTER THIRTEEN

Streamlining supply chain operations: A case study of bigbasket.com

S. Mahalakshmi[1], Anitha Nallasivam[1], Kavitha Desai[2], Sandeep Kautish[3,4] and Bharath H.[5]

[1]Faculty of Management Studies, CMS Business School, JAIN (Deemed-to-Be-University), Bangalore, Karnataka, India
[2]SVKM's Narsee Monjee Institute of Management Studies, Bangalore, Karnataka, India
[3]LBEF Campus (APU Malaysia) Kathmandu, Nepal
[4]Model Institute of Engineering and Technology, Jammu, Jammu and Kashmir, India
[5]CMS Business School, JAIN (Deemed-to-Be-University), Bangalore, Karnataka, India

13.1 Introduction and overview of the e-commerce market in India

The growth of e-commerce in India has been rapid; yet, it has been hindered by a comparatively low number of internet users. Nevertheless, the proliferation of technical accessibility, the emergence of 5G services, and the enhanced affordability of data plans and smartphones are collectively fostering a conducive milieu for the expansion of electronic commerce. To ensure the survival and success of firms in the sector, it is imperative to prioritize adaptability and customer-centric innovation, as the industry is characterized by fierce rivalry (Khosla and Kumar, 2017). The Indian online grocery retail sector has shown substantial expansion in recent years, mostly attributed to the rising prevalence of internet and mobile usage, evolving consumer preferences, and supportive governmental regulations. E-commerce is poised to emerge as a dominant entity within the realm of electronic business, providing unparalleled convenience and a vast worldwide reach. According to Jain et al. (2021), the advent of e-commerce brings advantages for both consumers and merchants, but it also presents obstacles for conventional enterprises. According to projections, the Indian online grocery retail sector is anticipated to attain a market value of $26.63 billion by the year 2027, exhibiting a compound annual growth rate (CAGR) of almost 33%. The growth

Computational Intelligence Techniques for Sustainable Supply Chain Management.
DOI: https://doi.org/10.1016/B978-0-443-18464-2.00001-7

observed in this context can be attributed to the rising customer demand for convenience and the opportunity to engage in grocery shopping from the confines of their residences. The future of e-commerce in India is promising, as there is an expected expansion in several sectors such as travel, electronics, hardware, retail, and fashion. The critical determinants of success encompass several variables such as the provision of guarantees, the integration of mobile commerce capabilities, the availability of different payment choices, the implementation of quality assurance measures, and the establishment of robust customer support services. According to Chanana and Goele (2012), the efficient implementation of fundamental components presents opportunities for both businesses and individuals. Bigbasket, Dunzo, Amazon Pantry, Blinkit, Swiggy Instamart, Zepto, Geomart, and Grofers are prominent participants in the online grocery retail sector in India. These organizations provide a diverse array of products, encompassing fresh produce and home needs while employing sophisticated technologies, including big data and artificial intelligence (AI), to enhance operational efficiency and enhance the overall consumer experience. The online food retail industry in India encounters various obstacles, such as inadequate logistics infrastructure, restricted internet accessibility in rural regions, and a lack of consumer trust, despite its notable expansion. The Indian market is transforming due to the influence of e-commerce, which has ample opportunities for expansion considering that just 19% of the population now participates in online purchasing and selling activities. Nevertheless, it is imperative to address difficulties such as inadequate legislation pertaining to cyber security. The realization of a prosperous e-commerce future is contingent upon the effective resolution of these concerns, the augmentation of accessibility, and the establishment of a legal structure that safeguards the rights of consumers and their privacy (Battase and Shankar, 2020). Furthermore, the presence of fierce competition from both domestic and international entities poses significant challenges for smaller enterprises in terms of maintaining competitiveness. In general, the online grocery retail market in India presents substantial prospects for expansion and advancement, as enterprises endeavor to address the evolving demands of customers and enhance their supply chain activities. By implementing effective strategies and making substantial investments in technology, organizations operating in this particular area are poised to achieve substantial growth in the forthcoming years. Numerous instances arise within the highly competitive domain of online grocery delivery (Fig. 13.1).

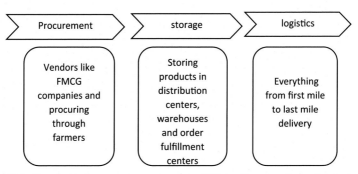

Figure 13.1 Supply chain management at http://Bigbasket.com. *From Original.*

13.2 The grocery industry in India

In the context of the Indian online grocery sector, two prominent business models have emerged as noteworthy: the inventory-based model and the hyperlocal model. The previous approach, characterized by a significant investment in warehouse facilities, has inherent benefits. In contrast, the hyperlocal model, which establishes collaborations with nearby retail establishments, represents a more cost-effective option (Carpenter and Moore, 2006). It is worth noting that organizations in India have consistently customized these models to suit their requirements. For example, BigBasket first implemented a strategy centered around order fulfillment but later shifted to an inventory-centric model by establishing a specialized warehouse. ZopNow.com initiated an inventory-centric approach before transitioning to a hyperlocal business model (Gupta and Chintagunta, 1994). In contrast, AaramShop Private Limited collaborated extensively with local vendors using its platform and implemented deliveries by forming partnerships with nearby grocery stores, distinguishing its delivery method from the conventional approach.

13.3 About Bigbasket.com

Bigbasket.com is an e-commerce platform that operates in the grocery and household goods sector. Bigbasket.com is an online grocery shop of significant magnitude in India. It was established in 2011 by a group of individuals including Hari Menon, V. S. Sudhakar, Vipul Parekh, and

Abhinay Choudhari. The establishment provides a diverse selection of recently harvested fruits and vegetables, essential food products, animal-derived protein sources, aquatic edibles, and many domestic necessities. The e-commerce platform, Bigbasket.com, currently operates in more than 30 locations around India, facilitating the delivery of over 1 million orders every month. According to a source, Bigbasket achieved the designation of a unicorn in 2019, denoting its classification as a privately held startup with a valuation of over $1 billion. Based on a recent analysis conducted by RedSeer Consulting, it has been determined that Bigbasket.com holds the dominant position in the online grocery sector within India, with a market share above 30%. According to a report by RedSeer strategic consultant in March 2017, the company boasts a user base of more than 20 million registered users, with over 10 million of them being active users. As of December 2022, the valuation of Bigbasket stood at approximately $2.7 billion. Bigbasket is a subsidiary of Innovative Retail Concepts Pvt Ltd, which operates under Supermarket Grocery Supplies, a holding entity controlled by Tata. The range of products offered by the organization encompasses approximately 100,000 items, including fruits, vegetables, spices, pulses, juices, hygiene, household basics, baby products, and various others (OECD, November 15, 2022). In addition to offering a diverse range of products and ensuring efficient and punctual delivery, the favorable return policy and affordable pricing have consistently positioned Bigbasket.com as the favored delivery service among its clientele (Thomas et al., 2017). Furthermore, Bigbasket's Fresho stores, which serve as offline retail establishments, are characterized by their integration of technology. This integration enables customers to conveniently place orders online and then collect their purchases in-store. Additionally, consumers have the opportunity to engage in traditional shopping experiences within the store premises, where they can independently complete the payment process. The establishment primarily offers a variety of fruits, vegetables, groceries, and everyday necessities for sale. Fresho's self-billing system incorporates AI technologies, similar to Amazon Go. This system allows consumers to independently choose and retrieve desired items, as well as weigh them at the checkout counter. Utilizing computer vision technology, the system automatically generates an accurate bill based on the selected items and their respective weights (Bhandari, 2016). These outcomes can be achieved through the utilization of a highly effective inventory management system and the establishment of strong relationships with suppliers. Additionally, they prioritize the promotion and sale of their exclusive brand. One notable advantage is that clients can choose their desired delivery time within a designated period. According to Ratna (2018), Bigbasket demonstrates a commendable order fulfillment rate of 99.5%, ensuring that a significant majority of customer orders are successfully processed. Additionally, the company achieves a 99% on-time delivery rate, indicating its commitment to timely and reliable service. Furthermore, Bigbasket maintains a substantial inventory availability rate of 96%, suggesting that the majority of requested items are readily accessible to customers. As of March 2023, the company's valuation stands at over $3.0 billion. It operates in more than sixty cities, providing a wide range of over one hundred thousand products and handling nearly 1.5 million orders daily. Bigbasket exhibits an impressive 99.5% rate of timely deliveries, providing a substantial advantage to its customer base, which exceeds 10 million individuals. Bigbasket.com recorded revenues above INR 10,000 crore for the fiscal year 2022−23. Bigbasket, a prominent online grocery supermarket in India, holds a significant market share of 30%. According to a report by The Economic Times in 2023, Bigbasket.com secured the 7th position in the ranking of the "Top 10 E-commerce Companies in India" (Appendix 13.2).

13.4 Business model of Bigbasket.com

The inception of Bigbasket was initiated by entrepreneurs who possessed practical expertise within the sector. At the outset, complimentary delivery was provided for orders above Rs. 600. According to Dr. Aparajita et al. (2021), Bigbasket provides a guarantee of on-time delivery, and in the event of failure to meet this guarantee, the company agrees to refund 10% of the bill value. The company implemented a straightforward and adaptable refund policy, which allowed customers to request refunds without being subjected to any inquiries. Customers were given a 48-hour timeframe to report their refund requests. The refunded amount would be credited back to the original payment method within 7−10 business days or instantaneously to the Bigbasket wallet. One possible option available to clients is the implementation of a slotted delivery system, which allows them to choose a specific time slot for their delivery within the designated timeframe of 7 am to 10 pm (Kumar, 2020).

The notion of Express delivery, which entails the delivery of items within a 90-minute timeframe, is currently limited to tier-1 cities. To facilitate this expedited service, Bigbasket has established partnerships with local vendors to ensure superfast delivery (Appendix 13.1). Additionally, during the pandemic, Bigbasket formed a special collaboration with local outlets such as bakeries and sweet shops, thereby enhancing the visibility and significance of its BB Super specialty store (Choudhary, 2020, March 2). The package will be delivered within a time frame of 90 minutes. Another alternative entails providing goods in large quantities to educational institutions, universities, workplaces, and businesses at a competitive rate.

The business model of the company is predicated upon the incorporation of the subsequent fundamental components:

E-commerce platform: Bigbasket is an e-commerce platform that facilitates a diverse range of product offerings across many categories, including groceries, household essentials, personal care items, and other related merchandise. Customers have the convenient ability to peruse and get things via the company's website or mobile application (Appendix 13.4) and most of this traffic comes from the direct site, search engine, and websites and a few also come from the inorganic or paid search (Appendix 13.5).

Supply chain management: Bigbasket possesses a very efficient supply chain management system, enabling the company to effectively procure products from suppliers and promptly deliver them to its clients. According to Choudhary (2020), the organization has made strategic investments in both technology and logistics infrastructure to facilitate the efficient and effective delivery of superior products to its consumer base.

Customer experience: Bigbasket places a high value on providing a great client experience. To ensure client pleasure, the organization provides rapid and dependable delivery, a user-friendly website and mobile app, and prompt customer assistance (Jain, 2019, September).

Data-driven approach: Bigbasket bases its business decisions on data and analytics, which include product selection, pricing, marketing, and customer service. This helps the organization to continuously develop its offerings and better satisfy its consumers' expectations (Kritika Kumar, 2020).

Own brands: Bigbasket also sells its private label items on its site. These in-house brands help Bigbasket stand out from the competition and increase profits (Singhal, 2020).

Overall, Bigbasket's business approach is centered on providing clients with a convenient, efficient, and high-quality online grocery and household goods purchasing experience. Bigbasket has built itself as one of the leading online grocery retailers in India by leveraging its e-commerce platform, supply chain management, customer experience, data-driven approach, and own brands (Kumar, 2020).

13.5 Supply chain management at Bigbasket

Although much emphasis has been placed on generating demand in the E-Commerce sector, it is now critical that we transfer our focus to supporting and interfering with the supply side of the market (Nougarahiya et al., 2021). Bigbasket offers a supply chain management system that allows it to buy directly from vendors such as P&G and HUL. Their quick delivery is entirely based on the hyperlocal concept of dark storefronts in all of the places where they operate. Vans and bikes are employed for carrying merchandise between warehouses and dark stores, as well as for end delivery (Choudhary, 2020, March 2).

Key practices that contribute to Bigbasket's supply chain efficiency include:

- **Regional warehousing:** To shorten lead times and increase the freshness of products delivered to clients, Bigbasket has developed regional warehouses in important areas. To shorten lead times and increase the freshness of products delivered to clients, Bigbasket has developed regional warehouses in important areas.
- **Inventory management:** To maintain ideal stock levels and avoid waste, the organization employs advanced inventory management techniques. This helps to cut expenses while also ensuring that products are always available when customers need them.
- **Direct sourcing:** Bigbasket sources fresh produce, dairy products, and other perishable things directly from farmers and producers, decreasing the need for intermediaries and boosting product quality.
- **Last-mile delivery:** To ensure rapid and dependable delivery of orders to customers, the company uses a combination of its delivery fleet and agreements with local transportation providers. To reduce

delivery times and costs, real-time tracking and route optimization technologies are used. A delivery app is used by the delivery people to enable the delivery.

- **Customer service:** Bigbasket lays a high focus on customer service, providing a variety of customer support alternatives such as a contact center, email, and social media outlets. In addition, the organization has a highly attentive customer care crew that has been trained to address customer problems and complaints swiftly and effectively.

Choudhary (2020, March 2) continues to invest in new technology and procedures to improve supply chain efficiency and preserve market competitiveness.

13.6 IT in supply chain management at Bigbasket

We are living in the e-commerce 2.0 age. E-commerce is mostly utilized for better and more effective supply chain management, with an emphasis on having less inventory, inventory-focused, and aggregator-based. Bigbasket began with a zero–inventory and zero–inventory approach, with Bigbasket executives picking up from local stores and delivering (Bhandari 2016). Later, they gradually relied on the just-in-time (JIT) strategy, in which deliverables arrived just before delivery, and they gradually deviated from the JIT approach. Bigbasket's supply chain management activities rely heavily on information technology (IT) (Wilson, 2015).

Some of the primary ways that IT is used in this regard are as follows:

Inventory management: Bigbasket tracks inventory levels in real-time using their systems, allowing them to make informed decisions about ordering and refilling products, eliminating waste, and maintaining product availability for clients. Traditional inventory models are used to replenish inventory, with algorithms that consider vendor criteria such as frequency of supply, quality, and timeliness (Raman et al., 2018).

Order management: To ensure timely and accurate delivery to consumers, the company's IT systems assist in the efficient processing of customer orders, including order tracking and fulfillment.

Fulfillment and logistics management: IT systems are utilized to manage and optimize activities such as route-based picking, route planning, delivery tracking, and vehicle dispatch quality assessment. This helps to ensure that deliveries are performed on schedule and within budget.

Supply chain analytics: Bigbasket makes informed judgments by leveraging data and analytics to acquire insights into supply chain performance. Data on inventory levels, order fulfillment rates, delivery durations, and other relevant parameters are included (Bhandari, 2016).

Customer relationship management (CRM): IT systems are used to handle customer contacts and feedback, revealing important information about client preferences and behavior. This data can help with product selection, pricing, and marketing tactics.

Bigbasket's supply chain management is supported by IT, making it more efficient. There are numerous data analytics technologies available for planning and anticipating demand. Bigbasket maintains an internet presence via the website Bigbasket.com (Appendix 13.3). It is available for both Android and iOS. It makes use of predictive analytics. It offers features such as a smart basket, which predicts the customer's needs based on previous purchases, minimizing the time to order and reorder (Meta Group, 2001). Bigbasket's specialized software optimizes the route by assigning delivery vans to specific routes based on the orders received. The smart algorithm optimizes the route, making delivery easier. Unlike all other firms, Bigbasket has its end-to-end e-commerce platform. It all began with the apps. Despite having a web presence, Bigbasket has developed apps for Android and iOS. An app dedicated to placing orders, picking delivery slots, and accepting multiple payment ways to make the process run more smoothly (Zohra Ennaji et al., 2016). Aside from that, Bigbasket depends significantly on AI and ML, as well as a personalized recommendation engine, to understand client purchasing patterns. It also contains a smart basket, which is used to determine preference for a specific product or brand. Bigbasket now uses an inventory-led strategy in which products are purchased directly from farmers and manufacturers (Hansen, 2008). As the number of orders increased, it became more difficult for the delivery executives to ensure quick and timely delivery. Bigbasket also offers a B2B operation, delivering to roughly 10,000 grocery stores across the country via a tech-enabled network. Bigbasket's farm fresh fruits and veggies were purchased directly from farmers, eliminating the intermediary and resulting in a win-win situation for both the farmers and Bigbasket customers. In addition to its relationship with farmers, Bigbasket employs big data, AI, and ML to improve supply chain operations and consumer experience (Munson et al., 2017). By leveraging these technologies, Bigbasket can optimize its operations, minimize waste, and deliver a high-quality customer experience (Patrick and Paul, 2001).

13.7 Bigbasket's supply chain management challenges

Bigbasket, like many other online grocery sellers, confronts several supply chain issues. Among these are (den Hertog, 2000):

Logistics management: In highly populated areas, where traffic delays are more common, it might be more difficult to guarantee punctual and effective product delivery to clients. Warehouses, delivery vehicles, and a network of delivery staff are all essential pieces of Bigbasket's logistics infrastructure that need to be in place to support the company's operations. Keeping delivery drivers on the job is a problem in many places and is therefore receiving attention (Lovelock et al., 2013).

Inventory management: Maintaining precise inventory levels is crucial for ensuring that products are available to customers when they need them. Furthermore, product assortment requirements can vary among cities. This is not always easy to do, especially when considering the local assortment and how it should interact with local vendors and suppliers. It's difficult to make accurate predictions about perishable goods because of their short shelf life. Bigbasket needs efficient inventory management solutions to cut down on waste and keep products in stock at all times (Mazumder, 2015).

Supply chain visibility: Bigbasket relies on being able to see all links in its supply chain to make educated decisions and find places for growth. However, this might be difficult in situations when there are several suppliers and delivery points involved in the supply chain (Vankipuram & Nandy, 2020).

Product quality: Maintaining a high standard of quality is essential to gaining and keeping loyal customers. Because of the importance of delivering high-quality goods on time, Bigbasket must implement stringent quality control measures (Richa Chhabra, 2016).

Competition: There is an increasing number of competitors in the online food retail business, making it intensely competitive. To maintain its competitive edge and grow its customer base, Bigbasket must always innovate and set itself apart from the competition (Sharma, 2016).

These are a few instances of the difficulties encountered by Bigbasket's supply chain. Bigbasket can maintain its position as the industry leader in online grocery retail by responding to these threats and constantly enhancing its operations.

13.8 Personal interview with the expert at Bigbasket.com

To enhance understanding of the case and the research subject, an interview was conducted with Mr. KB Nagaraju, the Director at Innovative Retail Concepts Pvt Ltd (Bigbasket), utilizing both a Zoom meeting and an email correspondence (Nagaraju, 2023). During the virtual conference that spanned 30 minutes, Mr. KB Nagaraju provided an overview of Bigbasket and offered insights into the supply chain management strategies employed by the company. Mr. KB Nagaraju is a highly experienced senior supply chain professional with comprehensive expertise in several aspects of the supply chain, including planning, import procurement, manufacturing, client order management, and logistics. The individual also possesses extensive experience in General Management, having served as the Logistics Business Head for a multinational third-party logistics (3PL) organization. Furthermore, they have demonstrated proficiency in establishing manufacturing processes, and developing systems, and have accumulated a wealth of experience in managing high-tech electronics supply chains. His area of expertise lies in the field of mergers and team integration, capacity and infrastructure development, project management, team training and motivation, supply chain management, and education.

In this interview, the questions were included as text within the electronic message and subsequently transmitted to Mr. KB Nagaraju. The electronic mail interview was conducted on March 22, 2023. After conducting comprehensive research, the interview questions were formulated to explore the intricacies of supply chain management at Bigbasket. The focus of the inquiry centered on understanding the company's adoption of best practices, as well as its utilization of technology and sustainability approaches. The interview consisted of a total of ten questions. The primary objective of this interview was to obtain a comprehensive understanding and acquire more profound perspectives on the innovative and technology-driven supply chain processes implemented by Bigbasket. Given this, a list of ten questions was developed and later used in a private email interview. The questions are listed below.

1. Can you describe your experience in managing supply chain operations for a large-scale e-commerce platform like Bigbasket?

2. How do you ensure efficient management of inventory and stock levels to meet customer demands while avoiding overstocking or stockouts?

3. What are some of the biggest challenges you face in managing supply chain operations for Bigbasket, and how do you address them?

4. Can you discuss any innovative supply chain strategies or initiatives that you have implemented at Bigbasket to optimize operations and reduce costs?

5. How do you maintain strong relationships with suppliers and vendors to ensure timely delivery of goods and services?

6. How do you leverage technology and data analytics to improve supply chain operations and enhance the customer experience at Bigbasket?

7. How do you ensure compliance with regulations and standards related to supply chain management, such as food safety regulations and labor laws?

8. Can you describe your approach to managing risk in the supply chain, such as disruptions in logistics, natural disasters, or political unrest?

9. How do you collaborate with other departments at BigBasket, such as marketing and finance, to align supply chain operations with broader business objectives?

10. How do you ensure sustainability and ethical practices in the supply chain, such as reducing carbon emissions, promoting fair labor practices, and reducing waste?

Mr. KB Nagaraju responded that it has been an exciting yet turbulent experience overseeing supply chain operations for a major e-commerce platform such as Bigbasket when questioned about his overall experience in the role. He said that Bigbasket's operations are extensive, dispersed throughout several cities, and that the company sources and stores its merchandise locally. Bigbasket has always prioritized important KPIs such as cost stack, customer complaints, fulfillment rate, availability, and on-time delivery. Additionally, he stated that the goal is to continuously enhance all aspects of the business, including planning, acquiring, receiving, picking, and delivery processes. To satisfy consumer demand and prevent overstocking or stockouts, the interviewer questioned Bigbasket about how it controls its inventory and stock levels. "Inventory management is a traditional concept that is backed by science and well understood principles," was the reply, in his own words. But at Bigbasket, the emphasis has always been on providing the appropriate assortment for local needs, working closely with farmers and suppliers, and having a proprietary inventory replenishment algorithm that takes into account all important factors like seasonality, perishability, vendor metrics, and so forth to make inventory management more scalable and straightforward. The following query focused on the main difficulties Bigbasket has encountered in overseeing its supply chain activities and how those difficulties were resolved. In response, he stated, "Bigbasket faced unique and local challenges due to its extensive operations across multiple cities and business lines." The first major problem for Bigbasket was making sure it had an assortment that was appropriate for a given city, but this challenge was solved by using some trial-and-error tactics. The second major challenge was making sure it had enough manpower, particularly for the blue-collar workers. This became problematic, particularly during the epidemic when managing migrant laborers took precedence. Acquiring excellent warehouse space in consumer catchment areas became difficult at the same time that the company was moving toward becoming more hyperlocal with the emergence of quick-commerce.

During our conversation, we also looked into any creative supply chain initiatives or tactics that Bigbasket has used to streamline operations and cut costs. Mr. KB Nagaraju emphasized that innovation has always been at the core of Bigbasket. Bigbasket has made extensive use of innovation to address problems in the supply chain. Bigbasket defines innovation as ideas or best practices that originate from all levels of the organization, not only the use of technology. A specialist team works on these concepts to produce workable solutions that lessen suffering, cut expenses, or improve and sustain operations. In keeping with that, he adds that using electric vehicles (EVs) for delivery is another cutting-edge and best practice used at Bigbasket. However, it is important to note that BigBasket has made a commitment to sustainability initiatives and lowering its carbon footprint. In this regard, they announced plans in 2019 to switch to EVs for last-mile deliveries in several Indian cities, and in 2020, they partnered with manufacturers of electric vehicles to add more than 1000 EVs to their delivery fleet. Furthermore, Bigbasket used electric cars since they were both economical and environmentally friendly. Not only that, but he also claims that Bigbasket discovered creative ways to increase pick-up truck productivity, including automation. The Bigbasket's relationship equation with its vendors and suppliers for the prompt delivery of goods and services was the subject of the following inquiry. In response, he said that Bigbasket has always been modest enough to acknowledge that its farmers, merchants, and workers—especially the blue-collar workers—are the key components of their success. Mr. KB Nagaraju stated that it is important to highlight that Bigbasket has policies that are fair and farmer-friendly, including electronic weighing, 48-hour payment terms, and agronomist assistance in choosing seeds, fertilizer, farming supplies, and government programs. In the same vein, "all other vendors are treated with respect in their daily interactions as well as in the policies." At their warehouses, all of our carriers and merchants have access to the canteen, he stated. The next question to be posed to Mr. KB Nagaraju concerned how Bigbasket is using technology and data analytics to improve supply chain operations and improve customer experience. Bigbasket is keeping up with the times, with businesses leveraging analytics to improve business and overall customer experience. In response, Mr. MB Nagaraju mentioned their extensive operations, presence in 60 cities, quick-commerce, B2C, and B2B services, as well as their retail presence, millions of registered customers, and more than 80% of regular buyers each month—all of which generate enormous amounts of data. Additionally, he stated that Bigbasket offers over 100,000 SKUs and that, presumably, data analytics is used to identify trends and patterns as well as to optimize inventory levels and prices—all of which are crucial for attracting new consumers and keeping hold of current ones. Additionally, he claims that its recommendation engine assists in recommending products to customers' baskets and in evaluating their lifecycle value. Data analytics is also utilized to increase delivery productivity, which lowers the cost per order.

We also questioned Bigbasket about its adherence to standards and regulations pertaining to supply chain management, such as labor laws and food safety regulations. He responded that Bigbasket has embraced these laws and regulations as a brand and has been in compliance with them, whether they are labor laws or related to food safety (FSSAI). There is no compromising, and we do frequent internal and external audits to make sure they stay compliant despite difficult local circumstances and shifting regulatory environments. Furthermore, we were informed that Bigbasket uses both proactive and contingency planning tactics to manage risks associated with its supply chain, such as disruptions in logistics, natural disasters, or political upheaval. In essence, they guard and maintain the protection and safety of their workers, farmers, and vendors when they are on our property. He added that throughout the epidemic, every precaution was taken for safety and cleanliness, and masks, gloves, medications, and emergency medical attention were all provided at no cost in the event of severe instances. Additionally, they provided Covid care for their blue-collar workforce. We generously extended our leave to allow them to heal. We also have backup plans to halt or scale back operations during natural disasters or disturbances, but we always notify our clients in advance.

Our next query concerns how Bigbasket works with other divisions inside the company, including finance and marketing, to synchronize supply chain activities with overarching business goals. He informed me that Bigbasket is an online retailer that was founded less than ten years ago and does not have any legacy problems. We do not operate with a "silo" mentality, which refers to a lofty tower used to store grains. Every time our marketing team does unique TV commercials, promotions, or events, the supply chain function is always involved. In a similar vein, supply chain provides a great deal of valuable client feedback on our app, pertaining to both product quality and pricing. Supply chain and finance collaborate closely to fund projects, pay farmers and vendors on time, and reconcile all funds collected during delivery. In conclusion, Bigbasket expressed pride in its foundation of ethical business practices, as evidenced by its response to our final question regarding sustainability and ethical practices in the supply chain, which included lowering carbon emissions, promoting fair labor practices, and cutting waste. It is said that Bigbasket instills these values in all of its employees. At Bigbasket, all labor laws are complied with, and workers receive excellent treatment even in situations when it is not required by law. In case some of the products are about to expire or lose their freshness, they kindly donate to those in need and claim to have creative ways to eliminate waste in their operations. To cut carbon emissions, they also employ electric cars or EVs. Above all, solar plants have been put on top of all of their warehouses and large dark stores to lessen reliance on grid electricity and to utilize fewer generators, making their operations sustainable.

We appreciate Mr. KB Nagaraju, Director of Innovative Retail Concepts Pvt Ltd (Bigbasket), and Bigbasket's time in explaining the company's supply chain methods. They were open and honest about their process, answering all of our questions with clarity and patience, which facilitated the development of this case study.

13.9 Way forward for Bigbasket.com

Bigbasket's supply chain relies heavily on space minimization, process simplification, and service level agreements (service level agreements). Bigbasket and its suppliers have established SLAs to ensure the timely and dependable delivery of products. In addition to reducing delivery delays, these SLAs ensure that products are constantly available to customers. Bigbasket is always evaluating its supply chain operations to make them more efficient and less wasteful. This requires adjusting distribution patterns, improving order processing, and adjusting storage strategies. Bigbasket combines state-of-the-art technologies and warehouse management systems to maximize storage capacity. This not only aids in reducing costs and ensuring that things are handled and stored efficiently, but it also aids in reducing waste. Bigbasket can improve the effectiveness and efficiency of its supply chain operations, providing a better customer experience and boosting its overall competitiveness in the online grocery retail sector, by concentrating on SLA, reducing procedures, and optimizing space.

13.10 Conclusion

This chapter discusses the current state of India's online grocery shopping and e-tail markets, before turning its attention to bigbasket. com and its supply chain methods and how it has leveraged IT to optimize its supply chain. The supply chain leader at Bigbasket.com talks about the difficulties and new approaches to embracing a tech-enabled supply chain. Equal treatment of merchants and careful consideration of local assortment planning are also emphasized. Key tactics for managing an optimized supply chain, including data analytics, regulatory compliance, and proactive risk management, are also highlighted. Supply chain strength is attributed to the coordinated efforts of finance, marketing, and operations. Ethical and sustainable methods, such as using solar energy and electric vehicles from Bigbasket.com to cut down on carbon emissions and waste, are also highlighted in this chapter. The company will be the best version of itself when it comes to supply chain management and will serve as a model for others to follow by adopting technology-enabled supply chain practices that improve efficiency, service quality, adherence to service legal agreements, process streamlining, space optimization, and faster delivery times.

Appendix 13.1

Fig. 13.A1

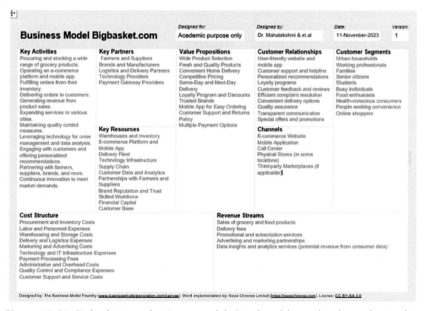

Figure 13.A1 Bigbasket.com business model. *Developed by author by understanding the business process of Bigbasket.com.*

Appendix 13.2

Fig. 13.A2

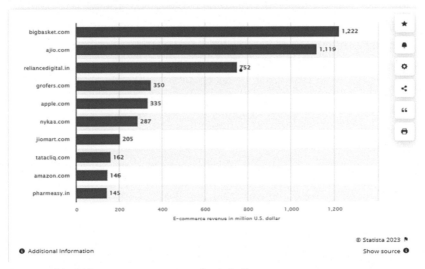

Figure 13.A2 Online e-commerce net sales in India.

Appendix 13.3

Fig. 13.A3

Internet penetration rate in India from 2007 to 2022

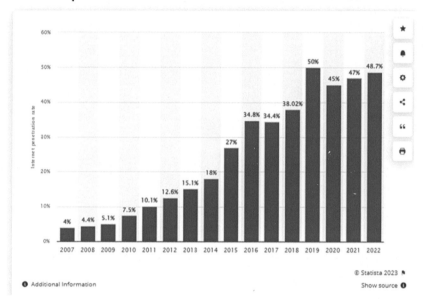

Figure 13.A3 Last 10 years Internet penetration rate in India.

Appendix 13.4

Fig. 13.A4

bigbasket.com Traffic Share by Device

Quickly understand where a website's traffic comes from and what devices visitors prefer to use. On bigbasket.com, desktops drive 13.08% of visits, while 86.92% of visitors come from mobile devices.

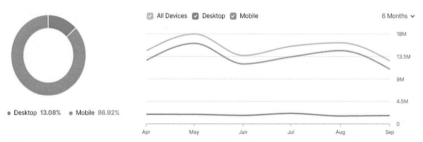

Figure 13.A4 Bigbasket device wise online traffic share. *From seoanalyzer.com.*

Appendix 13.5

Fig. 13.A5

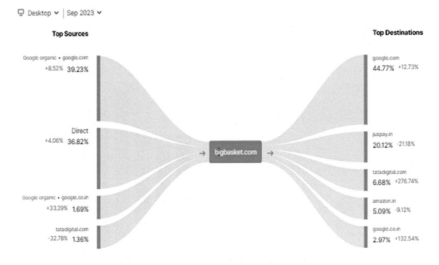

On bigbasket.com, visitors mainly come from google.com__Google organic (39.23% of traffic), followed by Direct (36.82%). In most cases, after visiting bigbasket.com, users go to google.com and juspay.in.

Figure 13.A5 Traffic source for Bigbasket.com.

References

Aparajita, A. D., Chawla, M., & Tulpule, D (2021). Story of big basket-pre and post pandemic. *Asian Journal of Science and Technology*, *12*(2), 11548–11552.

Battase, D. S. T., & Shankar, S. S. (2020). E-commerce in indian-an overview. *International Research Journal of Marketing & Economics*, *7*(2), 302–310, ISSN (2348-9766).

Bhandari, R. (2016). Impact of Technology on Logistics and Supply Chain Management, IOSR Journal of Business and Management (IOSR-JBM) e-ISSN: 2278-487X, p-ISSN: 2319-7668 PP 19–24.

Carpenter, J., & Moore, M. (2006). Consumer demographics, store attributes, and retail format choice in the US grocery market. *International Journal of Retail & Distribution Management*, *34*(6), 434–452.

Chanana, N., & Goele, S. (2012). Future of e-commerce in India. *International Journal of Computing & Business Research*, *8*(1).

Choudhary, M. D. (2020). How Big Basket is leveraging emerging tech for streamlining delivery. Available at: https://www.expresscomputer.in/news/how-big-basket-islever-aging-emerging-tech-for-streamlining-delivery/50150.

Gupta, S., & Chintagunta, P. K. (1994). On using demographic variables to determine segment membership in logit mixture models. *Journal of Marketing Research*, *31* (February), 128–136.

Hansen, T. (2008). Consumer values, the theory of planned behaviour and online grocery shopping. *International Journal of Consumer Studies*, *32*(2), 128−137.

den Hertog, P. (2000). Knowledge-intensive business services as co-producers of innovation (PDF). *International Journal of Innovation Management*.

Jain, V., Malviya, B., & Arya, S. (2021). An overview of electronic commerce (e-Commerce). *Journal of Contemporary Issues in Business and Government*, *27*(3), 665−670.

Jain, V. (2019). Bigbasket to have 3,000 e- vehicles by 2020. ET Retail.com https://retail.economictimes.indiatimes.com/news/food-entertainment/grocery/bigbasket-to-have-3000-e-vehicles-by-2020/70995283 Jain Varun (2019, September 05) Bigbasket to have 3,000 e- vehicles by 2020. ET Retail.com.

Khosla, M., & Kumar, H. (2017). Growth of e-commerce in India: An analytical review of literature. *IOSR Journal of Business and Management (IOSR-JBM)*, *19*(6), 91−95.

Kartika Kumar, K. (2020). Online Retail Grocery Market To Reach $3 Billion Valuation In 2020, Studies Reveal. Available at: https://www.theindianwire.com/business/onlineretail-grocery-market-to-reach-3-billion-valuation-in-2020-studies-reveal-297541/.

Kumar, P. (2020). How 'silent' e-commerce is already underway in India. Available at: https://www.financialexpress.com/bandwagon/writers-alley/how-silent-e-commerce-is-already-underway-in-india/2138551/.

Lovelock, C., Wirtz, J., & Chatterjee, J. (2013). Services Marketing People, Technology, Strategy: A South Asian Perspective, 7th Edition, Pearson Education India.

Mazumder, S. (2015). Can the 'Hyperlocal' supply chain model scale up? India.

Meta Group. (2001). Integration: Critical Issues for Implementation of CRM Solutions, Report by Meta Group Inc., February 15. Moriarty, Rowland T. and Moran, Ursula (1990), "Managing Hybrid Marketing Systems," Harvard Business Review, 68 (November/December), pp. 146−157.

Munson, J., Tiropanis, T., & Lowe, M. (2017). Online grocery shopping: Identifying change in consumption practices. In International Conference on Internet Science (pp. 192−211). Springer, Cham.

Nagaraju, K. B. (2023). Personal communication, March 22 2023.

Nougarahiya, S., Shetty, G., & Mandloi, D. (2021). A review of e−commerce in india: The past, present, and the future. *Research Review International Journal of Multidisciplinary*, *6*(3), 12−22.

Patrick, S., & Paul, M. (2001). "A Strategic Framework for CRM," Available at: http://crm-forum.com. (accessed April 13, 2002).

Raman, S., Patwa, N., Niranjan, I., Ranjan, U., Moorthy, K., & Mehta, A. (2018). Impact of big data on supply chain management. *International Journal of Logistics Research and Applications*, *21*(6), 579−596.

Ratna, T. (2018). Retrieval methods in M commerce applications: Challenges and prospects. *IJETSR*, *5*(1), 466−473.

Richa Chhabra, N. S. (2016). A Review on Hyper Local E-Commerce. India.

Sharma, R. (2016). Why it is Critical for Hyperlocal and eCommerce Companies in India to Get Last Mile Delivery Right. India.

Singhal, M. (2020). Online Grocery: What is the next frontier of growth? Available at: https://www.moneycontrol.com/news/opinion/online-grocery-what-is-the-next-frontier-of-growth-6068471.html.

Thomas, S., Abhishek, S., Vataywala, S., & Sinha, P. (2017). http://BigBasket.Com: Redefining the Business Model. Available from: https://doi.org/10.4135/9781526487513.

Vankipuram, M., & Nandy, M. (2020). Deal with Walmart, Flipkart will create retail-supply chain synergy, efficiency: Ninjacart co-founder. LiveMint. https://www.livemint.com/news/india/deal-with-walmartflipkart-will-create-retail-supply-chain-synergy-efficiencyninjacart-co-founder-11580186775510.html.

Wilson, M. N. (2015). Effects of information technology on performance of logistics firms in nairobi county. *International Journal of Scientific and Research Publications*, *5*(4), 1—26.

Zohra Ennaji, F., El Fazziki, A., El Alaoui El Abdallaoui, H., Sadiq, A., Sadgil, M., & Benslimane, D. (2016). Multi-agent framework for social CRM: Extracting and analyzing opinions. In: 2016 IEEE/ACS 13th International Conference of Computer Systems and Applications (AICCSA).

CHAPTER FOURTEEN

Applications of artificial intelligence in Echo Global Logistics

Amrita Chaurasia[1], Bhakti Parashar[2] and Sandeep Kautish[3,4]
[1]School of Commerce, Finance and Accountancy Christ (Deemed to be University), Ghaziabad, Uttar Pradesh, India
[2]VIT-Bhopal University, Bhopal, Madhya Pradesh, India
[3]LBEF Campus (APU Malaysia) Kathmandu, Nepal
[4]Model Institute of Engineering and Technology, Jammu, Jammu and Kashmir, India

14.1 Introduction

The ability to manage entire supply chain processes as a linked system of operations is a fundamental tenet of supply chain management. According to Akkermans, Zott, and Amit (2010), having a system perspective across the operations chain is crucial to prevent missing system linkages or focusing too much on a single activity. Furthermore, Teece (2018), emphasizes in his book "Industrial Dynamics" how decisions at any level have an impact on the circumstances that led to the decision in the first place. This is because industrial activities are closed-loop information systems. Effective applications of these ideas have also been shown in the automotive sector.

The design phase creates the products to be manufactured and distributed within the supply chain, thus determining a large portion of supply chain costs. Christopher and Towill (2001) have observed that a distribution system of cascaded inventories and ordering procedures seems to amplify small disturbances that occur at the retail level, based on the structure and policies within a multistage distribution system and a problem associated with the company's management's understanding of current strategic transitions, which implies changes to the supply chain design and/or product offerings.

The dynamics of global commerce are undergoing a seismic shift, driven by a confluence of technological advancements, consumer expectations,

Computational Intelligence Techniques for Sustainable Supply Chain Management.
DOI: https://doi.org/10.1016/B978-0-443-18464-2.00009-1

405

and a hyper-competitive business environment Bakos and Treacy (1986). At the heart of this transformation lies the intricate interplay between Artificial Intelligence (AI). Chen et al. (2012) mentioned the complex web of supply chain practices. The relentless quest for operational efficiency, cost reduction, and improved service quality has spurred companies across industries to explore AI as a transformative force in supply chain management. Pearson (2019) embarks on a journey of discovery, unveiling the profound impact of AI on Echo Global Logistics, a logistics and supply chain management powerhouse Elsevier Science (2010). With an operational footprint spanning the globe, Echo Global Logistics has emerged as a pioneer in integrating AI technologies to not only optimize existing practices but also to create innovative, value-driven solutions for its clients Dwivedi et al. (2021). Delving into the realms of AI and supply chain practices at this firm they unravel a narrative of evolution and adaptation and explore the motivation and challenges behind their strategic adoption of AI, and how this integration has propelled the company to redefine the essence of service delivery.

Bartlett et al. (2007) examine AI-driven demand forecasting, route optimization, inventory management, and proactive risk mitigation services to provide a base for scrutinizing the transformative potential of organizations that will shape the future of logistics.

14.1.1 Purpose and scope of the chapter

The purpose of this case study is to provide a comprehensive exploration of AI in supply chain practices at Echo Global Logistics. This chapter provides an understanding of the adoption of AI in the Supply chain and investigates the underlying motivations that led Echo Global Logistics to embrace AI technologies in their supply chain operations and ways to find the strategic imperatives driving the company's AI adoption.

The primary goal of the case study is to identify and analyze the challenges and obstacles encountered during the process of integrating AI into existing supply chain practices at Echo Global Logistics by gaining insights, the researcher in this case study will provide more understanding about supply chain operations by measuring the influence of AI on cost reduction, efficiency gains, service quality, and customer satisfaction. As a result, the goal of this chapter is to raise awareness about AI in supply chain logistics while also providing an overview of key areas where supply chain firms can save money and energy. Furthermore, the author investigates a formal approach to the applications of AI in supply chain firms.

14.1.2 Background of echo global logistics

Echo Global Logistics was founded in 2005 to simplify transportation operations. This firm connects businesses in need of transporting their products with carriers who can move goods efficiently, securely, and cost-effectively using various transportation methods, from coast to coast and port to port, Thomas and Uminsky (2022) explain the fundamental challenges faced by logistics firms and their solutions by streamlining the transportation operations, Thomas work is focus on the core activities of logistics firms and their clients, carrier partners, and team members all experience exceptional service, built on the principles which are supported by Biloslavo et al. (2018). This firm employed a set of skilled employees in logistics, they are experts and among the best in the field, dedicated to providing the best possible customer service. the team members ensure that clients' goods are handled with care and offer excellent content through their foundational service and improved technology is extremely adaptable, scalable, and user-friendly for customers, partners, and vendors, through seamlessly integrated portals, the employees of the firm ensure enhanced data collection and transmission, easy communication, comprehensive reporting, and real-time visibility.

14.1.3 Significance of artificial intelligence in supply chain

The network of people, businesses, organizations, resources, and processes involved in developing and providing a good or service to final consumers is referred to as the supply chain market. Buunk and Van Der Werf (2019) cover the entire spectrum, including the procurement of raw materials, manufacture, shipping, warehousing, distribution, and customer support. Long-time supply chain participants understand the value of building business relationships, whether they are business-to-business or business-to-customer. It has become increasingly complex in recent years due to globalization, technological advancements, and changing consumer preferences Yang et al. (2021). Companies must navigate multiple supply chain partners, deal with supply chain disruptions, manage inventory levels, and respond quickly to market demands. Supply chain friction at UK borders is replicated globally Sharma et al. (2021). Many companies are turning to supply chain technologies like automation, data analytics, and AI to tackle these challenges. These technologies can help streamline operations, reduce costs, and improve efficiency, as per Baryannis et al. (2018) challenges are from the standpoint of developing trust in supply chains and trust ecosystems.

The supply chain market is highly competitive, with a wide range of providers offering different products and services. Strange and Zucchella (2017) categorize the companies that specialize in specific areas of the supply chain, such as transportation or warehousing, while others offer end-to-end solutions. The supply chain market plays a crucial role in the global economy. Its importance is only expected to grow as companies continue to seek ways to improve their supply chain operations Grover and Dresner (2022) explained that the supply chain market to the global economy is likely to increase as businesses look for new ways to enhance their supply chain operations.

The relevance of the supply chain market to the global economy is likely to increase as businesses AI has the potential to alter how businesses distribute and sell their products significantly Samson (2020) Automating and streamlining distribution and sales operations with the aid of AI technologies like ML, natural language processing, and computer vision this will increase the productivity, lower costs, and enhance customer experiences for businesses. AI algorithms can accurately estimate future demand by analyzing previous sales data, consumer behavior, and other relevant variables. Frankenberger et al. (2013) enterprise that, trades may decrease waste, increase efficiency, and guarantee that their products are available when customers need them, AI can analyze information such as traffic patterns, delivery schedules, and weather conditions to make deliveries as efficient as possible, which will speed up deliveries and lower the cost of shipping. AI algorithms that examine client information, including purchase history, preferences, and behavior, can make personalized product suggestions which could raise client happiness and boost revenue. AI-powered chatbots can enhance the customer experience and boost revenue by offering customers real-time support and assistance.

14.2 Echo global logistics

Echo Global Logistics is a leading provider of technology-enabled transportation and supply chain operation services for all major modes of transportation, including expedited, intermodal, LTL, and partial truckload (TL).

The company operates through a unique web-based technology platform that gathers and analyzes data from its network of over 40,000 transportation providers to assist clients from various industries and simplify complex transportation operation tasks as mentioned in the report. Echo's priority is exceptional service in every customer interaction. In the firm, the employees understand that there is no one-size-fits-all solution, so they always strive to assess transportation needs, identify opportunities, and provide the most value for their client's requirements and business objectives.

14.2.1 Company overview

A user-friendly shipping platform ensures that clients receive the most suitable transportation solutions for both TL and less-than-truckload (LTL) shipments. The company's intuitive design eliminates the complex interfaces which helps offer the efficiency and functionality required for LTL shipping, including quoting, reservations, tracking, and invoice management. Clients can request quotes and book shipments from anywhere in the world, 24 hours a day, seven days a week. This firm provides real-time updates on load status through its Shipment Board, enabling clients to track their shipments, upload documents, and receive updates on specific loads of interest. Company Green Team is a dedicated group of customer service representatives who provide exceptional service and have been recognized with inbound logistics excellence awards for five consecutive years.

The straightforward and user-friendly design of the firm ship minimizes repetitive tasks, allowing clients to focus on their core activities. It is a well-established player in the logistics industry and recognizes the need to adapt to an evolving landscape. This case study delves into their remarkable journey to address a real-life problem that was undermining their efficiency and profitability, and the extraordinary solution they adopted to navigate the ever-evolving logistics terrain. fuel costs were soaring due to unnecessarily long routes and traffic bottlenecks, resulting in increased operational expenses.

14.2.2 History

This company simplifies transportation operations, it connects businesses in need of transporting their products with carriers who can move goods efficiently, securely, and cost-effectively using various transportation methods, from coast to coast and port to port. company green logistics solutions streamline transportation operations, allowing clients to focus on their core activities. clients, carrier partners, and team members all experience exceptional service, built on the principles of the Echo way. The team of skilled logistics experts is among the best in the field, dedicated to providing the best possible customer service, the employees of the organization ensure that clients' goods are handled with care and offer excellent content through their foundational service and improved technology, this technology is extremely adaptable, scalable, and user-friendly for customers, partners, and vendors. Through seamlessly integrated portals, firms improved data collection and transmission, easy communication, comprehensive reporting, and real-time visibility. In this company, they use the Poole bracket approach which was employed to provide a better overview. It is a leading provider of technology-enabled transportation and supply chain operation services for all major modes of transportation, including expedited, intermodal, LTL, and partial TL. The company operates through a unique web-based technology platform that gathers and analyses data from its network of over 40,000 transportation providers to assist clients from various industries and simplify complex transportation operation tasks. priority is exceptional service in every customer interaction in this firm. employees understand that there is no one-size-fits-all solution, so they always strive to assess transportation needs, identify opportunities, and provide the most value for their client's requirements and business objectives.

14.2.3 Benefits

Echo Global Logistics, often referred to as Echo, is a leading provider of technology-enabled transportation and supply chain management services. The company specializes in freight brokerage and managed transportation solutions, catering to a diverse range of clients across various industries it's a prominent American transportation and logistics company based in Chicago, Illinois, and offers several key benefits that showcase its strengths and contributions within the industry such as:

- Enhanced supply chain visibility and transparency.
- Efficient and cost-effective freight management solutions.
- Access to a wide network of carriers and transportation options.
- Streamlined logistics processes through technology-driven platforms.
- Dedicated customer support and personalized service.

Some of the key benefits and contributions of Echo Global Logistics are given in Fig. 14.1.

- Technology-Driven Solutions: Echo utilizes advanced technology and digital platforms to provide innovative and efficient transportation and logistics solutions. This includes automated systems for freight management, real-time tracking, and supply chain optimization.
- Diverse Service Offerings: Echo's comprehensive service portfolio encompasses freight brokerage, managed transportation, intermodal, and supply chain management solutions. This diverse range of services allows the company to cater to the unique needs of various industries and clients.
- Cost-Effective Operations: Echo's strategic approach to logistics and transportation management enables cost savings for its clients. Through effective route planning, carrier optimization, and competitive pricing, Echo helps businesses achieve greater efficiency and reduced transportation costs.

Figure 14.1 Echo's strengths and contributions within the industry. Source: *Authors own.*

- Extensive Carrier Network: Echo's strong relationships with a wide network of carriers contribute to its ability to offer reliable and flexible transportation solutions. This extensive network ensures that clients have access to a diverse range of transportation options, leading to improved service levels and timely deliveries.
- Customer-Centric Approach: Echo is known for its commitment to customer satisfaction and personalized service. The company emphasizes building strong client relationships through dedicated support, transparent communication, and tailored solutions that meet specific business requirements.

14.2.4 Range of services provided

Customer satisfaction ratings were plummeting as late deliveries became a recurring issue. Echo Global Logistics offers a diverse range of services designed to streamline transportation management and enhance supply chain operations for businesses across various industries. These services are founded on advanced technology and industry expertise, making Echo Global Logistics a comprehensive solution provider in the logistics sector with the following services:

- **Transportation Management:** Echo Global Logistics provides end-to-end transportation management services, encompassing everything from planning and execution to tracking and reporting. Their technology-driven approach helps businesses optimize their transportation processes for cost savings and efficiency.
- **Managed Transportation Services:** Echo's managed transportation services offer a more hands-on approach, where the company's experts take the lead in managing a client's transportation operations. This can include network design, carrier selection, and continuous optimization.
- **Intermodal Services:** Echo Global Logistics facilitates the integration of intermodal transportation solutions, combining the flexibility of trucking with the cost-effectiveness of rail. Intermodal services are ideal for long-haul shipping and can provide sustainability benefits.
- **Truckload Services:** Echo manages and arranges TL shipments, offering solutions that cater to the specific needs of each client. This includes full truckload (FTL) and LTL options.
- **LTL and Partial Truckload:** For businesses with shipments that do not require an FTL, Echo offers LTL and partial truckload services.

This allows for cost-effective and efficient transport of smaller shipments.

- **International Shipping:** Echo Global Logistics supports businesses with their international shipping requirements. This includes cross-border logistics, customs compliance, and international trade facilitation.
- **Expedited Shipping:** In situations where time-sensitive deliveries are crucial, Echo provides expedited shipping options to ensure that shipments reach their destinations within tight deadlines.

14.2.5 Real-time tracking and visibility

Echo's technology platform offers real-time tracking and visibility into shipments. This allows clients to monitor the status of their shipments and make data-driven decisions based on real-time information in the subsequent ways by which the company provides services for the delivery of perishable commodities, pharmaceuticals, and other temperature-controlled products to sectors that require temperature-sensitive shipping, they provide account managers to clients which guarantee that they receive individualized support and customized solutions that address their specific supply chain requirements, works together with clients to comprehend their goals and supply chain difficulties. After that, they offer their clients professional advice and insights to assist in making defensible selections. This company integrates environmentally friendly modes of transportation into its offerings, cutting carbon emissions and encouraging sustainable supply chain methods, the company is a comprehensive suite of logistics and supply chain services that cater to the diverse needs of businesses across different industries, with a strong emphasis on technology, data-driven decision-making, and customer-centricity, the company is committed to providing innovative and efficient solutions for optimizing transportation management and enhancing supply chain operations. This range of services positions the company as a valuable partner for businesses looking to navigate the complexities of modern logistics and transportation.

14.2.6 Industry position and client base

Echo Global Logistics holds a significant position in the logistics and supply chain management industry, making it a prominent player in a highly competitive and dynamic sector. Its industry position is a reflection of its growth, capabilities, and commitment to delivering technology-driven logistics solutions in various methods given below:

- **Market Presence:** Echo Global Logistics has established a strong market presence, both in the United States and on a global scale. The company's reputation and expansive network of operations contribute to its prominent position in the logistics industry.
- **Technology Leadership:** Echo's focus on technology and data-driven decision-making sets it apart as a leader in the logistics industry. The company's proprietary technology platform empowers businesses with tools to optimize transportation management and enhance supply chain operations.
- **Innovation:** Echo Global Logistics continues to drive innovation within the industry. Its commitment to leveraging emerging technologies, such as AI and ML, positions the company at the forefront of transformative logistics solutions.
- **Client–Centric Approach:** The company's client-centric philosophy has fostered long-lasting relationships with a diverse range of businesses. By understanding and addressing the unique needs of its clients, Echo Global Logistics has earned a reputation for exceptional service and support.
- **Collaborations and Partnerships:** Echo collaborates with a network of carriers, suppliers, and technology providers, enhancing its industry position. These partnerships strengthen its service offerings and extend its reach within the logistics ecosystem.

14.3 Artificial intelligence and its role in supply chain transformation

AI embodies the ability of a machine to emulate human abilities like thinking, learning, planning, and creativity. It enables technological systems to perceive their environment, react to visual information, troubleshoot problems, and take action to accomplish specific goals. By ingesting pre-processed or self-collected data through sensors like cameras, the computer interprets and processes this information to generate appropriate responses (Chaurasia et al., 2023). Demand planning and forecasting using computers are nothing new. It is built on a set of algorithms that are designed to forecast using different data sets throughout time, such as manufacturing data, ordering patterns, shipment data, and product life cycle data. The AI-enabled system, however, is aware of the best possible combinations of data sets and algorithms to take into account when making an accurate prediction. And above all, AI is assisting companies in:

1. Forecasting and projecting client demand with nearly 100% accuracy.
2. Maximizing their R&D to boost production at a reduced cost and with better quality.
3. Supporting them in the promotion (determining the demographics, target market, price, and design of appropriate message, etc).
4. Giving their clients a superior experience.

These four sectors generate value important for gaining a competitive advantage. AI plays a pivotal role in the ongoing transformation of supply chain management. The integration of AI technologies is revolutionizing the way businesses plan, execute, and optimize their supply chain operations. floor waste. The lean-green approach improved employee morale and quality.

14.3.1 The evolving landscape of supply chain management

Pricing solutions can now monitor purchasing patterns and calculate more affordable product prices thanks to AI. Several industries, including consumer goods, fashion, hospitality, and transportation, have used AI-driven pricing software. Businesses will eventually move from absolute, or static, pricing to dynamic pricing, which will allow them to provide clients with varying prices according to their unique purchasing behaviors and external circumstances. The foundation of dynamic pricing is the aggregate pricing data that is available from multiple sources, such as the internet, rival websites, and prices that are offered in different areas. Dynamic pricing algorithms use a variety of parameters to calculate the price at which customers are ready to pay for a given good or service, including pricing strategies used by competitors, consumer behavior, geography, time of day, and seasonality. As businesses expand into global markets, the supply chain becomes more complex and interconnected.

Understanding and adapting to the changing global landscape is essential for ensuring efficient and cost-effective operations. Rapid advancements in technology, such as the Internet of Things (IoT), AI, blockchain, and data analytics, have revolutionized supply chain management. These technologies enable better visibility, traceability, and decision-making within the supply chain, leading to increased efficiency and responsiveness. Today's consumers expect faster delivery, accurate tracking, and transparency in the supply chain. Meeting these expectations is crucial for customer satisfaction and retention.

The evolving landscape of supply chain management helps companies align their strategies with changing consumer demands. Sustainable supply chain management has become a critical focus due to environmental concerns. Businesses are under pressure to reduce their carbon footprint, minimize waste, and adopt eco-friendly practices. An evolving supply chain landscape is essential to address these sustainability challenges effectively. Recent global events, such as the COVID-19 pandemic and natural disasters, have highlighted the need for resilient supply chains. Companies are now reevaluating their supply chain strategies to better manage risks and disruptions.

14.3.2 Background literature

The network of people, businesses, organizations, resources, and processes involved in developing and providing a good or service to final consumers is referred to as the supply chain market. This covers the entire spectrum, including the procurement of raw materials, manufacture, shipping, warehousing, distribution, and customer support. Long-time supply chain participants understand the value of building business relationships, whether they are business-to-business or business-to-customer. Supply chain management has become increasingly complex in recent years due to globalization, technological advancements, and changing consumer preferences. Companies must navigate multiple supply chain partners, deal with supply chain disruptions, manage inventory levels, and respond quickly to market demands. Supply chain friction at UK borders is replicated globally.

Many companies are turning to supply chain technologies like automation, data analytics, and AI to tackle these challenges. These technologies can help streamline operations, reduce costs, and improve efficiency. The challenge is from the standpoint of developing trust in supply chains and trust ecosystems. The supply chain market is highly competitive, with a wide range of providers offering different products and services. Some companies specialize in specific areas of the supply chain, such as transportation or warehousing, while others offer end-to-end solutions. Thus, it plays a crucial role in the sglobal economy, and its importance is only expected to grow as companies continue to seek ways to improve their supply chain operations (Fig. 14.2).

Figure 14.2 Strategic alignment for sustainable superior performance.

The relevance of the supply chain market to the global economy is likely to increase as businesses look for new ways to enhance their supply chain operations. The four components that need to be in alignment with one another to produce consistently better performance are combined into the Strategic Alignment model in this model an organization needs to be in harmony with its surroundings for operation. The utility Illustrates the relationship between clients' requirements, the formulation of suitable strategic replies, as well as the triumphant application of these tactics by forming the essential interior capacities as well as matching types of leadership. the basic requirements and purchasing actions that eventually motivate, s The relevance of the supply chain market to the global economy is likely to increase as businesses AI has the potential to alter the business's distribution and sell their products significantly (Tina, 2023). Automating and streamlining distribution and sales operations with the aid of AI technologies like ML, natural language processing, and computer vision may increase productivity, lower costs, and enhance customer experiences for businesses.

AI algorithms can accurately estimate future demand by analyzing previous sales data, consumer behavior, and other relevant variables. By doing this, businesses may decrease waste, increase efficiency, and guarantee that their products are available when customers need them. AI can analyze information such as traffic patterns, delivery schedules, and weather conditions to make deliveries as efficient as possible. This can speed up deliveries and lower the cost of shipping. AI algorithms that examine client information, including purchase history, preferences, and behavior, can make personalized product suggestions. This could raise client happiness and boost revenue. AI-powered chatbots can enhance the customer experience and boost revenue by offering customers real-time support and assistance. The given figure explains the future of AI in the retail market.

AI can completely change how businesses approach distribution and sales, allowing them to work more productively, for less money, and with happier customers. Businesses are being made more efficient, and the consumer experience is being improved thanks to AI technology, The retail industry faces intense competition from digital platforms, leading retailers to adopt AI technologies like ML and deep learning to improve customer shopping experiences and stay relevant. However, reluctance to adopt these advancements and the need for a competent workforce may hinder market expansion. AI can be used for real-time marketing adjustments and inventory management. The integration of AI and shared mobility has made it possible to provide personalized customer experiences to users. The future of shared mobility is believed to be driverless cars, which employ AI and sensor technologies. The driverless fleets are more lucrative, and by 2030, they may replace 6.2 million drivers worldwide.

14.3.3 The business model of echo global logistics

Echo Global Logistics' business model revolves around using technology to streamline the transportation of goods and provide efficient and cost-effective logistics solutions. They specialize in TL and LTL transportation. Shippers can utilize Echo's transportation management services for carrier selection, routing, scheduling, real-time tracking, and visibility of products in transit. The company leverages its broad carrier network and technology platform to automate logistics procedures, reducing costs and increasing efficiency. The graphical representation of this model is given below (Fig. 14.3).

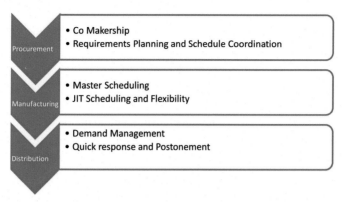

Figure 14.3 Echo Global Logistics. Business model. Source: *Authors Own.*

This model is a technology-driven transportation and supply chain management company that operates through a digital platform connecting shippers and carriers to optimize goods transportation across various transportation modes. The company gathers and analyses data from a multimodal network of transportation providers to meet its clients' transportation and logistics needs. Echo generally arranges transportation with TL and LTL carriers and also offers intermodal (rail and truck shipping), small parcel, domestic air, expedited, and international transportation. Echo's technology platform, consisting of web-based software applications and a database, allows it to identify redundant transportation capacity, negotiate competitive rates, and execute hundreds of shipments daily.

14.3.4 Supply chain management at echo global logistics

Supply chain management is a critical aspect of Echo Global Logistics' business model. The company offers a range of supply chain services to help its clients optimize their logistics operations and reduce costs in procurement, Manufacturing, and distribution. Some of the key aspects of Echo's supply chain management include Transportation Planning: Echo provides transportation planning services to help its clients optimize their logistics operations. The company uses advanced analytics and modeling to determine the most efficient transportation routes and modes for each shipment, helping clients reduce transportation costs and improve delivery times. Echo manages a large network of carriers, allowing it to provide its clients with a wide selection of transportation options.

The company matches shipments with the most suitable carriers based on factors such as capacity, location, and cost, using its proprietary technology platform. Echo acts as a freight broker, connecting shippers with carriers to transport their goods. The company leverages its extensive carrier network and advanced technology to find the most cost-effective and efficient shipping solutions for its clients. Echo offers warehousing and distribution services to assist clients in managing their inventory and streamlining their supply chain operations. Clients are served by the company's network of warehouses, which provide storage, fulfillment, and distribution services. Echo provides real-time tracking and visibility of products in transit, allowing clients to monitor their shipments and make informed decisions about their supply chain operations. The company also offers comprehensive reporting and analytics solutions to help clients track performance metrics, identify trends, and make data-driven decisions.

Here, supply chain management services are designed to help its clients optimize their logistics operations and reduce costs. By leveraging its advanced technology platform, extensive carrier network, and a range of supply chain services, Echo provides a comprehensive logistics solution that enables clients to focus on their core business operations. It generates profit by charging a fee for its transportation operation and freight brokerage services. The company also earns profit from its technology platform through subscription fees and sales-based fees.

This competitive advantage is based on its technology platform and extensive carrier network. The platform allows the company to provide more efficient and cost-effective logistics services to its clients, while its carrier network offers access to a wide range of transportation options. Echo Global Logistics' business concept is centered around using technology to streamline goods movement and provide clients with more efficient and cost-effective logistics solutions. It offers a comprehensive set of services, ranging from transportation planning to carrier management and warehousing, all underpinned by advanced technology and a commitment to exceptional customer service.

14.3.5 Environmental sustainability

Management is dedicated to responsible environmental management, sustainability, and natural environment protection in our workplace, which includes maintaining an Integrated Management System with ISO 14001 certification. The management team commits to upholding sustainable development, environmental stewardship, and environmental protection at work. This involves maintaining an ISO 14001-certified integrated management system, reducing waste through recycling techniques and innovative work practices, minimizing the environmental impact of our products, processes, materials, and services, and increasing the use of environmentally friendly equipment, technology, and materials.

The company also encourages our vendors to adhere to ethical environmental standards and actively promote environmental awareness among employees, clients, and the general public. Management bears the ultimate responsibility for ensuring environmental sustainability and compliance with environmental regulations, setting measurable goals to eliminate waste, pollution, and environmental degradation. Our employees are also responsible for adhering to all environmental legislation and guidelines, recognizing and reporting environmental hazards, and contributing creative ideas.

14.3.6 AI and predictive analytics at echo global logistics

AI and ML play a crucial role in Echo Global Logistics' fulfillment and logistics management operations. The company leverages these technologies to optimize logistics, reduce costs, and enhance the customer experience. Predictive analytics is also employed, allowing Echo to analyze vast amounts of data and anticipate future developments accurately. These techniques assist in identifying potential security or fraud issues and improving decision-making, operational efficiency, and customer satisfaction. By proactively addressing security concerns and minimizing risks, Echo gains a competitive advantage.

14.4 Challenges

Along with benefits, the company faces some challenges as well. Supply chain management poses significant challenges, and Echo Global Logistics is no exception. These challenges include dependence on third-party vendors, the potential impact of natural disasters and transportation delays, inventory management difficulties, and the need for seamless technology integration. To address these challenges, Echo Global Logistics can create contingency plans, invest in workforce development, and strengthen its supply chain management capabilities.

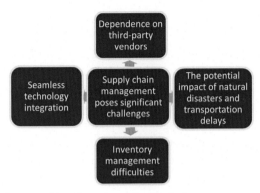

Figure 14.4 Challenges with echo green. Source: *Authors own.*

In the chart given below some of the major challenges are mentioned (Fig. 14.4).

Echo Global Logistics specializes in TL and LTL transportation. Over the years, the company has expanded through acquisitions, purchasing companies like Trailer Transport Systems, Plum Logistics, Shipper Direct Logistics, Sharp Freight Systems, and Open Mile Inc. Echo generates revenue by charging fees for transportation management and freight brokerage services. They also utilize subscription fees and transaction-based fees for their technological platform. Echo's competitive advantage lies in its technology platform and extensive carrier network, enabling it to provide efficient and cost-effective logistics services.

14.5 Research methodology

The interview with Echo Global Logistics CEO Doug Waggoner was conducted through email, with the questions sent as text within the email. The interview questions were crafted after extensive research, focusing on supply chain management, best practices, and technology and sustainability approaches at Echo's digital platform.

The purpose of this interview was to gain a comprehensive understanding of Echo's digital platform's unique and tech-enabled supply chain practices. The ten questions asked during the interview are designed to explore various aspects of AI and supply chain management. The email interview was carried out. The interview questions were developed after significant research, and they focused on understanding supply chain management at Echo's digital platform, best practices, and the technology and sustainability approaches that are adapted and implemented at Echo's digital platform.

Some questions were asked by the interviewer. Finally, the purpose of this interview was to obtain a full grasp of Echo's digital platform's unique and tech-enabled supply chain practices. In this sense, the 10 questions listed below were designed, and a personal e-mail interview was done. The discussion is predicated on understanding the following features of AI and supply chain management, as well as the impact of AI and efficient technology utilization:

1. What is the impact of the pandemic on the industry, including market predictions for 2030.
2. Capacity trends, all of which will be dependent on information flows.
3. Which materials or orders will we investigate in terms of AI use in the supply chain industry?
4. How does AI affect corporate distribution and sales?
5. How will changes in management policies affect the firm's internal supply chain management?
6. How does the system leverage AI to produce distribution amplification?
7. Does AI support on a small- or large-scale affect sales?
8. What will such a system do in reaction to technological advances?
9. What are the effectively profitable supply chain management changes?

We will discover that these questions can be explored by analyzing Echo Global Logistics' distribution system, which connects businesses that need to distribute their products with carriers that move items promptly, securely, and cost-effectively.

"I am looking forward to working with Echo and Roadtex as we endeavor to cultivate both companies," Hurst said of his overall experience managing force chain operations for a large-scale e-commerce platform similar to Echo Global Logistics. "These associations have constantly proven to be outstanding leaders in developing LTL results for shippers, and I am recognized to now be a part of that trouble."

As he stated, "I am thrilled to be joining the Roadtex platoon and working with founders Bruno Ciccarelli and Bob Kelly to make on their vision and grow the company," said Recendez. "As Roadtex pursues ongoing invention for our guests at Echo Global Logistics, we have a fantastic occasion to work Echo's technology and coffers." As he stated, "I am thrilled to be joining the Roadtex platoon and working with co-founders Bruno Ciccarelli and Bob Kelly to make on their vision and grow the company," said Recendez. "At Echo Global Logistics, we have a fantastic occasion to work Echo's technology and coffers as we pursue ongoing invention at Roadtex for our guests." The thing is to continuously enhance operations, beginning with planning, procurement, entering, picking, and shipping.

14.6 Proposed solution

The interviewer inquired about the impact of the pandemic on the industry and market predictions for 2030. They also asked how capacity trends would be influenced by information flows. Echo's CEO responded, emphasizing that Echo's digital platform, which employs AI, ML, and predictive analytics, is integral to its business model. As he stated, "I am thrilled to be joining the Roadtex platoon and working with founders Bruno Ciccarelli and Bob Kelly to make on their vision and grow the company," said Recendez. "As Roadtex pursues ongoing invention for our guests at Echo Global Logistics, we have a fantastic occasion to work Echo's technology and coffers." As he stated, "I am thrilled to be joining the Roadtex platoon and working with co-founders Bruno Ciccarelli and Bob Kelly to make on their vision and grow the company," said Recendez. "At Echo Global Logistics, we have a fantastic occasion to work Echo's technology and coffers as we pursue ongoing invention at Roadtex for our guests." The thing is to continuously enhance operations, beginning with planning, procurement, entering, picking, and shipping.

14.6.1 Industry impact and capacity trends

The canvasser inquired as to how capacity trends will be grounded on information overflows, to which he responded, "Echo's digital platform is the backbone of its business model." The platform employs AI, machine literacy, and prophetic analytics to match shippers with applicable carriers, optimize transportation routes, and give real-time shadowing and reporting. The question that follows focuses on the most serious difficulties with accouterments or orders that we will dissect in terms of AI deployment in the force chain- grounded assiduity.

14.6.2 Tie up with third-party merchandisers

Echo Global Logistics will take the step of erecting strong working ties with third-party merchandisers. Echo Global Logistics will also ameliorate collaboration and communication by establishing strong hookups with carriers and other third-party merchandisers. We also moved on to bandy any innovative force chain strategies or enterprise that Echo Global Logistics has enforced for optimizing operations and lowering costs, with Mr. Doug Waggoner emphasizing that invention has been the backbone at Echo Global Logistics, which has reckoned heavily on invention to break force chain issues.

14.6.3 Electric vehicles

Likewise, he argues that another unique and stylish practice at Echo is the operation of electric vehicles (EVs) for deliveries and that AI is changing the establishment's distribution and deals. still, it's worth noting that Echo has been committed to sustainability enterprise and reducing its carbon footmark; they blazoned plans to transition to using EVs for last- afar delivery operations in 32 service centers across the country; and Roadtex leads the request in temperature-controlled warehousing and exact time defined LTL service. Using a vast network in confluence with technology streamlines shipping to mass merchandisers and big box stores, furnishing peace of mind that shipments will arrive on time, complete, and damage-free.

14.6.4 AI in supply chain and collaboration

The interview delved into the materials or orders that AI will impact in the supply chain industry and how AI affects corporate distribution and sales. It also discussed how changes in management policies affect the firm's internal supply chain management. The questions explored how AI leverages distribution amplification and whether AI impacts sales on a small or large scale. The question that followed asked how changes in operation programs would prompt the establishment's internal force chain operation. "At Echo, they're happy to mate with Frank and Phil as Bruno and I continue to support Roadtex guests and work with our devoted workers," stated Bob Kelly, Co-Founder and Head of Deals at Roadtex. Working with Echo to strengthen our long-standing customer connections through our devoted transportation network and public storehouse footmark remains an important part of our vision. Various methods can

Figure 14.5 Methods for AI in Supply Chain and Collaboration Practices. Source: *Authors own.*

be used for AI in supply chain and collaboration practices, some of them are mentioned below (Fig. 14.5).

The coming question for Echo CEO Doug Waggoner is regarding the AI-powered distribution system. Mr. Doug Waggoner replied by pressing their expansive businesses as well as In response, they mentioned their huge operations as well as the use of AI technology for logistics and force chain duties. "Echo was innovated on the idea that by using disruptive technology, we could simplify transportation and bring edge to the massive and largely fractured transportation request," stated Doug Waggoner, Chairman of the Board of Directors and CEO of Echo. "We first introduced an online shipping gate 12 times agone, and our new personal Echo Accelerator platform, along with Echo Drive and Echo Ship, marks the rearmost advancement in Echo's ongoing digital elaboration."

14.7 Findings and conclusion

The interview covered innovative supply chain strategies and practices at Echo Global Logistics for optimizing operations and reducing costs. It highlighted Echo's use of electric vehicles (EVs) for deliveries and how AI is transforming the company's distribution and deals. It's worth noting that Echo is committed to sustainability efforts and transitioning to EVs for last-mile delivery operations. The discussion shifted to changes in operational programs and how they impact the company's internal supply chain operations. Echo's CEO emphasized their commitment to strengthening customer connections and leveraging AI technology for logistics and supply chain duties. We also enquired as to how efficiently economic advancements in force chain operation are subordinated to regular internal and external check-ups to maintain compliance, indeed in a changing nonsupervisory terrain.

Echo Drive allows carriers to bespeak, manage, and be paid for hauling products for Echo. The cargo operation tool and document upload capabilities of the platform help dispatchers stay organized and motorists get back on the road briskly. Echo Drive guests can search for, see, and shoot on available goods in real time, saving time and trouble in choosing goods for their available outfits.

Its CEO provided insights into its AI-powered distribution system, including the use of disruptive technology to simplify transportation and improve the efficiency of transportation requests. Echo's digital platforms, including Echo Accelerator, Echo Drive, and Echo Ship, play a significant role in their ongoing digital development. The content mentioned that Echo Global Logistics' distribution system connects businesses with carriers for efficient and cost-effective product distribution. Echo Ship was highlighted for its user-friendly design and time-saving benefits, supported by Echo's 24/7 customer service. The interviewer inquired about how economic advancements in supply chain operations are aligned with regular internal and external inspections to maintain compliance, especially in a changing regulatory landscape. Echo Drive was discussed as a tool that allows carriers to request, manage, and get paid for hauling products for Echo. It also provides benefits for dispatchers and drivers. Echo Ship, known for its user-friendly design, simplifies the shipping process, saving time for logistics personnel.

References

Bakos, J. Y., & Treacy, M. E. (1986). Information technology and corporate strategy: A research perspective. *Management Information Systems Quarterly*, *10*(2), 107. Available from https://doi.org/10.2307/249029.

Bartlett, P., Julien, D., & Baines, T. (2007). Improving supply chain performance through improved visibility. *The International Journal of Logistics Management*, *18*(2), 294–313. Available from https://doi.org/10.1108/09574090710816986.

Baryannis, G., Validi, S., Dani, S., & Antoniou, G. (2018). Supply chain risk management and artificial intelligence: State of the art and future research directions. *International Journal of Production Research*, *57*(7), 2179–2202. Available from https://doi.org/10.1080/00207543.2018.1530476.

Biloslavo, R., Bagnoli, C., & Edgar, D. (2018). An eco-critical perspective on business models: The value triangle as an approach to closing the sustainability gap. *Journal of Cleaner Production*, *174*, 746–762. Available from https://doi.org/10.1016/j.jclepro.2017.10.281.

Buunk, E., & Van Der Werf, E. (2019). Adopters versus non-adopters of the green key ecolabel in the Dutch Accommodation Sector. *Sustainability*, *11*(13), 3563. Available from https://doi.org/10.3390/su11133563.

Chen, H., Chiang, R. H. L., & Storey, V. C. (2012). Business intelligence and analytics: From big data to big impact. *Management Information Systems Quarterly*, *36*(4), 1165. Available from https://doi.org/10.2307/41703503.

Chaurasia, A., Parashar, B., & Kautish, S. (2023). Artificial intelligence and automation for industry 4.0. In *Disruptive Technologies and Digital Transformations for Society 5.0* (pp. 357−373). https://doi.org/10.1007/978-981-99-5354-7_18.

Christopher, M., & Towill, D. R. (2001). An integrated model for the design of agile supply chains. *International Journal of Physical Distribution & Logistics Management*, *31*(4), 235−246. Available from https://doi.org/10.1108/09600030110394914.

Dwivedi, Y. K., Hughes, L., Ismagilova, E., Aarts, G., Coombs, C., Crick, T., Duan, Y., Dwivedi, R., Edwards, J. S., Eirug, A., Galanos, V., Ilavarasan, P. V., Janssen, M., Jones, P., Kar, A. K., Kizgin, H., Kronemann, B., Lal, B., Lucini, B., ... Williams, M. D. (2021). Artificial Intelligence (AI): Multidisciplinary perspectives on emerging challenges, opportunities, and agenda for research, practice, and policy. *International Journal of Information Management*, *57*, 101994. Available from https://doi.org/10.1016/j.ijinfomgt.2019.08.002.

Frankenberger, K., Weiblen, T., & Gassmann, O. (2013). Network configuration, customer centricity, and performance of open business models: A solution provider perspective. *Industrial Marketing Management*, *42*(5), 671−682. Available from https://doi.org/10.1016/j.indmarman.2013.05.004.

Grover, A. K., & Dresner, M. (2022). A theoretical model on how firms can leverage political resources to align with supply chain strategy for competitive advantage. *Journal of Supply Chain Management*, *58*(2), 48−65. Available from https://doi.org/10.1111/jscm.12284.

Pearson, A.W. (2019). *The A.I. marketer*. Intelligencia.

Samson, D. (2020). Operations/supply chain management in a new world context. *Operations Management Research*, *13*(1−2), 1−3. Available from https://doi.org/10.1007/s12063-020-00157-w.

Sharma, M., Joshi, S. K., Luthra, S., & Kumar, A. (2021). Managing disruptions and risks amidst COVID-19 outbreaks: Role of blockchain technology in developing resilient food supply chains. *Operations Management Research*, *15*(1−2), 268−281. Available from https://doi.org/10.1007/s12063-021-00198-9.

Strange, R., & Zucchella, A. (2017). Industry 4.0, global value chains, and international business. *The Multinational Business Review*, *25*(3), 174−184. Available from https://doi.org/10.1108/mbr-05-2017-0028.

Teece, D. J. (2018). Profiting from innovation in the digital economy: Enabling technologies, standards, and licensing models in the wireless world. *Research Policy*, *47*(8), 1367−1387. Available from https://doi.org/10.1016/j.respol.2017.01.015.

Thomas, R. L., & Uminsky, D. (2022). Reliance on metrics is a fundamental challenge for AI. *Patterns*, *3*(5), 100476. Available from https://doi.org/10.1016/j.patter.2022.100476.

Tina. (2023, October 25). Artificial intelligence (AI) in supply chain and logistics. *ThroughPut Inc.* https://throughput.world/blog/ai-in-supply-chain-and-logistics/.

Yang, M., Fu, M., & Zhang, Z. (2021). The adoption of digital technologies in supply chains: Drivers, process and impact. *Technological Forecasting and Social Change*, *169*, 120795. Available from https://doi.org/10.1016/j.techfore.2021.120795.

Zott, C., & Amit, R. (2010). Business model design: An activity system perspective. *Long Range Planning*, *43*(2−3), 216−226. Available from https://doi.org/10.1016/j.lrp.2009.07.004.

Index

Note: Page numbers followed by "*f*" and "*t*" refer to figures and tables, respectively.